Birendra Prasad (Editor)

CAD/CAM Robotics and Factories of the Future

Volume I: Integration of Design, Analysis and Manufacturing

3rd International Conference on CAD/CAM
Robotics and Factories of the Future
(CARS and FOF '88) Proceedings

With 223 Figures

Springer-Verlag Berlin Heidelberg New York
London Paris Tokyo Hong Kong

ISBN 978-3-642-52322-9 · ISBN 978-3-642-52320-5 (eBook)
DOI 10.1007/978-3-642-52320-5

Offsetprinting: Color-Druck Dorfi GmbH, Berlin; Binding: Lüderitz & Bauer, Berlin
2161/3020 543210 – Printed on acid-free paper

Conference Objectives

Improving cost competitiveness and remaining abreast in high technology are some of the challenges that are faced by a developing enterprise in the modern times. In this context, the roles of engineering, manufacturing and plant automation are becoming important factors to enhance productivity and profitability, and thereby increase market share and product quality. The commuter automobile, actively controlled car, the U.S. space station, the unmanned platform, and commercial space ventures are all real life examples of a few explorations now being undertaken on earth and space - requiring a greater dependence by people on machines. Complete shop floor automation - a "lights out" plant may be unrealistic to many but automating and integrating the engineering and manufacturing process, where it makes sense from a cost/benefit stand point, are certainly viable undertakings.

Hence, the objective of the Third International Conference on CAD/CAM, Robotics and Factories of the Future (FOF) is to bring together researchers and practitioners from government, industries and academia interested in the multi-disciplinary and inter-organizational productivity aspects of advanced manufacturing systems utilizing CAD/CAM, CAE, CIM, Parametric Technology, AI, Robotics, AGV technology, etc.. It also addresses productivity enhancement issues of other hybrid automated systems that combine machine skills and human intelligence in both manufacturing (aerospace, automotive, civil, electrical, mechanical, industrial, computer, chemical, etc.) and in non-manufacturing (such as forestry, mining, service and leisure, process industry, medicine and rehabilitation) areas of application. Such an exchange is expected to significantly contribute to a better understanding of the available technology, its potential opportunities and challenges, and how it can be exploited to foster the changing needs of the industries and the marketplace.

Conference Scope

The conference included the following areas of active research and application:

CAED: CAD, CAT, FEM, Kinematics, Dynamics, Simulation, Analysis, Computer Graphics, Off-line Programming

CIM: CAD/CAM, CNC/DNC, FMS, AGV, Integration of CNC, Interactions between Robotics, Control, Vision, AI, Machine Intelligence, and other Automation Equipments, and Communications Standards

Design/Build Automation: Parametric Programming, Design, Sensitivity, Optimization, Variational Geometry, Generic Modeling, Identification, Design Automation, Value Engineering,, Art to Part, Quality, Cost & Producibility

Knowledge Automation: Artificial Intelligence, Expert Systems

Robotics: Mechanical Design, Control, Trajectory Planning, Mobility, End Effecters, Maintenance, Sensory Devices, Work Cells, Applications, Testing and Standardization

Factory of the Future: Planning of Automation, Management, Organization, Accounting, Plant Design, Informative Systems, Productivity Issues, Socioeconomic Issues, Education, Seminars and Training.

Conference Theme

The theme of the 3rd International Conference was:

C4 (CAD/CAM/CAE/CIM) Integration, Robotics, and Factory Automation for improved productivity and cost containment.

Conference Organization

Committee Chairpersons

Conference General Chairperson: Dr. Biren Prasad,
Electronic Data Systems, GM, USA

Program Chairpersons: Dr. Suren N. Dwivedi,
UWV, USA ; William R. Tanner,
Cresap Manufacturing Cons., USA ;
Doug Owen, EDS, USA

Technical Chairpersons: Rakesh Mahajan,
Deneb Robotics, Inc., USA ;
Dr. Jean M. Mallan, EDS, USA

International Chairpersons: Dr. Ario Romiti,
Politechnico di Torino, ITALY ;
Dr. Marcel Staroswiecki, Universite De Lille,
FRANCE ; Dr. Jon Trevelyan,
Computational Mechanics Institute, UK

Panel Session Chairpersons: Dr. Frank Bliss,
EDS, USA ; Dr. Subra Ganesan,
Oakland University, USA

Workshops Chairperson: Dr. Pradeep K. Khosla,
Carnegie Mellon University, USA

Video/Tech Display Chairperson: Dr. Addagatla
J. G. Babu, University of South Florida, USA

Student Session Chairperson: Dr. Hamid R. Parsaei,
University of Louisville, USA

Exhibits Chairpersons: Jon Keith Parmentier,
Tektronix Inc., USA ; Forrest D.
Brummett, GM, USA

Receptions Chairperson: Umesh B. Rohatgi, Charles
S. Davis Associates Inc., USA ;
Dr. Bhagwan D. Dashairya, Inventors Council of
Michigan, Ann Arbor, MI, USA

Administration Chairperson: Dr. Prakash C.
Shrivastava, GM, USA

Conference Directory: Dr. Yogi Anand, Consultant,
Rochester Hills, MI, USA

Committees' Roster

INTERNATIONAL ORGANIZING COMMITTEE

BELGIUM

M. Becquet (Brussels)

R. Gobin (Leuven)

J. Peters (Leuven)

R. Snoeys (Leuven)

H. Van Brussel (Leuven)

P Vanherck (Leuven)

BRAZIL

Maria Emilia Camargo (Santa Maria)

Edger Pereira (Porto Alegre)

CANADA

B. Manas Das (Calgary)

Mark B. Zaremba (Hull)

DENMARK

Finn Fabricius (Lyngby)

FRANCE

Bourjault Alain (Besancon)

Phillipe Pract (Besancon)

Marcel Staroswiecki (Villeneuve-D'Ascq)

Claude Viebet (Evry)

INDIA

C. Amarnath (Bombay)

P. C. Pandey (Roorkee)

Rakesh Sagar (Delhi)

V. Singh (Varanasi)

N. Viswanadham (Bangalore)

ITALY

Ario Romiti (Torino)

JAPAN

Yoshiaki Ichikawa (Hitachi)

T. Yamashita (Tobata)

NETHERLANDS

J. A. M. Willenborg (Utrecht)

PEOPLES REPUBLIC OF CHINA

Qixian Zhang (Beijing)

POLAND

A. Morecki (Warsaw)

REPUBLIC OF CHINA

Shui-Shong Lu (Taipei)

ROMANIA

Voicu N. Chioreanu (Sighetu Marmatiei)

Mircea Ivanescu (Craiova)

SPAIN

R. Ceres (Madrid)

THAILAND

R. Sadananda (Bangkok)

UNITED KINGDOM

John Billingsley (Portsmouth)

Carlos A. Brebbia (Southampton)

M. A. Dorgham (Milton Keynes)

David G. Hughes (Plymouth)

David Paul Stoten (Bristol)

Letter from the President, ISPE

Dear Participants and Guests;

1987-1988 was the best and the most fruitful year in the history of ISPE. With your continued support and co-operation, ISPE has seen considerable growth and popularity. You will agree that our focus is very much mainstream and activities are clearly aimed towards bringing all the pertinent issues found in technological, business, socio-economic, and organizational horizons for discussion and resolution.

After successful sponsorship of three conferences in the USA, ISPE is now sponsoring the Fourth International Conference at I.I.T. Delhi, India during December 19-22, 1989. I hope, with your active participation and support, the fourth conference is bound to be a success.

We would like you to know that your continued technical input, written to share constructive ideas and innovative development strategies have been our backbone. your involvement has been the key to our success but our continued growth requires more efforts. The society is constantly in need of creative ideas and experienced hands. So far, we have been carrying out the responsibilities with sustained contributions from a limited number of members. Now, we are requesting your cooperation and help.

With this letter, I extend a personal invitation to each of you to come up with fresh ideas and new ways of thinking - a partnership that can strengthen ISPE technical and financial foundations so that we could be more aggressive in promoting yours interests and improving the quality of life to which ISPE stands.

With good wishes,

Dr. Suren N. Dwivedi
West Virginia University
Morgantown, West Virginia
USA

ISPE Conference Mission

ISPE was founded in 1984 with the goal to accelerate the international exchange of ideas and scientific knowledge with absolutely no barriers of disciplines or fields of technological applications. The main objective of ISPE is to foster cross-fertilization of technology, strategy and 4M resources (manpower, machine, money and management) to enhance productivity - to increase profitability and competitiveness, and thereby improve the quality of life on land, sea, air and space. One of the aims of the Society is to provide opportunities for contact between members through national and international conferences, seminars, training courses and workshops. The Society also aims to create a channel of communication between academic researchers, entrepreneurs, industrial users and corporate managers.

ISPE embraces both the traditional and non-traditional fields of engineering, manufacturing and plant automation, all areas of computer technologies, strategic planning, business and control. Equal emphasis is being placed on the cross-fertilization of emerging technologies and effective utilization of the above 4M resources.

Acknowledgements

The Third International Conference on CAD/CAM, Robotics and Facto-ries of the Future (CARS & FOF '88) was hosted by the International Society for Productivity Enhancement (ISPE) and was endorsed by more than 18 societies, associations and international organizations. The conference was held in Southfield, Michigan at Southfield Hilton Hotel during August 14-17, 1988. Over 450 people from 12 foreign countries attended. People from industries, universities, and government were all represented. Over 250 technical presentations organized into 11 forums (panels), 61 specialty sessions, 3 plenary sessions and 4 workshops were conducted during the four days program. Six major symposia were concurrently held.

I wish to acknowledge with many thanks the contributions of all the authors who presented their work at the conference and submitted the manuscripts for publication. It is also my pleasure to acknowledge the role of keynote, banquet, and plenary sessions speakers whose contributions added greatly to the success of the conference. My sincere thanks to all sessions chairmen and sessions organizers. I believe that the series of the International Conferences on CAD/CAM, Robotics and Factories of the 'Future which emphasizes on cross-fertilization of technology, strategy and 4M resources (manpower, machine, money and management) will have a major impact on the correct use of productivity means - to increase profitability and competitive-ness, and thereby improve the quality of life on land, sea, air and space.

I acknowledge with gratitude the help and the guidance received from the various organizing committees. I also wish to extend my gratitude to the sponsoring organizations. Grateful appreciations are due to student volunteers from Oakland University, Wayne State University, University of Detroit and University of Michigan for their enthusias-tic participation and help in organizing this conference. Thanks are also due to all my colleagues, friends, and family members who extend-ed their help in organizing this conference and making it a success. In particular, I acknowledge the help and cooperation extended by Electronic Data Systems (EDS) without which this would not have been possible.

I would like to appreciate the excellent work done by Springer-Verlag in publishing this proceedings.

B. Prasad
Conference Chairman and Chief Editor

Conference Proceedings

The papers included in this volume were presented at the Third International Conference on CAD/CAM, Robotics and Factories of the Future (CARS & FOF '88) held in Southfield, Michigan, USA during August 14-17, 1988.

CARS & FOF '88 featured 11 panels, 6 symposia and 4 workshops. The symposia covered six specific themes of productivity tracks (representing foundations of connectivity) in "The Look of the Future in Automated Factories". Under each symposium, several key sessions were planned, focussing both on the opportunities and challenges of new or emerging technologies and the applications. Over 250 papers from over 12 countries covering a wide spectrum of topics were presented in the following six symposia:

Symposium   I: CAED - Product & Process Design

Symposium II: CIM & Manufacturing Automation

Symposium III: Design/Build Automation

Symposium IV: AI & Knowledge Automation

Symposium   V: Robotics & Machine Automation

Symposium VI: Plant Automation & FOF

The conference proceedings are published in three bound volumes by Springer-Verlag. The three Volumes are:

Volume   I: Integration of Design, Analysis and Manufacturing

Volume II: Automation of Design, Analysis and Manufacturing

Volume III: Robotics and Plant Automation

Volume I includes papers from Symposia I and II, Volume II includes papers from Symposia III and IV, and Volume III includes papers from Symposia V and VI. The papers presented in the panel sessions and plenary sessions are distributed to the Volumes based upon the subject matters. The complete list of papers for all volumes are included at the end of each Volume.

Preface

The total integration of the process of designing, manufacturing, and supporting a product from the earliest conceptual phase to the time it is removed from service remains an unfulfilled dream. Yet, when we look at the enormity of the process of integration even for the most simply conceived and manufactured items, we can recognize that substantial progress has been and is being made. It is our nature to be dissatisfied with near term progress, but when we realize how short a time the tools to do that integration have been available, the progress is clearly noteworthy - considering the multitudes of subjects we have to deal with. Most of the integration problems we confront today are multidisciplinary in nature. They require not only the knowledge and experience in a variety of fields but also good cooperation from different disciplined organizations to adequately comprehend and solve such problems. In Volume I we have many examples that reflect the current state of the art in integration of engineering and production processes.

The papers for Volume I have been arranged in a more or less logical order of conceptual design, computer-based modeling, analysis, production, and manufacturing. Chapter I is devoted to those with a design and geometric modeling emphasis; Chapter II is devoted to an engineering analysis emphasis; and Chapter III to a production/manufacturing emphasis. It is a measure of the times and the multidisciplinary environments for the source of the problems that we encountered a significant number of papers which fit into more than a single category. A fictitious paper entitled "CAD-Based Computer Simulation of the Dynamic Behavior of a Flexible Fixture System Using the Finite Element Method for the Purpose of Monitoring Tool Wear to Develop Preventive Maintenance Schedules" only mildly exaggerates the breadth of categories often encountered in a single paper. This also illustrates the multidisciplinary nature of the problem dealt with.

The conference contained six specific themes of productivity tracks. Though a paper may fit into more than one track, it is arranged according to its emphasis on a subject matter. In Volume I, the various topics are arranged in a manner that, while providing bits and pieces of current research emphasis, reflects their relation to the larger picture - the integration of design, analysis and manufacturing. With such arrangement, the editors have provided a certain amount of direction and flow as the reader proceeds from paper to paper.

Contents

CHAPTER I: Integrated Design and Geometric Modeling

Invited Lectures

Keynote Speech:
  Eric Mittelstadt,
  President and Chief Executive Officer, GMF Robotics
  Auburn Hills, MI, USA
Banquet Speech:
  Senator Carl Levin,
  Chairman, Senate Small Business Sub Committee on
  Innovation, Technology and Productivity,
  US Senate, Washington, DC, USA
Plenary Sessions:
A Case for Computer Integrated Manufacturing
  J. Tracy O'Rourke,
  President and Chief Executive Officer,
  Allen Bradley Co., Rockwell International, Pittsburgh, PA, USA
Future Trends in AI/Robotics - A Pragmatic View
  Randall P. Shumaker,
  Director, Navy Center for Applied Research in AI,
  Washington, DC, USA
Future of Engineering Design Practice
  Kenneth M. Ragsdell,
  Director, Design Productivity Center,
  University of Missouri, Columbia, MO, USA
A New Departure in Programmable Robotic Design
  G.N. Sandor,
  Research Professor and Director, M.E. Design and
  Rotordynamics Labs, University of Florida,
  Gainesville, FL, USA
Cost Management as the Criterion for Integrated Design and
Manufacturing
  Ali Seireg,
  Mechanical Engineering Department, University of Wisconsin,
  Madison, WI, USA
Earth Observing Satellite System
  Gerald A. Soffen,
  Director, NASA Program Planning, Goddard Space Flight Center,
  Greenbelt, MD, USA
Rapid Response to Competition
  Raj Reddy,
  University Professor of Computer Science and Director
  Robotics Institute, Carnegie Mellon University,
  Pittsburgh, PA, USA
Engineering Research Centers - A Vision for the 90's
  Howard Moraff,
  Program Director, Cross-Disciplinary Research,
  National Science Foundation,
  Washington, DC, USA
Robots Beyond the Factory
  W.L. Whittaker,
  Robotic Institute, Carnegie Mellon University,
  Pittsburgh, PA, USA

CHAPTER I:
Integrated Design and Geometric Modeling

Introduction

The papers in Chapter I have a substantial interest in the design process. While many of these papers illustrate the integration of design and analysis, and design and manufacture, others are devoted to more closely defined topics. The chapter begins in Section I.1 with papers on developing and organizing the conceptual design phase. It proceeds in Section I.2 to examine several aspects of integrated CAD/CAM. The next Section, I.3, more specifically treats certain aspects of geometric modeling for display, finite element preprocessing, and numerical control. The chapter continues in Section I.4 on mechanism design and in Section I.5 on several aspects of simulation as a design tool. It concludes in Section I.6 with papers on design for manufacture.

This selection of topics calls attention to the fact that computer-based tools for design are much further advanced for dealing with detailed design functions but remain largely a promise at the conceptual and preliminary design level. It is also apparent that design for manufacture is an emerging field. It will be interesting to compare this volume with the proceedings of future conferences as more attention is paid to both the early design and the later manufacturing phases.

Conceptual Design

Structured Planning: A Method for Developing Design Concepts

C. L. Owen

Design Processes Laboratory
Institute of Design
Illinois Institute of Technology

Summary

Structured Planning is a process devoted to the conceptual part of a design project. It presumes that a product development project is divided into two phases: the development of a comprehensive concept for the product, and the detailed design of its elements. In the Structured Planning phase the goal is the production of a plan thorough enough to become a proper specification for the detailed design phase.

The process involves producing an extensive information base which associates "Functions" which must be performed by the product (or system) with qualitative information shedding insight on what occurs when those Functions are performed. The insights, frequently involving actions of the product's users, suggest speculative ideas (or "Speculations") for how the product can be designed to resolve the problems noted or take advantage of the phenomena observed. Insights and Speculations are documented in the information base as "Design Factors".

Three computer programs are used to create the information base and establish relationships among the Functions: INFORM, RELATN and VTCON. INFORM helps the user to enter information and builds the information base; RELATN establishes links between Functions on the basis of how Speculations support or obstruct their fulfillment; and VTCON creates a hierarchical information structure from the graph of links thus formed.

Developing Information

Artifacts exist in time. They cannot be designed without regard for the way they operate and are used. Any product can be viewed as a system operating with a user or users in different ways that are appropriate for different modes of existence. For effective design, these modes of existence must be identified, the activities that occur in them must be understood, and the Functions that the system must perform (or the user must perform for it) within each activity have to be recognized -- along with insights about how they may be accommodated. In Structured Planning, this information is gathered as "Functions", "Design Factors" (insights) and "Speculations" (speculative solution ideas) -- an extended discussion and a sample use are given in [1] and [2].

To assist a design team using this information assembling process, a computer program called INFORM is called upon. Information collected on paper is typed into computer files using prompts and form-fill-out procedures. Once entered, Functions and Design Factors can be retrieved with search processes, read in browsing mode, edited and printed. But the major use of the program is in helping the design team to confirm or modify ideas as they are evolved during the process of developing an overall system concept. In conjunction with two other programs, RELATN and VTCON, the INFORM program establishes links between Functions so that those Functions which ought to be considered together are organized as Related Functions in the information base.

Establishing Relationships among Functions

How should Functions be organized? The controlling factor for whether two Functions are related from the design standpoint is not whether they are "alike", but whether they share potential solutions -- or, put more correctly, whether a significant number of their potential solutions are "of concern" to both Functions. This includes, in a sense, whether they are "unalike" because of their potential solutions. If the designer can consider those Functions together that have a number of potential solutions in common -- that is, a solution for one Function also (in some way) fulfills the requirements of a second Function -- there is an excellent chance that he will be able to fine-tune one or a few solutions so that they will meet the requirements of the Functions under consideration. In the second case, if he can see Functions together that have potential problems of conflict because of some of their potential solutions (a solution for one Function, if accepted for the overall system concept, aggravates or prevents meeting the needs of a second Function), he has the opportunity early-on to select, modify or devise solutions which will avoid the difficulties.

The RELATN program [3], embodies this concept to establish links between Functions based on the Speculations given for a project. How it does this can be illustrated with two diagrams. In the first diagram (Figure 1A), the "bull's-eye" represents a two-part abstract space in which exist all the speculative solutions for a project that in some way are of concern to a Function (consider it Function 1). The diagram has two regions because some of the Speculations help to fulfill the Function (+), and some -- if they are chosen for the project -- will make it difficult to fulfill the Function (-). Both kinds of Speculations are obviously of concern. There are, of course, other Speculations in the collection for the whole project; they are represented in this diagram as being outside the bull's-eye (0), because they have no bearing on Function 1 -- they neither support nor obstruct its fulfillment.

In the diagram of Figure 1B, a similar bull's-eye for Function 2 is combined with that for Function 1. The intersection of the two creates regions with all the possible combinations of the characteristics from the two original bull's-eye diagrams. The pairings of plus, minus and zero values indicate the support or obstruction the Speculations in each region exhibit for the Functions, left position for Function 1, right for Function 2. The five regions of importance are those which contain the "positive" Speculations, in other words, all solutions that might be selected to fulfill either of the two Functions. These five regions determine the degree to which the two Functions are related -- the amount of "interaction" between the two Functions.

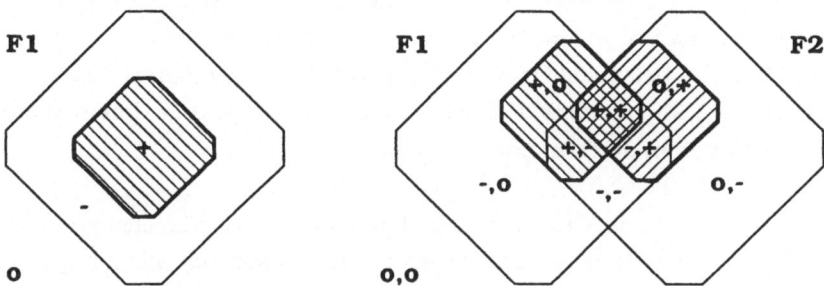

Fig. 1A (left). Solution space for Function 1. Fig. 1B (right). Interaction between Functions 1 and 2 is defined by solutions spaces.

Speculations that fulfill both Functions are in the (+,+) region. These are the "elegant" solutions because each fulfills both Functions at once. The (+,0) and (0,+) regions also contain Speculations that might be used with confidence. Two Speculations, one from each of these regions, would create a total solution for the two-Function system. While not as elegant, this set of choices at least does not introduce difficulties. The two remaining regions, (+,-) and (-,+), are troublesome. A Speculation chosen from either will create a situation in which it will be difficult to successfully fulfill the Function for which the (-) value was given. Based on the effect they have on the two Functions, the five regions are labeled: reinforcement (+,+); independence (+,0) and (0,+); and conflict (+,-) and (-,+).

The concept of interaction can be drawn intuitively from the diagram. Assuming that the reason two Functions should interact (or be linked) is that they have potential solutions of concern in common, the amount of interaction should be proportional to the number of Speculations in the "common" regions of reinforcement and conflict relative to those in the five regions including those and the two independence regions.

None of the other regions is relevant because no Speculation would be chosen from them to fulfill either Function. Thus, in its simplest form, a measure for interaction is the ratio of the number of reinforcing and conflicting Speculations to the sum of those plus the independent Speculations.

In the RELATN program, the interaction concept is extended with three additions. First, instead of a count for the Speculations in a region, the program accepts scaled evaluations for how much a Speculation supports or obstructs fulfillment of a Function. Second, weights for the Speculations are allowed to increase or reduce the role of any Speculation in its effect on the amount of interaction. Weights typically reflect the likelihood that a Speculation will be used in the final solution -- some ideas are more practical than others, or some may be heavily favored by constraints on the project. Finally, a balancing factor is incorporated for the problem that some Functions have more Speculations of concern than others and, thus, interaction may appear differently from different points of view.

The result of operations with the RELATN program is a nondirected graph in which Functions are the vertices. Links between Functions indicate which Functions have enough interaction to warrant being considered together in any conceptual development activity. For many purposes, this level of organization is sufficient, but for most design projects, further structuring is valuable.

Another program, VTCON [4], finds clusters of Functions (vertices) in the graph. The clusters represent primary groupings of Functions. Once the clustering organization is established, the designer can choose a Function at will and know which other Functions are of direct concern. Functions linked to others outside their primary clusters provide the basis for higher level, broader-reaching clustering, and VTCON uses these links to create a "condensation" hierarchy. Higher level clusters are formed from the information on common Functions and links between Functions across clusters. Levels of hierarchy grow with smaller numbers of larger clusters at each succeeding level until the entire graph is recomposed. In form, the hierarchical structure is a semi-lattice rather than a tree because Functions can be in more than one cluster and clusters can be members of more than one higher level cluster. This is the most general hierarchical form and the one most appropriate for design -- where it is natural to find that the same Function is performed in more than one activity.

Using Information -- the Conjectural/Evaluative Process

Generally speaking, two schools of thought exist on the structure of the design process as it proceeds into the stage of generating concepts. In the simplest formulation of the traditional model, the process flows from analysis to synthesis to evaluation. More complex versions break down the three phases into substeps and introduce feedback loops, but the procedural dependence remains intact -- analysis is done before synthesis, and synthesis is done before evaluation.

The conjectural/evaluative model challenges the lockstep relationship of the phases. In this version, ideas are synthesized and evaluated as they take form. The advantage is that mistakes can be detected earlier. In a large project, this may mean avoiding massive redesign. To use this approach, however, there must be an effective means of evaluation. For the process to work, information must be explicit, available in detail and richly cross-related.

Not coincidentally, the Structured Planning process provides the means to take advantage of the conjectural/evaluative approach. First, there must be a way of knowing what to work on: the information base of Functions provides that. Second, there must be a way to know whether an idea is contributing to a good solution: the Design Factors in the information base provide that at a local level; constraints, objectives and other goal statements provide it at a global level. Third, there must be a mechanism to ensure that piecemeal solutions are not "climbing the wrong hill"; the structuring induced with the RELATN and VTCON programs reduces that danger significantly by tying together those Functions which ought to be considered concurrently.

References

1. Owen, C.L., *"Structured Planning: A Computer-Supported Process for the Development of Design Concepts"*, **Design Processes Newsletter** (1987), Vol. 2, No. 1: 1-5 and Vol. 2, No. 2: 1-5, Chicago: Design Processes Laboratory, Institute of Design, Illinois Institute of Technology.

2. Owen, *"Space Station Project -- Systems Design"*, **Design Processes Newsletter** (1986), Vol. 1, No. 4: 2,7-8, Chicago: Design Processes Laboratory, Institute of Design, Illinois Institute of Technology.

3. Sato, K. and Owen, C.L., ., *"A Prestructuring Model for System Arrangement Problems"*, **Design Studies** (April, 1981), Volume 2, Number 2: 67-76.

4. Owen, C.L., *"An Algorithm for the Decomposition of Non-Directed Graphs"*, in **Emerging Methods in Environmental Design and Planning**, 133-146, Moore, G.T. (ed.). Cambridge, Massachusetts: MIT Press 1970.

Hypertext for Organizing Engineering Design

Rollin C. Dix
Illinois Institute of Technology
Chicago, Illinois 60616

WHAT IS HYPERTEXT?

Hypertext is the name given to software systems which provide multi-media, "relational", interactive display to large quantities of diverse information through intuitive, user-controlled search. Vannevar Bush's 1945 Atlantic Monthly article "How We May Think" is recognized for describing the "Memex", the forerunner of an information storing method which provides this type of "associative" access. A memex is a device in which an individual stores all his books, records, and communications, and which is mechanized so that it may be consulted with exceeding speed and flexibility. Microfilm was the media that Bush expected to be used. Buttons and levers would provide rapid access to information by subject, but building trails of "automated button pushes" to associate sequences of text was considered essential.

In 1962 Douglas Engelbart wrote a scholarly report defining "augmentation of human intellect" by adapting Bush's ideas on the computer.* Original work was then performed at SRI International with Advanced Research Projects (ARPA) funding. Today, the U.S. government, especially the Air Force, is said to be the prime user of Engelbart's creation. A commercial version is now available through Tymshare under the name "Augment".

In the early 1960's another pioneer, Ted Nelson, began creation of his hypertext system, "Xanadu" - an electronic data pool of text, sound, graphics and other computer-storable information in multiple versions.**
The Xanadu system is today a multi-user, distributed data and processing, network-based system with capability to attach a variety of "frontends".

----

*Jeffrey S. Young, 'Hypermedia', *Macworld*, (March 1986)

** Roger Gregory, 'Xanadu - Hypertext from the Future', *Dr. Dobb's Journal*, Num. 75 (January 1983)workstations.

COMPUTERS IN ENGINEERING DESIGN

Computer Aided Engineering (CAE) tools development began in the 1960's with "systems engineering" studies of performance through dynamic simulations, finite element stress analysis, aerodynamic flow calculations and many other methods. In the 1970's Computer Aided Design (CAD) began to bring systemization to the production of engineering drawings. At the same time computers took over the operation of machine tools through Computer Aided Manufacturing (CAM).

In the 1980's CAD and CAM are combining to produce geometries that are effectively defined and accurately made. Some systems permit CAE analysis to be applied to the parts, but this is not yet the norm. Computer-based office and factory information systems provide information flow between management and workforce. Very recently, the application of Knowledge-based Expert Systems (KES) to assisting humans in decision making regarding carefully defined tasks of limited scope is growing rapidly.

HYPERTEST IN DESIGN

Were a hypertext CAD system to try to integrate all of the above, the task would be enormous and the resulting system, unlearnable. For proper design of products, it is primarily necessary that design engineers give adequate, informed attention to designing producibility, reliability, maintainability, cost and other factors as well as functionality into the product and that experiments and tests be scheduled throughout the life of the product to measure the success in these aspects of the design.

Therefore, the hypertext CAD system might be simply a checklist for continuously guiding engineers and electronic storage of decisions for intercommunication between them. Accumulated wisdom from many sources should be tranformed into more usable formats and incorporated in the processes.

COMPUTER HARDWARE ADVANCES

The latest computer hardware and software advances provide some very interesting possibilities for bringing this methodology into being and disseminating it to large numbers of engineering groups at reasonable cost. Personal computers with very large storage capacity provided by CD-ROMs or videodiscs have just become available. Graphics in color and with quite high resolution is also available. Processing speeds are increasing by a factor of two every other year. Access to very large quantities of text, graphic, sound and algorithmic information can be provided through hypertext.

APPLE AND IBM HYPERTEXT PRODUCTS

The personal computer user has recently obtained access to hypertext through two products introduced in the fall of 1987. In August, Apple Computer Company began selling Hypercard for the Macintosh computer and including it with every new Macintosh sold. In November, IBM announced at the Paris Auto Show that it is offering Hyper Document for the PS/2 Models 50 and 60 equipped with CD ROM drive, Windows, a mouse, and the 8514 monitor. IBM intends Hyper Document for industrial customers' technical documentation needs. At the show Renault announced it would install the system at its 9,000 dealerships worldwide after its release in January, 1988, in France. Renault wants to replace the current paper and microfiche system containing assembly nomenclature and parts lists with a database on CD ROM. Hyper Document is based on the Guide software of P. J. Brown. *

Although IBM's product is not yet available in this country, for the Apple Hypercard a number of public domain and commercial stacks are already on the market.

APPLE HYPERCARD TECHNICAL DETAILS

A detailed description of Apple's HyperCard* is included here to illustrate
the user's and the developer's tools for providing and for accessing information of the various types provided. HyperCard is not a classical object-oriented programming system, but it does do message passing and it has objects. The five types of objects in HyperCard are: buttons, fields, cards, backgrounds and stacks. The basic unit of information is the card. When the user looks at the Macintosh display, he sees a card with buttons
and fields on it. Sets of cards are called "stacks" which are Macintosh
files. Within a stack a subset of cards may all have the same general appearance because they all have the same "background". Actually,

----

* P. J. Brown, 'Interactive Documentation', *Software-Practice and Experiment,* vol. 16 , n. 3, 291-299 (March 1986)

information belonging to a specific card overlays the background when the card is displayed on the Macintosh screen. Figure 1 shows a card from a hypothetical stack for engineering design. This card displays three text areas and ten buttons. The text areas contain general concepts, specific checklist recommendations, current design decisions, and references. The "New" buttons create new cards for either a new checklist item or new information pertaining to a design decision in a category. The "Find" button does a rapid text matching search on a selected field for other cards and the "Export" button creates a report file. The arrows navigate to nearby cards.

The card and background both contain graphic information (pictures) and textual information (characters). HyperCard contains the McPaint tools for creating bitmapped graphics to display on the cards. Scanned pictures can also be imported into Hypercard as bitmaps. Buttons and fields belong either to cards or to backgrounds. Fields carry text: strings of characters. Buttons are action objects which cause something to happen when clicked by the user with the mouse. Each button has a script (written in "HyperTalk" by the developer) which determines what happens.

The objects in HyperCard have a natural hierarchy: buttons and fields are at the lowest level followed (in order) by card, background, stack, the Home Stack and HyperCard itself. These HyperCard objects communicate with each other, with the user, with HyperCard, and with the rest of the Macintosh equipment, by sending messages. Various objects receive, and act on messages, depending on their position in the hierarchy, the nature of the message, and other conditions. For example, mouse click messages (the fact that the mouse has been clicked) go first to the topmost button or field (if any) under the mouse pointer on the screen. If not handled there, the message goes to the card, the background, the stack, the Home stack, and finally to HyperCard as shown in Figure 2. How an object responds to a message depends on its HyperTalk script. Figure 3 shows the Hypertalk script for the "Find" button in Figure 1. HyperTalk command statements can also be thought of as messages - orders to do some particular thing, like add two numbers or display some other card, or execute an algorithm.

----

*'HyperCard Technical Reference Package, APDA# KMBHTL', Apple Programmer's and Developer's Association, August 11, 1987.

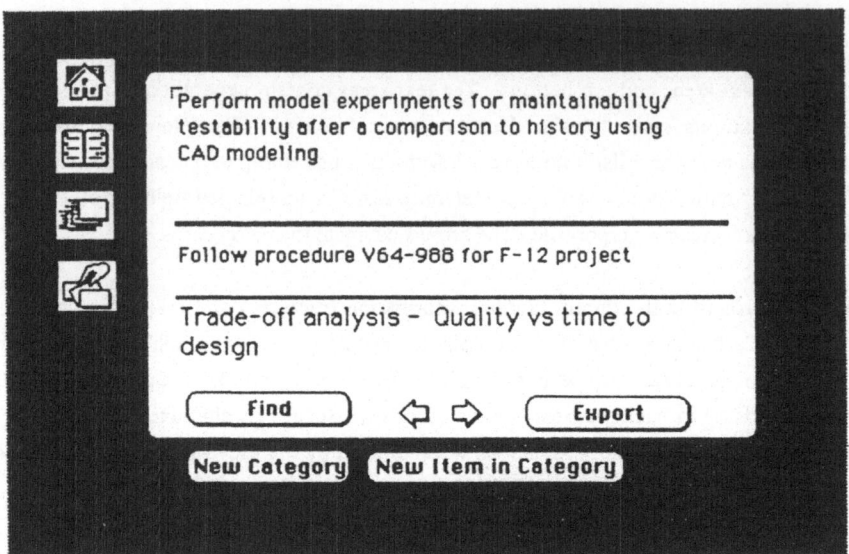

Figure 1. A card from a Hypercard stack for engineering
design. This card displays three text areas and ten buttons.
The text areas contain general concepts, specific checklist
recommendations, design decisions, and references. The "New"
buttons create new cards of either a new design concept
category or new item in a category. The "Find" button does a
rapid text matching search on all cards in the stack, the
"Export" button creates a report file, and the arrows
navigate to nearby cards. The left hand column buttons
either navigate, sort, or display cards. The action to be
performed when a button is clicked, is indicated by Hypertalk
script.

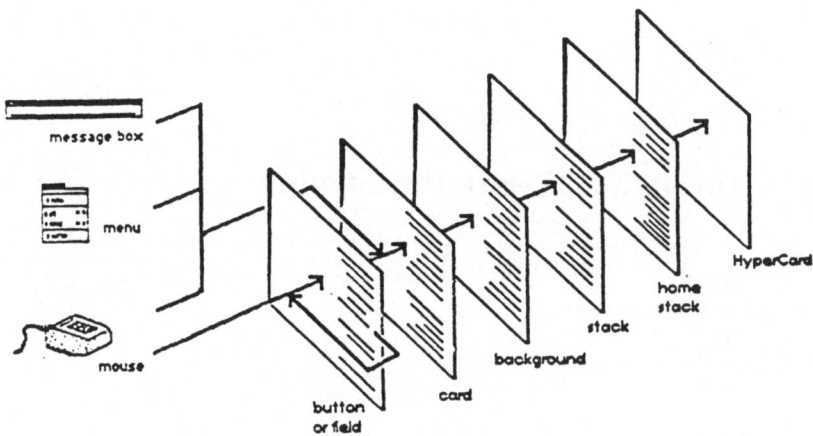

Figure 2. Button clicks are "handled" by Hypertalk script for
the button, or associated with the entire card, or associated
with all cards of a particular layout (background), or
associated with the stack, or the home stack or by Hypercard.
Each level is said to "inherit" the capabilities of the
higher levels.

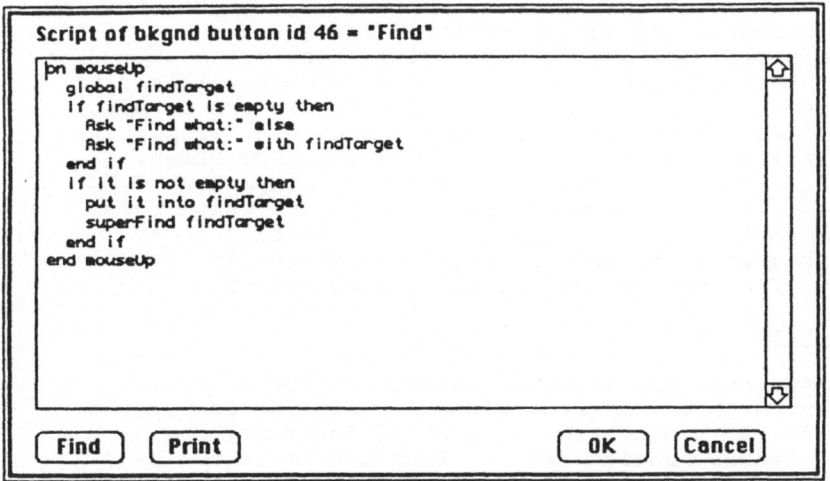

Figure 3. The Hypertalk script of the "Find" button in Figure
1. The "mouseUp" or click-action is recognized by this script
which then sends messages to the user asking for the text to
be matched. If the user's response is not empty, then the
message "superFind" is sent from this button's script to
another script at the stack level, where the search is
actually controlled.

Productivity Through Software Prototyping

Jack Horgan, Ph. D.
VP of Product Development
Aries Technology
600 Suffolk Street
Lowell, MA   01854

Abstract

The development process for mechanical products has been
characterized for centuries by its heavy reliance on
physical prototyping. This approach is lengthy, expensive,
and inflexible in adapting to new materials, innovations in
manufacturing processes, and changing market requirements.
An alternate approach based upon software prototyping
enables concept engineers to predict the behavior of their
design through visualization and simulation and also to
examine many more alternatives. By leveraging productivity
in the conceptual design phase firms can achieve
substantial savings during the verification phase where the
bulk of preproduction time and expense occurs. This paper
describes the needs and benefits of software prototyping
and also describes a desktop system capable of supporting
visualization, design, and analysis (mass properties,
clearance, interference, finite element, kinematic). The
paper stresses those aspects of this systems which make it
suitable to the concept engineer. An interface to a
sterolithography machine for producing three dimensional
hardcopy of conceptual models is also described.

Introduction

Every major news magazine has featured a cover story
seriously questioning the ability of American industry to
compete in the global arena. Much of the furor has
centered around manufacturing. Possible solutions have
been couched in terms of factory automation, jit
(just-in-time) inventory, and quality circles. Significant
progress is now being made in improving various aspects of
manufacturing. More recently, however, there has been an
increasing awareness of the strategic role that engineering
has to play in regaining America's rightful place in world
markets. Products have to be redesigned for automation and
to comply with emerging government regulations in the areas
of safety and environmental protection. Quality and
manufacturability must be designed into products and not be
left as a problem for manufacturing to solve. The
strategic goals of engineering have been extended to
include increased product quality, reduction in time to
market, and rapid design response to changing markets,
materials, and manufacturing processes. Over the next five

years engineering will be expected to develop twice as many product in half the time with half the resource. Engineering organizations must be prepared to meet this major challenge.

In seeking to improve the productivity of engineers it is tempting to investigate the tasks they are performing today and automate them. This would likely result in an electronic pencil for sketching, word processing, and online handbooks. The problem with this approach is that it focuses on local rather than global optimization. Consider the parallel of CAD/CAM as the automation of the drafting function. After nearly a decade of use and considerable claims for increased productivity, CAD/CAM remains an island of automation with limited impact outside of the drafting department. If the productivity of drafting magically became infinite would it lead to better products or significantly reduce the cost and time to design products? The answer is a resounding "No!". We need to step back and look at the overall process to discover where potential leverage exists.

MCAE Philosophy

The fundamental design decisions affecting cost, quality, appearence, manufacturability, maintainability, fit, and function are made during the concept engineering stage, the first step in the product development cycle. Unfortunately, the only way to validate these decisions is through a lengthy and costly process of building and testing physical prototypes. It could be months before a design problem is discovered. We refer to this time period as the knowledge gap, the time when engineering in effect is running blind. When there are problems time-to-market pressures frequently prohibit a significant redesign effort, so that the problem is simply bandaged over and passed along to manufacturing. Worse, of course, is a fundamental design flaw that goes undetected and reaches the marketplace. If engineering is to significantly shrink the time and cost associated with this process, to compress the knowledge gap, then they must be given tools to visualize their designs, predict the behavior of their designs through simulation, and to examine more design alternatives. This is where the opportunity exist to impact the overall process. Using software prototypes to evaluate design concepts before committing substantial time and funds to building and testing a prototype can reap substantial time and cost savings.

Partnership Program

In 1984 and 1985 as part of our Partnership Program [1] we conducted indepth interviews with engineers and engineering managers at over 40 Fortune 500 firms concerning their

requirements for a concept engineering system. These firms
spanned the automotive, aerospace, ag/construction, machine
tool, and consumer products industries. Despite
significant differences in industry segment, location, and
size there was considerable agreement on their basic
needs. They told us that a concept engineering system must
be desktop, easy to use, based upon solid modeling, capable
of supporting full range of analysis capabilities, provide
personal productivity tools, comply with standards, and be
compatible with existing systems. They also confirmed our
model of product development process.

Due to significant advances in relevant computer
technologies, dramatic hardware price reductions, and
increasing computer literacy among graduating engineers,
the development of such a system is now possible.

System Overview

The core of the system is a hybrid solid modeler [2]. A
facted boundary representation is used for rapid display of
graphics and a precise CSG representation is maintained in
support of certain applications and for communication of
boundary and surface data to downstream application on
foreign systems. Solid Modeling obviously provides
dramatic visualization capabilities in the form of shaded
images and hidden line removed views. These images are
important not only when styling and asthetics are involved
but more importantly in communicating design concepts and
problems to experts in the fields of manufacturability,
maintainability, and reliability. Using these images
consultants can quickly and easily grasp the problem to
solve without pouring over engineering drawings. This is a
significant step towards simultaneous or coengineering.

The solid modeler quickly calculates mass properties such
as weight, volume, and moments of inertia in any user
specified coordinate system. These calculations are
important in determining overall system stability and are
particularly valuable in design interations when weight
reduction is a major design goal. Interference and
clearance studies are also supported not only by visual
inspection but also through the determination of minimum
distance of separation.

Application Integration

Historically the major obstacle to the acceptance of solid
modeling has been the lack of application integration [3].
CAD/CAM companies have sewn solid modelers onto their
system in such a manner that existing application could
continue to function umodified with data derived from solid
models. The obvious problem is the difficulty of keeping

the two representations in synch. The remainder of this
paper gives a brief overview of three major application [4]
which operate directly on a shaded solid modeling data base
and function in a common user interface environment

The finite element application allows users to apply loads
and retraints directly upon the solid model. Graphical
symbols indicate their type, size, and orientation. A
powerful mesh generation capability greatly simplifies and
accelerates the process of creating nodes and elements.
Only the default element size need be specified. The
operator can optionally constrain the mesh density in areas
of interest. Figure 1 shows an engine valve cover which
took only six minutes of CPU to automatically mesh.
Traditional mapper mesh generation techniques would have
taken days to produce the same finite element model.
Automatic mesh generation enables the average engineer to
perform tasks previously restricted to the fem specialists.

The analysis can be performed on the workstation or on a
computational server in a network environment. The user
can post process the analysis results as deformed geometry
plots, stress contours, or a superposition of both. Figure
2 shows Von Mises stress contours for the engine valve
cover. The results at a given node or element can be
viewed in a pop-up window.

A mechanism application exists for studying the time
response of assemblies composed of rigid parts connected by
joints and allowed to move with respect to one another.
Figure 3 shows a slide of an assembly that transforms
rotary motion to linear reciprocating motion without the
use of a slider in a channel. The system first performs a
test to determine if the pieces can be assembled consistant
with the specified constraints. Then it determines the
motion of the linkage shown in Figure 4 as a superposition
of annimation frames. The user can also display
displacement, velocity, and acceleration vectors as well as
coupler curves.

Cross sectional data extracted from the solid modeler can
be used to drive a new and exciting process called
stereolithography. The system uses a lasar and
photocurable plastics to create plastic models in a matter
of hours. The advantage of this new process versus
conventional machining is the absence of tooling and
fixture leadtime and the limited inputs beyond the solid
modeling data needed to drive the process. No equivalent of
NC part programming is required. The model is built from
the bottom up layer by layer, cross section by cross
section as an elevator lowers the inprocess model in a vat
of photocurable plastic as shown schematically in Figure 5.

20

Conclusion

This paper has identified a significant opportunity for
increasing the productivity of engineering which is
strategically important to the competitiveness and even
possibly the survival of American industry. Industry can
capitalize on this opportunity through the adoption of the
tools and methodology of Mechanical Computer-Aided
Engineering (MCAE), i.e. the use of software prototype
rather than physical prototypes to evaluate design
alternatives. This paper also presents an overview of some
of the applications supported by MCAE.

References

1. Horgan, J.; MCAE: Market Needs Dictate Industry
    Standards. MICAD 88 Proceedings of 7th European
    Conference on CAD/CAM & Computer Graphics, Paris.

2. Rothstein, S.; Geometry and Topology in the Aries
    ConceptStation. Fourth Annual State-of-the-Art
    Conference on Solid Modeling, Boston, 1987.

3. Horgan, J.; Schmidtberg R.; Yerry, M.; Marriage Between
    Solid Modeling and Finite Element Technologies. Fourth
    Annual State-of-the-Art Conference on Solid Modeling,
    Boston, 1987.

4. Horgan J.; MCAE in Action. NCGA '88 Conference
    Proceedings, Anaheim, California.

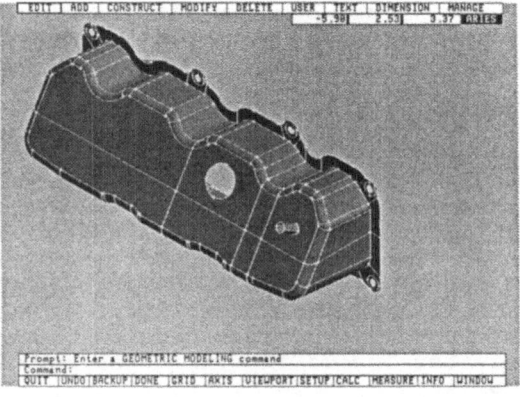

FIGURE 1:
Solid Model of
Engine Valve Cover

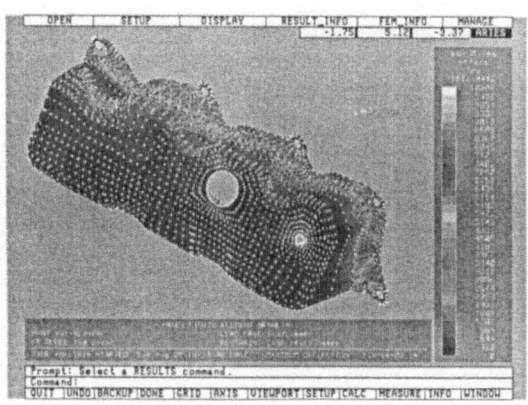

FIGURE 2:
Von Mises Stress Contours
for Engine Valve Cover

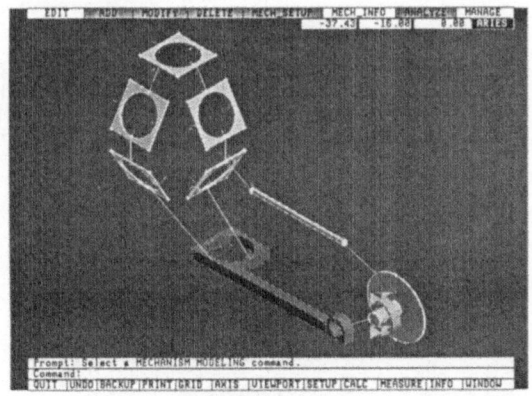

FIGURE 3:
Solid Model of Linkage
Assembly

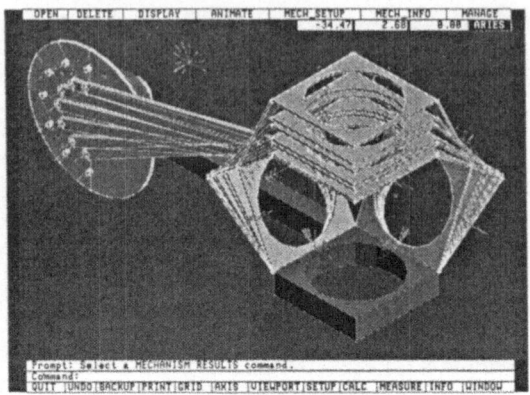

FIGURE 4:
Superposition of
Animation Frames

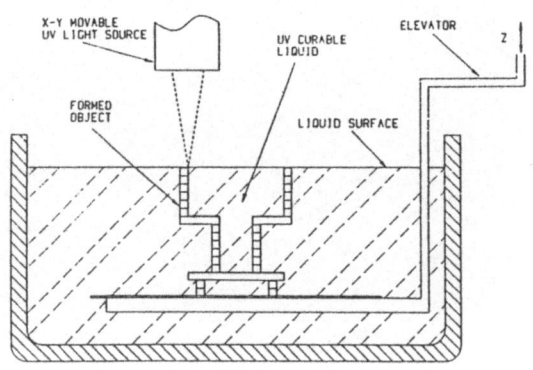

FIGURE 5:
Stereolithography Apparatus

Integrated CAD/CAM

Integration of CAD/CAM Into Engineering Sciences Education

Dr. NIR BERZAK

Mechanical Engineering Department
Rochester Institute of Technology
Rochester, New York

Abstract

As both engineering education and engineering practice increase the extent of CAD/CAM application, engineering science education must be restructured to make the application of CAD/CAM a natural step rather than a new approach. It is shown that a procedure for applying relevant principles to a class of problem and ordering mathematical operations in a general and logical sequence of steps can be defined. The emerged procedure is not a recipe but a description of logical orderly relationship between the defined variables. This systematic approach structured the body of knowledge in the same manner CAD/CAM utilized it. This approach is demonstrated with an example from the Mechanics of Materials course.

Introduction

Historically, the natural sciences and technology have developed along separate courses. The early practice of technology was not based on an understanding of natural laws but on accumulated information generated through experience. During this period, engineering was learned through apprenticeship. The major change in engineering education came with the fusion of engineering knowledge with natural sciences. It triggered the continuing process of formulating engineering sciences based on research and on scientific analyses of engineering problems. The accumulated experience was classified and formulated into various disciplines. The change precipitated an increase in the share that the natural sciences held in engineering education. The industry viewed this process favorably, convinced that this was the only way to assure that future engineers would have the capacity to adapt to changing technology.

analytical approach was emphasized, even though complete
analytical solutions were possible for only very limited,
idealized problems. Numerical solutions were not considered
seriously. With the increase in the speed and accuracy of
computers, numerical solutions began to command more attention
and importance in engineering sciences. At the present time,
most of the engineering calculations are carried out by computer,
yet the overall approach to engineering science education remains
virtually unchanged.

As the industry revolutionizes its practice of engineering
through the rapid assimilation of CAD/CAM applications, and as
the engineering curriculum offers an ever increasing number of
CAD/CAM courses, the centrality of CAD/CAM to engineering today
has become predominated. To ease the student's adaptation to
this new approach to engineering within the rigid time
constraints, it is not enough to offer additional courses in the
subject matter. The curriculum of engineering sciences must be
restructured to emphasize this reformalization of engineering
sciences body of knowledge. The student should be exposed to the
formulation of problems in more general terms through a procedure
resembling the thought process used in CAD/CAM, even though no
computer calculation may be utilized at this stage.

In this paper, two approaches to a problem taken from the
Mechanics of Materials course are presented. One solution was
obtained using the classical approach, while the second solution
employs the new approach. The example illustrates that a
equivalent knowledge of engineering sciences is required to
obtain the correct solutions.

Objective

In engineering schools curriculum, natural and engineering
sciences have a two-fold objective: the mastery of scientific
principles and the successful implementation of these principles
to solve relevant engineering problems. The solutions to these
problems should be reached by employing a procedure which can be
applied effectively and efficiently to a particular class of
problems. The procedure should include problem definition,
analysis, mathematical solution of the analysis, and evaluation
of the results. The mathematical operations used to obtain the

results rather than a subject by itself. In fact, at a later stage, this part of the procedure will be replaced by a computer code. This approach will eliminate the need for intuition, insight, or ingenuity in solving engineering problems.

Methodology

The objective can be attained by either of two methods: 1) by studying the scientific principles and then their applications, which is the standard method in natural science education, or 2) by studying a class of problems with their associated scientific principles, as is done in engineering science education. In the second approach, it is important to adequately define the class of problem and relevant scientific and engineering principles involved. The analysis of both the problem and the solution should evaluate the effect of the various principles qualitatively and quantitatively, and be arranged in a logical, ordered sequence of steps. It is desirable to describe this procedure in the most general possible form in order to facilitate the eventual substitution by a CAD program.

Implementation

The above Objective and Methodology can be implemented in engineering science courses with minimal effort and expense; it is, in essence, a change in philosophy. In the following example, the method was utilized to approach a problem taken from the Mechanics of Materials course. In essence, mechanics of Materials utilizes the principles of statics, strength of materials, and material engineering. The design objective is to select, for a structure or an element, the appropriate dimensions and material to withstand the external forces and moments which will be applied. In this class of problem, Newton's first law is employed to calculate the forces in each section, following which the various stresses in the selected sections of the structure are calculated and compared to the allowable stresses of the selected material. This paper suggests that this logical and ordered sequence of steps toward the solution be maintained. Most of the computer codes which calculate stress use this same sequence of steps. In the long run, it is very beneficial for the student to use this sequence of steps, since it will always lead him to the solution and will enable him to readily utilize carried out.

Example

The following example [1] is a classical problem in the Mechanics of Materials course.

For a given geometry with ultimate stresses ($_u$, $_u$) and a desired factor of safety, the maximum force, P, should be calculated for the structure in Fig. 1.

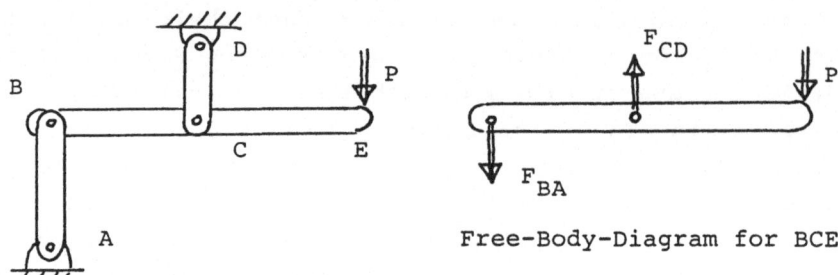

Fig.1. The analyzed structure.

In the classical approach, the maximum allowable force, P, is found by replacing the pin D or A with the maximum allowable force, determined by the ultimate shear stress and the associated factor of safety. The advantages of this approach are that P is obtained by direct calculation without the need to solve a set of algebraic equations and that there is an absence of inequalities in the procedure.

In the suggested approach, the engineering method of solution will always start from the static equilibrium; in other words, satisfying the Newton's first law. In this example, the equilibrium condition for the Free-Body-Diagram of link BCE will result in two equations with three variables F_{BA}, F_{CD}, and P. At this point in the teaching process, it is important to review a dependent solution since there are more variables than equations.

As presented by many CAD codes for static equilibrium and structural design, the input should include, in addition to the geometry, sufficient known variables, while leaving the choice of the unknown variables to the designer. In this example, the forces, F_{BA} and F_{CD}, can be expressed as a function of P.

The second step of the solution is the calculation of stresses.

stresses in the links, CD and BA, are calculated. Since the forces are expressed as a function of P the stresses will also be expressed this way.

The third step is the comparison of the calculated stresses with the allowable stresses. This step produces a set of inequalities based on the requirement that the calculated stresses not exceed the allowable stresses. In this example, four inequalities were derived. The lowest allowable P is determined by the inequality of the shear stress at pin C.

Finally, as the steps of the solution are laid out, a discussion of the method for improving the design by increasing the allowable P should take place. It is obvious that the shear stress at pin C is the limiting factor and that modifying the design can increase the limit.

Conclusion

The classical approach which relies on obtaining the solutions by hand calculation searches for the simplest possible mathematical procedures. It is generally true that the classical solution employs simpler mathematical procedures. It may appear that solving the given example by the new approach was unnecessarily complicated. However, bearing in mind that students are going to solve a limited number of problems, and based on these experiences will develop a general engineering approach to problem solving, the benefit becomes obvious. In the work environment, the calculation will be carried out by computer codes. With this more general approach to problem solving, the student will concentrate on evaluating the results and improving the design by proper modification rather than being obstructed by the mathematical operations required.

References

1. Beer, F. P., Johnston, E. R., "Mechanics of Materials", McGraw-Hill, Inc., New York, 1981.

From Systems that Interface to Systems that Integrate CAD/CAM for Computer Integrated Product Realization

Charles A. Fritsch
Advanced Design Technology Group
AT&T Bell Laboratories
Columbus, Ohio

ABSTRACT

Industry is at the beginning of a second generation of computer aided design and manufacturing systems for computer aided product realization. The first generation linked "islands of automation" through systems that built interfaces to these islands scattered throughout the corporation. Additional efforts strove to unite the enterprise at a higher level by providing a common repository of information so that product design would "data-drive" manufacturing.[1] These views recognize an enterprise as being composed of inseparable parts united by a common mission needing access to consistent data.

The implementation of enterprise integration based on a common repository is difficult when design and manufacturing divisions are geographically separate. A half-step to this level of integration has been the building of systems that interface activities along the lines of organizational structures. This level views the customers for CAD and CAM systems as particular divisions of the corporation such as design engineering, drafting, manufacturing engineering, and the factory.

The second generation is computer integration that views the corporation as the customer. Instead of CAD and CAM systems, it speaks of a design (information) environment that feeds shop floor (execution) processes. Projects are managed and controlled not through the definition of data at organizational interfaces but rather through the precedence related activities in a product realization process.

---

1. Mike Kutcher, et al.,"Data-Driven Automation", IEEE Spectrum, May 1983.

In the building of such a second generation computer integrated product realization system, several major issues must be addressed:

- The demand for customized products in a short life cycle requires rapid throughput in design and manufacturing.

- The demand for high quality products requires that design and manufacturing errors be eliminated before any products are shipped and that high quality be maintained in the face of design changes.

These factors require a product realization environment that has not only a fast response time but also one that maintains data integrity while allowing the concurrent consideration of design and manufacturing engineering issues.

Here we note that, for such products, the concept of a common repository of information is unworkable in that the introduction of a design or manufacturing change can be efficiently managed only if the effects of such a change are held to a minimum. To address this concern, we introduce the following concepts:

- The generation of information is *precedence controlled*. That is, the process of information generation and maintenance is laid out as related tasks that have been strung together to achieve a particular output. Design changes are controlled through the selective reactivation of only those tasks affected by the change.

- Multiple sequences are set up as *streams of information flows* to allow parallel development to the extent possible.

- An *information generation process* is used to provide a minimally complex data structure. Data consistency is guaranteed in that the generation of each data item is the result of a specific task and only that task.

- In the process, *control flow is separated from data flow* for the management of iterations and changes. Thus, the iterative nature of design is supported in a straight forward way without a loss of efficiency.

A cross-organization integrated design environment is required where instabilities due to product design and manufacturing changes that affect information and the specification of parts are dealt with by both development and manufacturing engineers. The concept of concurrent engineering is used where the output of the information environment is a jointly approved product definition/manufacturing package. Manufacturing engineering activities are partitioned to deal with design changes in the information environment. Shop floor interruptions are minimized by loading only stable information on the shop floor processes.

The effects of such a partitioning of activity on present organizational structures will be extensive. Organizational structures that were set up to cause production through a hierarchical decomposition of tasks must be recognized as having limited value. That is, they often do not support laterally networked integration. Additionally, organizations that were chartered to develop interfacing CAD and CAM "systems" will have to be reoriented to support integrated computer aided product realization. Specifically, the data centered approach to product realization must be replaced by a goal centered process based realization. These changes must be made if we are to move from systems that interface to systems that integrate.

CAD/CAM: A Microcomputer-Based Integrated Approach

S. P. KOLLURI and A. A. TSENG

Department of Mechanical Engineering and Mechanics
Drexel University
Philadelphia, Pennsylvania, USA

Summary

A microcomputer based system has been developed to integrate the functions of design and manufacturing. The integrated approach facilitates smooth flow of information between all portions of the Computer-Aided Design and Manufacturing (CAD/CAM) system. This system uses graphics based software to strengthen the integrated approach by permitting efficient programming and allowing convenient access to product and tool information. The graphics based approach can not only be used for CNC programming but also enhance the manufacturing activities. As an example, the software system has been used to simulate some advanced controlling features. These features are especially useful to machine shops to upgrade their basic controllers without modifying the existing hardware.

Introduction

The application of microcomputers in design and manufacturing constitutes the most significant opportunity for substantial gains in industry, such as higher productivity, better quality, and lower cost [1]. Advances in microcomputer hardware and software have made it easier to link a Computer-Aided Design (CAD) system with the rest of the manufacturing operations, so that product design and manufacturing can be integrated into a single system [2]. A typical integrated Computer-Aided Design and Manufacturing (CAD/CAM) system includes capabilities for interactive engineering design, Computer Numerical Control (CNC) part programming, and process planning.

An integrated system essentially provides links between the different types of modules. Typical modules of a CAD/CAM system for machining include a CAD module, a CAM module, a tool data base, and a CNC machining center. The link between the CAM module and the machining center includes a post-processor and a communication package. The post-processor is required for the conversion of CAM software related codes, such as the tool path, into machine codes that can be interpreted by the machine controller. The communication package transfers the machine codes from the computer to the machine controller. With these modules and interfaces, the system is integrated [3].

This paper describes the development of a microcomputer based approach to the integration of CAD/CAM systems. This system uses graphics based software to strengthen the integrated approach by permitting efficient programming and allowing convenient access to product and tool information. The graphics based approach has not only been applied for CNC programming but also used to enhance the manufacturing activities. In the former, the graphics based CNC programming is explored as a tool for integration where the CNC programs have been generated by employing graphics instead of time-consuming programming statements. In the latter, as an example, the software system has been used to simulate some advanced controlling features. These features are especially useful to machine shops to upgrade their basic controllers without modifying existing hardware. The present approach represents the trend that more and more emphasis will be placed on increasing the capabilities of the microcomputer-based software, because of its flexibility and economics

System Description

The integrated system set up for graphics based CNC programming consists of two software packages, Cad-pack and SmartCAM. These are written for the MS-DOS Operating System. SmartCAM can also be used on Apple compatible microcomputers. The integrated system includes a vertical CNC milling machine with a simple controller having minimum capabilities, as shown in Figure 1, which is linked to the microcomputer.

Figure 1 : The Bridgeport milling center with the Bandit controller

Cad-pack is a drafting package that is a menu-oriented software package. This package makes extensive use of the function keys on the keyboard to access commands. Cad-pack offers the ability of data retrieval and storage and allows the drawings to be created and stored in either format A or B. Format A stores files in the ASCII format and format B stores files in the binary format. Format A is used for the integrated system developed.

SmartCAM is a graphics-based CNC programming package which is broken into modules, menus, commands, and options. SmartCAM is claimed to be the complete package that helps the user from the drawing stage to the generation of CNC code and its communication. SmartCAM allows itself to be linked to many CAD software packages through a module named CAM connection [4].

The controller considered in this study is the Mill Bandit CNC controller, level I, that runs a Bridgeport vertical mill. The Bandit CNC controller was manufactured by Summit Dana and uses memory image format to store commands and other information in its memory [5]. Six digits are allowed for a given entry on this controller. The numerical entries can be made with or without using a decimal point. The Bandit CNC has the capability of data transfer in ASCII format. An ASCII/memory image conversion occurs internally when input/output is performed. Consequently, the CNC code from the microcomputer is transferred to the Bandit CNC as ASCII code; the CNC controller upon receiving it converts the ASCII code into its own format [5].

Figure 2 : The part frame that is machined.

<u>System Implementation</u>

The part produced by this system is shown in Figure 2. Cad-pack is designed to generate two-dimensional engineering drawings of moderate complexity and to review and modify drawings created by other CAD systems. The drawing of the part is completed using the Cad-pack software. Geometric files are transferred between Cad-pack and SmartCAM through a conversion software called 'Cad-pack CAM Connection'.

SmartCAM allows the user to specify whether to cut inside or outside the geometry, where to begin and where to end cutting, direction of cut, and toolpart position before the geometry is resequenced. Another way to handle geometry transfer is to convert points and full circles to base points, which are used as center points for drilling, boring, tapping, and reaming. All drawings and tool paths are manipulated in the shape module of the SmartCAM system [4]. The drawing of the tool path as seen in the shape module is shown in Figure 3.

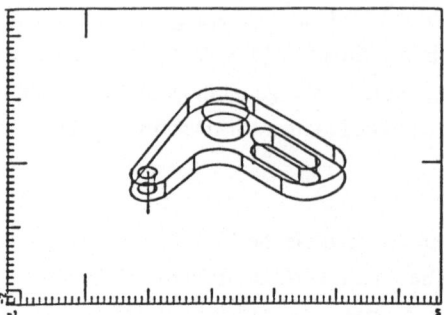

Figure 3 : The tool path in the SmartCAM system.

The machine and template files in the SmartCAM system are used to create the the post processor for the Bandit CNC controller. The machine define file uses specifications of the controller. The machine define file associated with the template file converts the SmartCAM tool path into appropriate CNC code. The CNC code is transferred to the controller using the communication package. The communication package can transmit data in the ASCII or in the EIA format at variable baud rates. The Bandit CNC controller requires the ASCII format. The CNC program can be tested for its validity using system graphics. Tool paths can also be visualized on the monitor to effect changes in tool path.

Simulation of CNC Controller Features

The integrated CAD/CAM system is capable of simulation of advanced controller features at the programming stage. The Bandit CNC controller that is used for this work does not have the more advanced controlling features such as, the tool offset and the canned cycle facilities [5]. This is a very time consuming task since the programmer has to take the tool offsets into account each time he defines a tool path and to alter the depths of the tool profile in the Shape module of the SmartCAM system. To illustrate this problem, consider the part shown in Figure 2. The outer profile has been machined using a 12.7 mm (1/2 in.) two-flute end milling cutter and the inner slot has been machined using a 6.35 mm (1/4 in.) end milling cutter. The drawing of the part is clear if the depths of the profiles are exactly as specified. If the tools are of different lengths (as they always are) and if the machine controller has not the tool offset facility, then the profiles appear displaced as shown in Figure 4. Also, if the controller has no canned cycle facility, then the drilling operation required cannot be performed in a single programming step. The mathematical capability and flexibility of the system is used to rectify these two problems.

Tool Offsets   The template file is used to manipulate the CNC code output. In the template

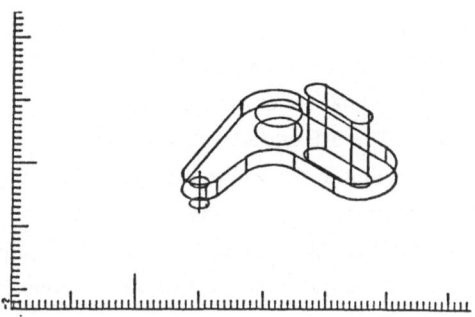

Figure 4 : Tool path without tool offset compensation.

file, as an example, to perform tool offset simulation, the following statement is introduced at the point where the Z-position of the tool is to be output:

#EVAL (#ZPOS = #ZPOS + #TLEN)

where #EVAL (-------) is used for evaluating a parameter. In this case the #ZPOS is evaluated. Let us say, that all tools, irrespective of their lengths, must move in the rapid mode to 25.4 mm (1 in.) above the part for a Z-Clear mode. This is not possible, unless the controller knows the lengths of the tool. This would have been an easy task, if the controller itself had a page of tool offsets. But now the post processor itself must simulate the tool offset. Therefore, the above statement indicates that the #ZPOS value must have the Z-Position value assigned in the Shape module. That Z-Position value must be added the length of the tool. The difference in lengths between various tools has been recorded as #TLEN in the job plan (tool data base). The values of #TLEN are picked up from the job plan. The values assigned in #TLEN are the difference in lengths between a standard tool and various other tools which will be used.

The #TLEN values in the job plan module of the SmartCAM system cannot be negative. To perform subtraction operation in in the above statement, if the selected tool is shorter than the standard tool, all negative values of the tool are preceded by 10. That is, if the tool happens to be 1.02 cm shorter than the standard tool, then the value of #TLEN in the job plan will be *10.102*. Therefore the following statement is introduced in the template file.

#IF(#TLEN > 10) <#EVAL (#TLEN = (10 - #TLEN) *10>

The above statement is a test to determine whether the selected tool is shorter or longer than the standard tool. A listing of the software developed can be found in reference 3.

Canned Cycles   A software has been developed for simulation of canned cycles using the template file. A canned cycle is a machine subroutine that groups several movements of the machine into one programming statement. A listing of this software for drilling operations can be found in reference 3. The developed software makes use of statements such as @FXD to address machine movements. The @FXD1 statement invokes a command such as G81 through the sub-statement <Z#ZDPTH #FXD>.

However, the Bandit CNC level I does not have the capability to interpret any of the G80 series of the G-functions. Therefore an alternative is to develop a series of commands combining the effects of G00 and G01s such that G80 series can be simulated. Accordingly, the @FXD1 statement or cycle is modified. The #IF and #EVAL statements explained in the previous section are added used in this section to determine the Z-position of the tool. The @FXD1 performs the following. After the correct Z-Clear and Z-Check have been achieved, the tool is moved to the correct X and Y position. Then, after assigning a feed value, the drill bit is taken to the assigned Z-Depth (based on the calculations made) and is then returned to its Z-Clear position as dictated by the /Z#ZPOS command. It can be seen that even though the CNC controller lacks advanced features, the flexibility of the SmartCAM system has allowed these desirable features be made available to the programmer.

Conclusion

A completely graphics-based, integrated CNC programming system has been developed and used to design and produce parts. The system was successful in reducing lead-time of operation considerably when compared to the traditional CNC operation. Designing the part on the CAD module and transferring it to the CAM module in an integrated environment has been a more efficient process than using these two as stand alone systems. The graphics-based integrated system facilitated programming for simulation. The simulation features, such as tool length offsets and canned cycles, that were not available on the basic Bandit CNC controller have been achieved. Thus the graphics-based system could be used with a basic controller. This result will be specially useful for those shops that have simple controllers. The simulation features have opened up the possibility of providing more emphasis in the development of systems which utilize microcomputers to program for machining.

References
1. Morgan, M. E.: Interactive computer graphics CAD/CAM interfaces to existing design and manufacturing systems. CAD/CAM Integration and Innovation, ed. by K. Taraman, pp. 99-111, Association of Computer and Automation Systems of SME, 1985.

2. Tseng, A.A.; Kolluri, S.P.; Radhakrishnan, P. : Design and construction of a CNC machining system for hardware and software development. Symposium on Product and Process Design, Vol. I, pp. 43 - 49, Proceedings of Manufacturing International '88, American Society of Mechanical Engineers, 1988.

3. Kolluri, S.P.: A microcomputer approach to integrated CAD/CAM systems. M.S. Thesis, Drexel University, March, 1988.

4. SmartCAM reference manual, pp. 1-6, Point Control Company, Eugene, Oregon, 1987.

5. Bandit CNC reference manual, pp. 1-10, Bendix Corporation, Detroit, MI, 1980.

CAD/CAM Integration for Injection Mold Design

SHUI-SHENG CHERN and ANIL SAIGAL

College of Engineering
Tufts University
Medford, MA 02155

Summary
Improving productivity is essential for competing in today's global market.
Availability of relatively low cost microcomputers with reasonable graphics
and calculation capabilities has revitalized the field of manufacturing
engineering. However, for substantial gains, it is essential to integrate
the concepts of interactive CAD and CAM rather than deal with them
separately. This paper shows how CAD and CAM can be successfully integrated
and implemented for mold design using the currently available hardware and
software.

Introduction

Plastics can be readily formed into many complex shapes to produce parts
which have excellent mechanical, chemical and electrical properties [1].
These parts may be difficult, if not impossible, to manufacture from other
materials.

Injection molding is an inherently efficient process for mass producing
intricate plastic parts to close tolerances at low costs. However, the
state-of-the-art of mold design today is still largely based on the experience and skill of the mold designer using a few rules of thumb or empirical
formulas [2]. This often results in large lead time and huge cost of
experimental molds which may have to be finally scrapped. With the increasing usage of plastic parts, particularly in mass production industries
such as automotive and appliances, it becomes essential to provide a
scientific basis for mold design.

The steps encountered in a typical mold-development cycle may include:
part and mold design, mold analysis, mold manufacture, etc. The advent of
CAD/CAE/CAM technology can speed up the part/mold design process. The
designer may try a possible geometry and analyze it to see if it has the
desired characteristics. If not, the geometry can then be altered and the
process repeated. The final version of the computer model is then used to

produce formal drawings, perform manufacturing process planning, generate NC tool path for cutting, and perform off-line robotic programming for assembly.

The successful users of CAD/CAM systems will be those companies that maximize the use of the electronic data base and communicate with many other systems. Integration is the key to the success in today's highly competitive and rapidly changing global market.

This paper illustrates how this integration process can be applied to mold design by using CADDS 4X, a CAD software from Computervision; ANSYS, a general purpose finite element software package from Swanson Analysis Systems, Inc.; EZ-Mill, a CAM package from Bridgeport Machines, Inc.; and a Bridgeport CNC Milling Machine.

Part/Mold Design

Parts can be poor in quality and wastefully produced unless a proper mold is designed. For a given part configuration, a mold designer has to make decisions on such things as the location and the number of gates to be used, the number of cavities and their arrangement in a single mold assembly, the size and location of runners, and the design of mold cooling system.

To illustrate how CAD/CAE/CAM integration may work we will consider the design and manufacture of a cam mold as an example. A cam is a mechanism device for transforming one motion into another [3]. This mechanical element has a curved or grooved surface which mates with and imparts motion to a follower. The motion of the cam (usually rotation) is transformed into an oscillation, translation or both. Practical applications for cam mechanisms are numerous including timing for automotive engines.

Cam design is a synthesis process in which one determines the required cam shape to meet a predetermined set of conditions on displacement, velocity and acceleration of the follower. Two important parameters which determine cam shape are follower type and follower displacement. Some commonly used follower types are point, roller, and flat-face. The follower displacement motion may be linear, parabola, harmonic or cycloidal.

Figure 1 shows a typical harmonic follower displacement which is to rise 1.5 inches during the first 180 degrees of the cam rotation and return to the initial position during the last 180 degrees of the cam rotation. The motion is to repeat this pattern with every camshaft revolution. Figure 2 shows the desired disk cam shape for a point follower which meets the displacement requirements. The cam profile is a B-spline curve which has been constructed by using CADDS 4X CAD system [4].

Once the cam shape has been determined, the mold can be designed. Two rectangular blocks of size 6 X 6 X 2 inches are selected as plates for male and female mold respectively. Cavities of 0.5 inch depth are then constructed within each plate by projecting the 2-dimensional cam profile along perpendicular direction. Gate position and runner size are provisionally set. Figure 3 shows the front section view of the mold assembly. Cylindrical cooling channels of radius 0.25 inches are drilled close to the cavity bed for achieving uniform and balanced cooling of part and shortening cycle times. Figure 4 shows the pictorial view of the female mold with cavity and cooling channels.

Mold Analysis
In the injection molding process, hot plastic is forced into a cold mold which can produce high cavity pressure and thermal stresses. A typical injection mold plate is under repeated mechanical and thermal stresses shot after shot which could result in fatigue failure. As a mold designer, it is important to understand the forces, deformation, and cooling behavior which are present during the injection molding process.

The basic analysis principle used here is the finite element method [5]. To conduct a finite element analysis, the mold is first discretized into mesh of small elements of simple shapes. To illustrate how finite element method may be applied to this mold design problem, we will perform the stress, deflection, and temperature analysis for the female mold as shown in Figure 4.

A section, as shown in Figure 5-a, has been sampled for further analysis because of the axi-symmetrical shape of female mold (approximately). Figure 5-b shows a mesh of 115 nodes and 80 elements which is generated based upon the geometrical data available in the CAD database by using CADDS 4X finite element modeling package. This section is assumed to be subjected to pressure of 5000 psi uniformly distributed along cavity wall during material filling and pressurization phase. The bottom face of the mold which is in direct contact with the machine platens is restricted from any deflection in the vertical direction. The temperature of polymer melt is 200°C at the time of fill; heat then flows into the mold through cavity wall and is then carried away by circulating water at 25°C through the cooling channels. The temperature on the other mold boundary surfaces is maintained at 20°C.

The mold is made of steel with Elastic Modulus of 3×10^7 psi, Poisson's ratio of 0.33, and Coefficient of Thermal Expansion 1.5×10^{-5} (1/°C).

Based upon the geometrical data, loading conditions, and material properties for mold, an input file for ANSYS finite element analysis package [6] can be generated. The input file is downloaded to IBM Personal System/2 Model 50 Personal Computer and then processed by ANSYS for obtaining the deformation, stress and temperature patterns within the mold. Figures 6,7, and 8 show the deflection pattern, equivalent stress contour lines, and temperature contour lines respectively. The design and analysis process can be repeated several times by changing the size and location of gate, runner and cooling channels until the mold satisfies the desired specifications.

Mold Manufacturing

Once the mold has been fully designed, the CAD database contains the geo- metrical information required for mold production. A typical computer aided manufacturing process may include process planning, tools and fixtures selection, tool path generation, and computer-controlled machining [7] etc. Based on a workpiece of 6 X 6 X 2 inches, pocketing (by end milling) and drilling may be required for producing the desired mold cavity and cooling channels respectively.

The CAD database was then transferred to the EZ-Mill package for tool path generation. EZ-Mill, a CAM package from Bridgeport Machines, Inc., was used in the MS-DOS Operating System environment on an IBM Personal System/2 Model 50 Personal Computer. EZ-Mill is also available for Apple compatible micro- computers. The EZ-Mill package is a menu driven software package. It allows the user to select various milling, drilling and tapping operations; the direction of the cut; type of cut (inside or outside the geometry) and the tool offset values. Once the tool path is generated, it can be pro- cessed using the POST module to obtain an ASCII text file for the CNC machine controller.

The CNC machine used for machining the mold is a Bridgeport Series I Inter- act 412 Vertical Maching Center with a Heidenheim TNC-151 Controller. The controller has 3100 blocks of storage memory and can be programmed via a plain language dialogue routine (M-code) or in standard G code (ISO) format. The TNC controller is equipped with standard data interface ports for input and output of program data and on-line operation with a host computer. One of the major advantages in integrating CAD and CAM is to obtain a better understanding of the concepts of designing for manufacturability. Figures

9 and 10 show the tool path from the EZ-Mill system and the final mold produced respectively.

Conclusions

An integrated CAD/CAM system has been implemented to design and manufacture mechanical parts. Using the geometry database from the CAD system for CAM helped in reducing the lead time for the manufacturing of the mold. This paper maximizes the use of the electronic database and communication among many systems.

References

1. Tadmor, Z.; Gogos, C.G.: Principles of Polymer Processing. New York,: John Wiley & Sons 1979.

2. Mock, J.N.: Plastics Engineering, January, pp. 15-21, 1978.

3. Levinson, I.J.: Machine Design. Reston Publishing 1978.

4. CADDS 4X Basic Geometry/Graphics Construction References, Computer-vision Corporation, Bedford, 1983.

5. Bathe, K.J.: Finite Element Procedures in Engineering Analysis. Prentice-Hall, Englewood Cliffs, N.J. 1982.

6. ANSYS PC/ED User's Manual, Swanson Analysis Systems, Inc., Houston, PA 1985.

7. Groover, M.P.; Zimmers, E.W. CAD/CAM Computer Aided Design and Manu-facturing, Prentice-Hall, N.J. 1984.

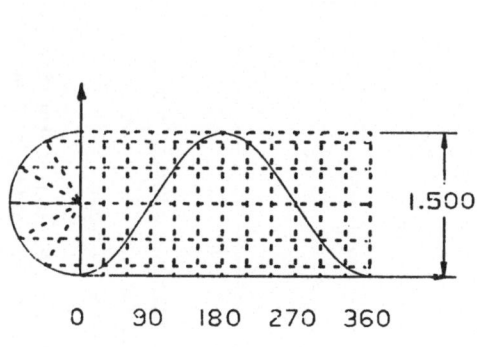

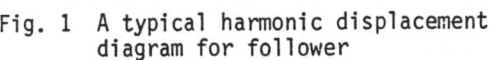

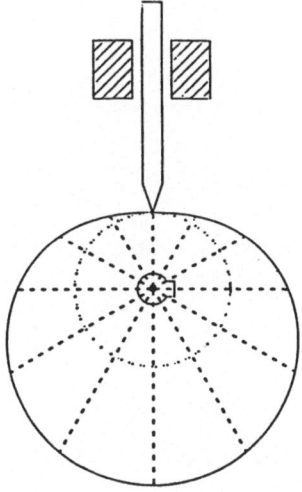

Fig. 1 A typical harmonic displacement diagram for follower

Fig.2 The harmonic cam

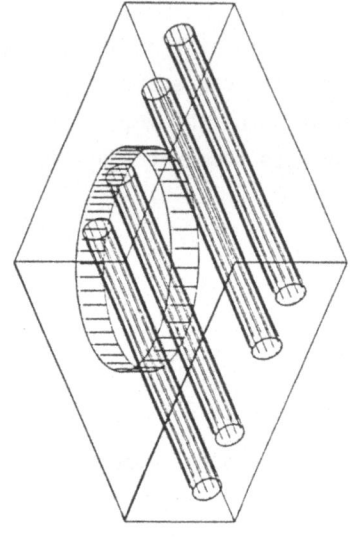

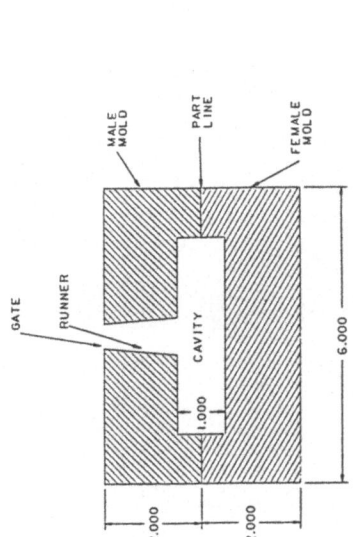

Fig. 3 A sectional view of the mold assembly

Fig.4 Pictorial view of female mold with cooling channels

115 NODES
80 ELEMENTS

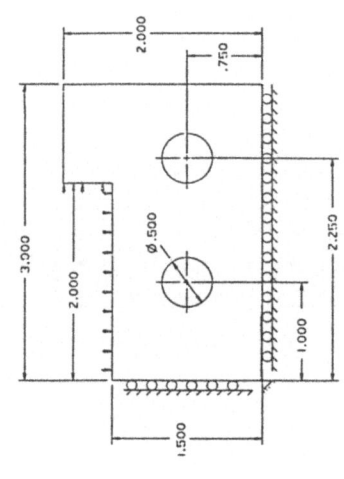

Fig. 5b Finite element mesh of the given cross section

Fig. 5a Cross section of the female mold for finite element analysis

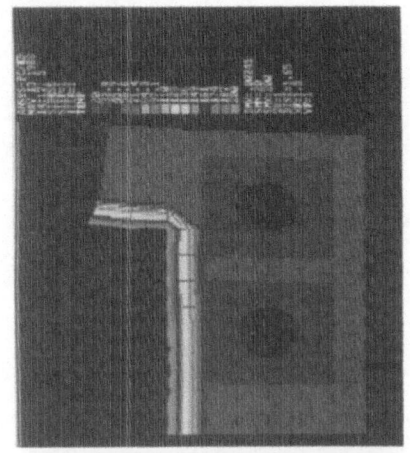

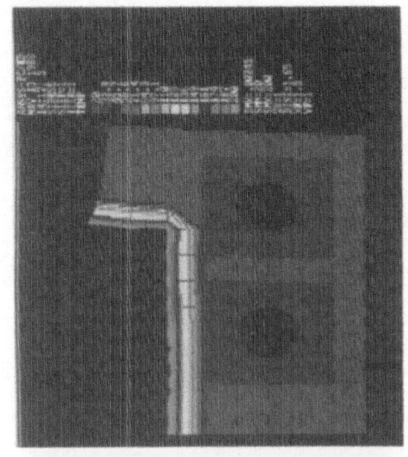

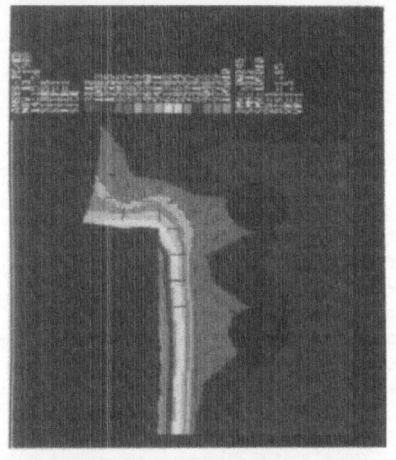

Fig. 6 The deformation pattern of the mold

Fig. 7 The equivalent stress contour lines within the mold

Fig. 8 The temperature distribution within the mold

Fig. 10 The machined mold

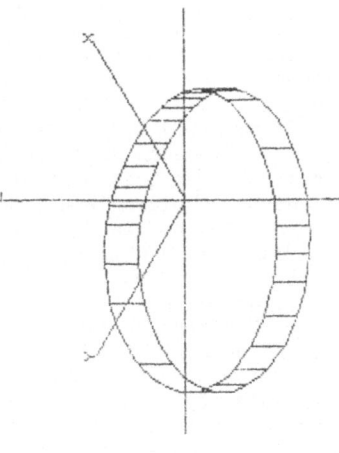

Fig.9 The tool path for pocketing mold cavity

Microcomputer Aided Tolerance Chart
for Squareness and Concentricity

R. C. LIN, E. A. LEHTIHET

Department of Industrial and Management Systems Engineering
The Pennsylvania State University
University Park, PA 16802

INTRODUCTION

Tolerance control in design and manufacturing represents one of the major activities that are undertaken to preserve the product functionality and facilitate manufacture. A computer module for tolerance control can be integrated with other CAD/CAM modules and serve as a post processor for a resident computer aided process planning system (CAPP), help engineering with the validation of tentative process plans and determine the capability required of production equipment. Such a module will also help manufacturing evaluate the significance of each operation appearing in the process sequence and pay particular attention to those critical operations playing dominant roles in the resultant tolerances of a process plan. Farmer and Gladman [1] discussed the requirements of computer based facilities for tolerance analysis and specification. Gadzala [2] and Wade [3] provided detailed treatments for the manual set up and analysis of tolerance charts. Ahluwalia and Karolin [4] described the implementation of a computer based module for the dimensional tolerance chart method treated by Wade. Fainguelernt et al. [5] implemented a microcomputer based module for the generation of a tolerance chart with automatic distribution of tolerances among operations.Most of the literature described above provides valuable modules for the treatment of linear dimensional tolerances in the context of a tolerance chart. In this paper, a modulirized, microcomputer based system for the control of concentricity and perpendicularity tolerances is described. This module interacts with a general purpose computer aided design software to extract design and tolerance information and enable the set up and solution of a tolerance chart procedure to control the two geometric tolerances under consideration.

TOLERANCE CHART FOR CONCENTRICITY

MATOL, Microcomputer Aided Tolerance Control System, is composed of three major modules. The first module is a geometric and dimensional information transformation module which extracts minimum design information from the engineering drawing created by AutoCAD, Autodesk, Inc. The second and the third modules allow users to input proposed process plans and construct the concentricity and perpendicularity tolerance charts. In order to differentiate between the basic geometric data created in cartesian coordinates, dimensional and tolerance specifications, and other auxiliary text information, the CAD design drawing is developed in a structured manner and divided into different layers; each layer will store different CAD information, such as features, horizontal/vertical tolerance specifications, datum locations, and concentricity/perpendicularity specifications.

The first module of MATOL, a post processor for the CAD data file, is then executed to filter all auxiliary drawing information, such as the default value of coordinates and angles, the text font style, line type, and so on. It will extract only required information and translate the dimension and tolerance specification data of each layer into a more comprehensive and compact style which can be directly processed by the other modules.

INFORMATION RETRIEVAL FROM CAD SOFTWARE

After the design drawing has been filtered and translated, the user can execute the concentricity tolerance module, to validate the resultant concentricity of a proposed process plan. The major procedure for creating a concentricity tolerance chart is discussed below.

Step 1: MATOL will generate a simplified side view of the original drawing automatically. This step is a key point in the construction of the concentricity chart. The drawing is distorted for ease of graphical representation.

Step 2: Input the proposed process plan sequentially.

Step 3: When all operations have been entered into the system, MATOL will calculate the resultant concentricity tolerance based on the trace-back method used in linear dimension tolerance charts and generate a temporary concentricity tolerance chart automatically. The basic notation in the concentricity tolerance chart is given below:

[*]: the center line of the cylinder to be measured
[x]: the center line of the datum cylinder based on
[>]: the surface to be machined
[*---*]: the resultant concentricity tolerance

The algorithm used in calculating the resultant concentricity is briefly described below.

Step 3-1: trace-back starts from the right hand side of the resultant tolerance bar which is at the bottom of the tolerance chart. An imaginary line (which will not show on the screen) will be drawn upwards until it hits a target axis [*] of the same operation.

Step 3-2: Move the dashed line to the datum axis [x] of the same operation.

Step 3-3: Repeat step 3-1 and step 3-2 until all operations have been examined.

Step 3-4: Use the left hand side axis [*] of the resultant tolerance bar as the other starting point. Repeat step 3-1 through step 3-3 until the two lines cross each other and fall into the same path (either at [*] or at [x].

Step 4: If two imaginary lines do not ultimately cross and fall into the same path, it is an indication that the proposed process plan can not be implemented. The user may either elect to modify the proposed process plan or start MATOL again; otherwise, the system will stop.

Step 5: Accumulate the concentricity tolerance of all critical operations that have been traced. The addition starts from the crossing point and down to the last critical operation.

Step 6: MATOL will display both resultant concentricity tolerance and the B/P value on the lower part of the screen.

Step 7: A final output list of the concentricity tolerance chart will be generated by executing the report generator module. The valid process plan can also be stored in a process plan data file for interfacing with a CAPP system.

TOLERANCE CHART FOR PERPENDICULARITY

The basic algorithm for constructing perpendicularity tolerance control charts is about the same as that of the concentricity chart. Both use the trace-back method. However, a simplified drawing with 90 degree rotation is generated instead of a side view for the concentricity chart; such transformation will make the horizontal center lines on the original drawing to appear vertical. The algorithm for constructing the perpendicularity tolerance is summarized below:

Step 1: MATOL generates a new drawing which is rotated 90 degrees relative to the original drawing at the same plane. The drawing is also distorted and the coaxial center lines are separated for purpose of clarity.

Step 2: Input the proposed process plan.

Step 3: Apply the trace-back method and repeat exactly the same procedure (step 3-1 to step 3-4) used for the concentricity tolerance chart to calculate the resultant perpendicularity tolerance.

Step 4: Generate the output list of perpendicularity tolerance control chart and save the proposed process plan to interface with CAPP system.

CONCLUSIONS

MÁTOL, Microcomputer Aided Tolerance Control System, has been successfully implemented on a personal computer and provides the process planner with a helpful tool for verifying the process plan with respect to concentricity and perpendicularity tolerances. MATOL is able to interface with AutoCAD and save the dimension and tolerance information in a modulized format. MATOL also provides self-diagnostic functions to detect contradictory process plans. Whenever resultant tolerances cannot be satisfied (tighter or looser than the B/P value), the user can elect to modify the proposed process plan or create a new plan. The final concentricity and perpendicularity chart can be printed out and the valid process plan can be saved and linked with a CAPP system. Besides its efficiency, MATOL provides an effective and user-friendly way to construct tolerance charts.

REFERENCES

1. Farmer, L. E., and Gladman, C. A., "Tolerance Technology - Computer Based Analysis", CIRP Annual Proceedings 1986, V35/1, 1986.

2. Gadzala, J. L., Dimensional Control in Precision Manufacturing, McGraw-Hill Inc., 1959.

3. Wade, O. R., Tolerance Control in Design and Manufacturing, 1967.

4. Ahluwalia, R. S. and Karolin, A. V., "CATC - A Computer Aided Tolerance Control System", Journal of Manufacturing Systems, Vol. 3, No. 2.

5. Fainguelernt, D., Weill, R. and Bourdet, P., "Computer Aided Tolerance and Dimensioning in Process Planning", CIRP Annual Proceedings, 1986, V. 35/1.

6. Foster, L. W., The Application of Geometric Tolerancing Techniques, Addison-Wesley, 1986.

7. Gladman, C. A. and Williams, R. A., "Research in Tolerance Technology - the Geometric Analysis of a Product", The Prod. Engr. Oct. 1974.

8. Luzadder, W. J., Basic Graphics, Prentice-Hall Inc., 1962

9. Spotts, M. F., Dimensioning and Tolerancing for Quantity Production, Prentice-Hall Inc., 1983.

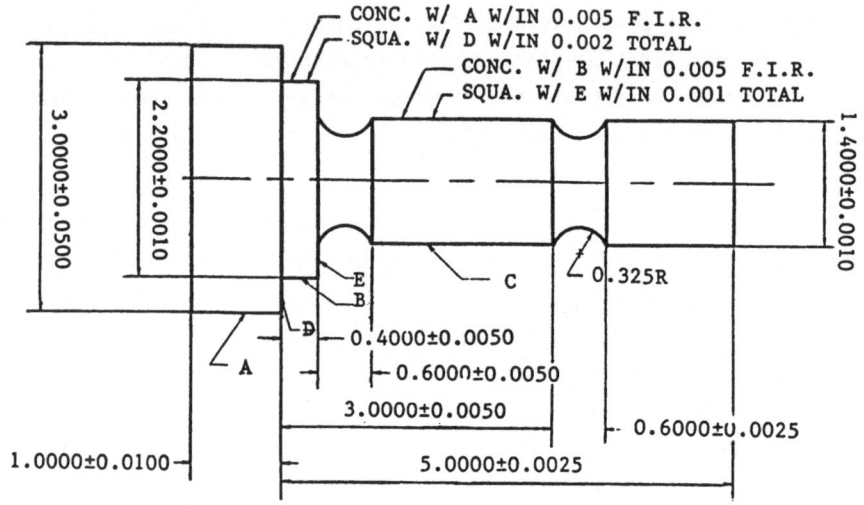

Figure 1: Sample Part

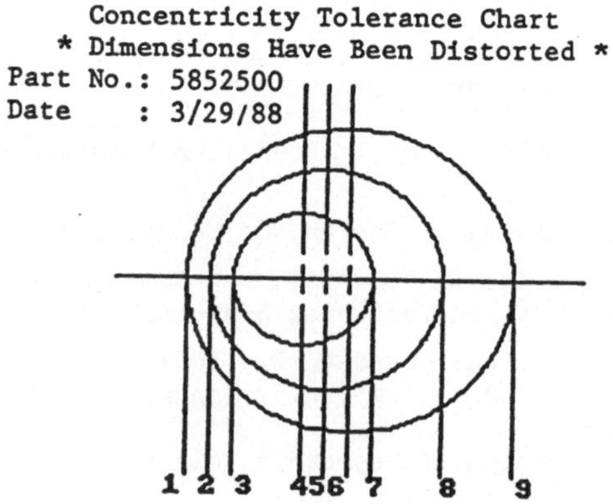

Concentricity Tolerance Chart
* Dimensions Have Been Distorted *
Part No.: 5852500
Date   : 3/29/88

Figure 2: MATOL Part Representation for
Concentricity Tolerance Chart

```
                    **** AUTOCAD DATA CONVERSION ****
                    ------------------------------------

                    DATA CONVERSION COMPLETED !!

        THE DRAWING INFORMATION NOW BEING STORED IN THE FOLLOWING DATA FILES :

                    BASIC DIMENSION     => BODY.DAT
                    HOLE                => HOLE.DAT
                    CENTER LINE         => CNTR.DAT
                    HOR. DIM. & TOL.    => HDIM.DAT
                    VER. DIM. & TOL.    => VDIM.DAT
                    REFERENCE DATUM     => DATU.DAT
                    CONCENTRICITY       => CONC.DAT
                    PAERPENDICULARITY   => SQUA.DAT

            PLEASE INPUT ANY KEY TO GO BACK TO MAIN MENU
```

Figure 3: AUTOCAD File Conversion Screen

```
          >>>>>>        PROPOSED PROCESS PLAN EDITING MENU      <<<<<<
SEQ. OP CODE  MACHINE  OPERATION DES.   MACHINING   CENTER  DATUM    WORK. TOL
                NO.                      SURFACE     LINE    SURFACE   F.I.R.
────────────────────────────────────────────────────────────────────────────
  1    R-TURN   CNC-TR  ROUGH-TRN           7          4       6      0.0500
  2    R-TURN   CNC-TR  ROUGH-TRN           8          5       6      0.0500
  3    F&C      CNC-TR  FACE & CNTR         9          6       5      0.0500
  4    R-GRD    GRD-NC  ROUGH-GRIND         7          4       6      0.0050
  5    R-GRD    GRD-NC  ROUGH-GRIND         8          5       6      0.0050
  6    F-GRD    CNC-GD  FINISH-GRD          8          5       4      0.0030
  7    F-GRD    CNC-GD  FINISH-GRD          9          6       4      0.0030
  8    SF-GRD   CNC-SG  SUP-FIN-GRD         7          4       6      0.0020
  9    SF-GRD   CNC-SG  SUP-FIN-GRD         8          5       6      0.0020

    I-INSERT    E-EDIT    D-DELETE    S-SAVE    Q-QUIT   INPUT (I/E/D/S) : ? S
```

Figure 4: Tentative Process Plan

```
        **** CONCENTRICITY TOLERANCE CHART FOR PART NO. 5852500    ****
 JOB| OP  | MACH.| DESCRIP  | WORKING |1    2    3    4*   5*   6*   7    8    9
 SEQ|CODE | USED |          | DIA.-FIR|

  1|R-TU|CNC-TR|ROUGH-TRN  | 0.0500  |                   *────x───>
  2|R-TU|CNC-TR|ROUGH-TRN  | 0.0500  |                  *──x──>
  3|F&C |CNC-TR|FACE & CNTR| 0.0500  |                x──*──>
  4|R-GR|GRD-NC|ROUGH-GRIND| 0.0050  |              *──────x─────>
  5|R-GR|GRD-NC|ROUGH-GRIND| 0.0050  |                 *──x──>
  6|F-GR|CNC-GD|FINISH-GRD | 0.0030  |          x───────────*─────>
  7|F-GR|CNC-GD|FINISH-GRD | 0.0030  |          x────────*─────>
  8|SF-G|CNC-SG|SUP-FIN-GRD| 0.0020  |             *──x──>
  9|SF-G|CNC-SG|SUP-FIN-GRD| 0.0020  |              *──x──>
B/P F.I.R.=.0050   RES. F.I.R.=.0020   TARGET:          *──*
B/P F.I.R =.0050   RES. F.I.R.=.0040   TARGET:    *────*
INPUT 1) ASSIGN OPTIMAL TOLERANCE 2) CHANGE PROCESS PL7N 3) MAIN MENU ? 1
```

Figure 5: Balanced Concentricity Tolerance Chart

Geometric Modeling

An Interactive CAD System for Styling Design

Mitsuo Inagaki

Toyota Motor Corporation

1, Toyota-cho, Toyota

Aichi, 471 Japan

Masatake Higashi

Toyota Technological Institute

2-12-1, Hisakata, Tempaku-ku, Nagoya

Aichi, 468 Japan

1. Introduction

In order to supply attractive cars to the market timely, automobile manufacturers are continuously improving body engineering processes. In the styling design field, the utilization of CAD systems is progressing at a high pace. However, conventional systems were only used to produce drawings after the body style had been decided on. We have developed and put into practical use a system called "A Styling CAD System". Fig.1 shows that the styling CAD system can assist not only the process of drawings but also modeling work of the entire processes for styling designer's creative activities.

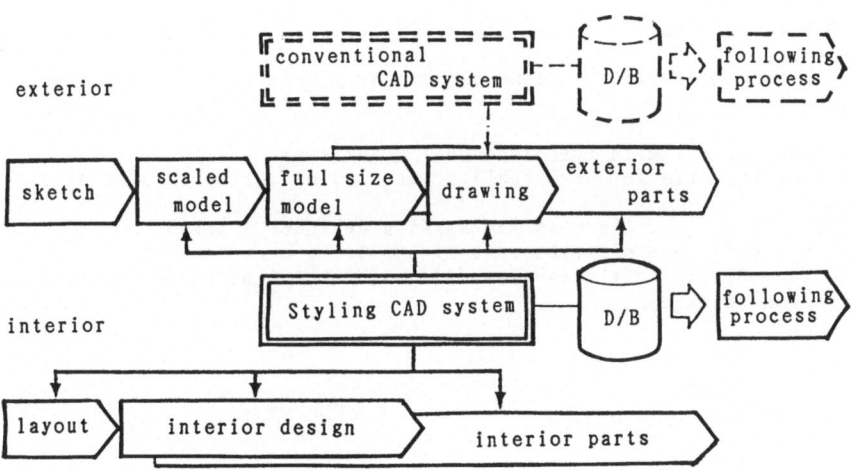

Fig. 1 Styling design process

2. CAD system in the styling design field

There are two objectives of the CAD system.

(1) Improving the efficiency of the modeling work and the design quality.
Fig. 2 shows roles of the CAD system in modeling work along with the conventional method. On a graphic display the styling designer can perform creatine activities rapidly. This allows more design variation and styling refinments resulting in higher quality designs.

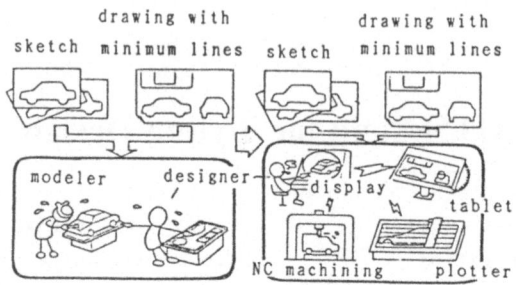

Fig. 2 Modeling by CAD system

(2) Improving the efficiency in the entire body engineering processes by supplying the high quality data of surfaces to the following processes.
It is very important to supply high quality data of surfaces timely to the CAD/CAM systems in the following processes, because structural parts are designed based on them, and stamping dies are designed and manufactured to realize the shape.

3. Technologies established
In order to achieve objectives , capabilities required by styling designers for the CAD system are as follows.
(1) To construct curves and surfaces freely into shapes as they desire.
(2) To evaluate curves and surfaces on a graphic display , based on the evaluation criteria.
(3) To allow them to use a graphic display friendly, from mathematical model.
(4) Quick verifications of three-dimensional models according to the final shapes.
This section describes technologies newly established to achieve the requirements.

3-1 Evaluation criteria for high quality surfaces
Evaluation criteria for high quality surfaces on a graphic display are as follows.
(1) Smooth highlight lines as the styling designer desires.
(2) Smooth curvature distributions of any section curve.
Fig. 3 and Fig. 4 show examples of highlight line check and curvature distribution check.

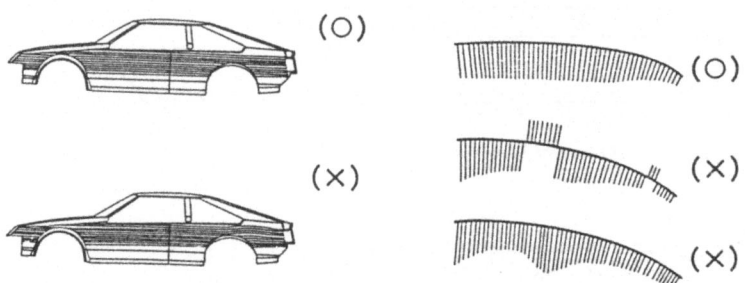

Fig. 3 Highlight line check   Fig. 4 Curvature distribution check

3-2 Establishing categorization of unit surfaces

It is imperative that surfaces reflecting faithfully the styling designer's intention can be created on the graphic display. Therefore, we categolized unit surfaces into three types for surface operation on a graphic display according to shapes after we analized the conventional drawing work and modeling work on a clay model. Fig. 5 shows the three types : namely a cross type, a "H" type, and a rectangular type. We call the section as "a base line", and the boundery as "a guide line". With cross type surfaces the base line moves along one guide line. With "H" type surfaces the base line moves along two guide lines. For rectangular type suefaces a surface with four boundery lines is stretched so that the boundery lines are bent inside.

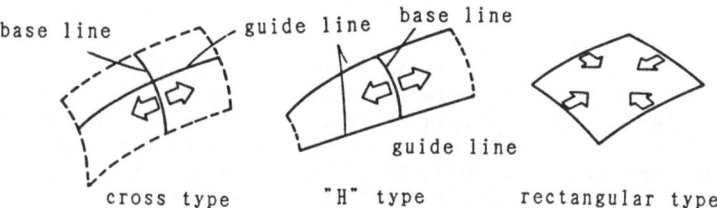

Fig. 5 Categorization of unit surfaces

We have classfied the above three types into more detailed categories according to moving constraints of the base line which would correspond to specific intention of the designer. Fig. 8 shows some examples of unit surfaces with hige light lines' property.

	control method	highlight line
cross type level fix	level	
"H" type one tangent fix		
"H" type cross point fix	fix	

Fig. 6 Example of unit surfaces

3-3 Establishing how to express high quality surfaces

In order to express the unit surfaces , we have adopted Hosaka's equation [1] and developed a method of its application to the unit surfaces. In the Hosaka's equation, B-spline control points are calculated with scale factor k i s from input points in the following equations. Fig. 7 shows the relation of a input point P and a control point Q.

58

$$\frac{Q_{i-1}+x_i(1+x_{i-1})Q_i}{1+x_i(1+x_{i-1})}+\frac{x_i(1+x_{i+1})Q_i+x^2{}_ix_{i+1}Q_{i+1}}{1+x_{i+1}(1+x_i)}$$
$$=(1+x_i)P_i \cdots\cdots\cdots\cdots\cdots\cdots\cdots\cdots\cdots (1)$$

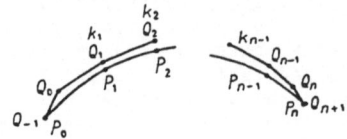

Pi:input point

Qi:control point

ki:ratio of tangent vector at Pi

Fig. 7 Relation of input point and control point

Usually, k i is determined as the ratio of the chord lengths, but to get smooth curveture distribution of the curve we decided k i according to the fifth degree Bezier curve determined by curveture values at the end points. In the case of surfaces , the locus of control points of the base line are represented in the Hosaka's equation. Here the scale factors of coresponding guide lines are made same values.

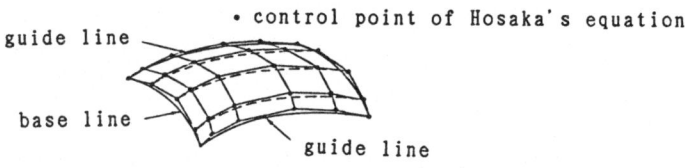

Fig. 8 Expression of surface by Hosaka's equation

In the case of corner surfaces with which both sides are designated, C^1 continuity between unit surfaces are assured by subdivision of patches.

3-4 Development of high speed NC machining method for three-dimensional models

It is indispensable to produce three-dimensional models in order to allow the overall evaluation of shapes created by this CAD system on a graphic display. This modeling work must be achieved within a cycle of the creative work by the CAD system. This means that it should be effective in as short time as possible. In this regard, we have developed appropriate technologie as follows.
· An efficient method to caliculate the cutting tool location.
· The system is capable of machining clay model (soft work material), in order to ease to modify and finish after NC machining.
To get NC tapes quickly and calculated the offdet points without calculation of interference, in this system local shapes which might cause some interferences with the tool are instructed to a computer by an operater, and the computer determines strictly the interference region at the local shapes. On the other hand we have developed new cutting tools and established conditions for NC machining of clay models to get easy finishing of the clay models.

4. Effects of the system applications

Fig. 9 shows examples of mathematical models and Photo. 1 shows an example of NC machining.

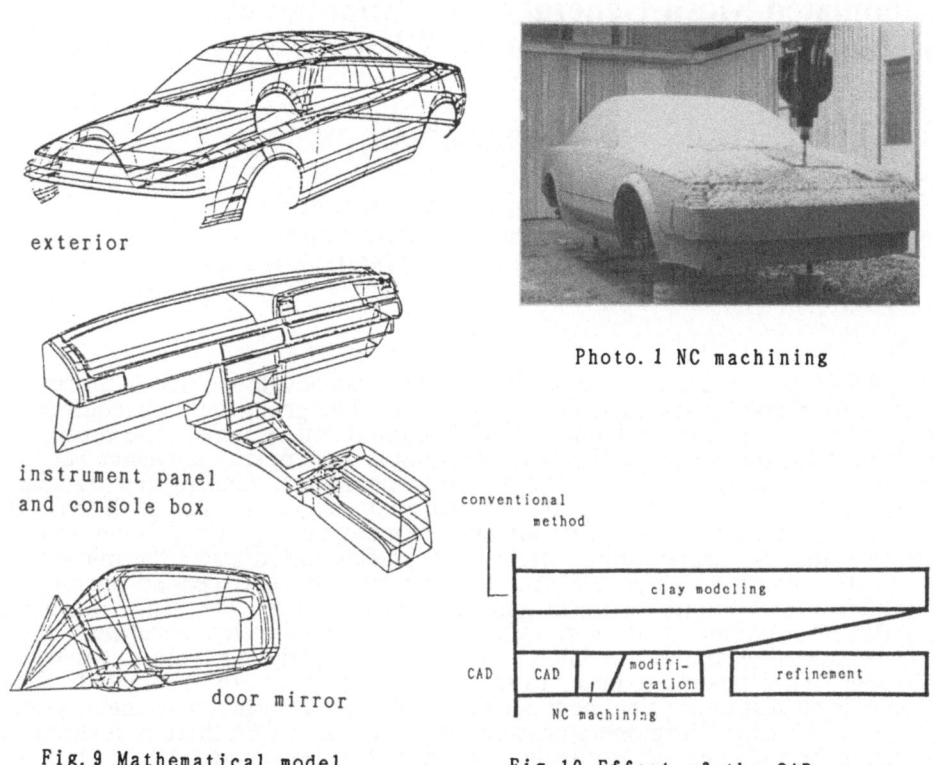

exterior

instrument panel
and console box

door mirror

Fig. 9 Mathematical model

Photo. 1 NC machining

Fig. 10 Effect of the CAD system

(1) Effects on the styling design

The time per cycle required in the modeling work has been reduced to half or less of the time of the conventional method as shown in Fig. 10. This reduction has allowed more design variation and styling refinements resulting in higher quality and improved products.

(2) Effects on the following processes

High quality data of surfaces have been provided to the following processes. It has become possible to use an integrated CAD/CAM system throughout the entire body engneering process. The styling CAD system has had a strong influence to substantially improve the efficiency of it.

Reference

[1] Hosaka, M. and Kimura, F.: A theory and Methods for 3 Dimensinal Free Form Shape Construction , Journal of Information Processing, Vol. 3, No. 3 1980

[2] Higashi, M., Kohzen, I., and Nagasaka, J.: An Interactive CAD System for Construction of Shape with High-quality Surface, CAPE '83, April, 1983.

[3] Nagasaka, J., Higashi, M., Sannokyo, H., Aoyama, N., Ogo, K., and Suzuki, T.: High Speed Machining System for Styling Design, ISATA Sept., 1984

Automated Mesh Generation for Simulations Exhibiting Extreme Geometric Change

JOHN M. SULLIVAN, JR.

Mechanical Engineering Department
Worcester Polytechnic Institute
Worcester, MA 01609

ABSTRACT: An automatic mesh generator was developed that conforms to arbitrarily shaped two dimensional geometries. The grid generator requires no user intervention and can handle multiple component systems. The routine is well suited to dynamic situations experiencing large geometrical change such as in forging, superplasticity or solidification processes. During the transient process the physical shape of the domain may change greatly such that the original topology (ie. node count and element connections) is no longer appropriate. An aspect ratio criterion is specified that dictates the maximum allowable distortion of an element. If this criterion is exceeded during the simulation the automatic mesh generator is invoked. Therein, a uniform equilateral triangular grid overlays the physical domain. Increased resolution can be specified in areas of interest or in areas undergoing large geometrical change. All elements exterior to the physical domain are eliminated. The elements closest to physical boundaries are adjusted to conform to the physical shape, and a smoothing operator is employed to assimilate these adjustments. Interior boundaries of multiple component systems are retained. An interpolation routine based on the previous finite element mesh provides nodal parameter estimates for the new mesh afterwhich the simulation resumes. This dynamic mesh generation strategy virtually eliminates user interaction during the simulation which reduces the actual time required for solution substantially.
    The adaptive finite element mesh generator was applied to several processes illustrating the generic nature of the routine. Herein, the multiple discretizations required during the iterative solution of optimum cooling line placements in a mold are rendered trivial. In this situation the automatic mesh generator was used within the iteration process for the placement of the cooling lines. Increased resolution was specified automatically about the cooling lines.

INTRODUCTION: Numerous computer-aided design packages (CAD) are available for displaying geometric objects. Using today's powerful personal computer (PC) systems, some of these CAD products have become fully operational at the PC level.[1] However, a subsequent operation is required that converts the CAD shape description to a valid numerical description compatible with traditional finite element or boundary element solvers. This usually involves construction of a numerical mesh which is a tedious, time consuming task without some form of automated mesh generation procedures. Grid generation codes exist for specific applications where the component shape is known apriori.[2] However, these techniques are not usually viable for arbitrarily shaped objects. As an additional complication, the initial CAD design may undergo several iterations, each requiring a different mesh, before a preliminary experimental design is finalized.

Recently, investigators have expanded the capabilities of mesh generation routines for arbitrarily shaped objects or geometries.[3-7] The majority of these algorithms use triangle and tetrahedral elements in two and three dimensional space, respectively. These elements are the simplex elements of their respective dimensional domains and have the ability to completely define geometries without overlap or voids. A highly successful triangulation procedure is based on constructing Voronoi regions followed by the Delaunay triangulation.[5,6] This process requires a previously defined set of numerical points and involves the calculation of localized circular or spherical domains about each nodal point. These radial calculations can be CPU intensive. The research presented herein avoids the intensive radial computations and provides a mesh generation preprocessor as an integral part of the main finite element program. This system coupled with a postprocessor for display of results provides the user with a vehicle for interactive steering of calculations.

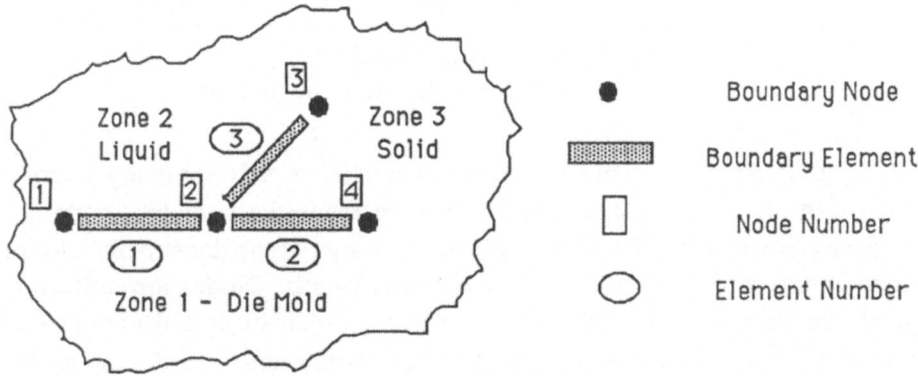

Figure 1 Unambiguous Boundary Definition via Linear Line Elements

RESULTS: Key to this mesh generation routine is the description of the geometric boundaries. A simple, yet unambiguous, boundary definition exists using linear boundary elements. Consider the situation where three distinct physical bodies meet as in Fig. 1. Node #2 is common to each zone which illustrates the difficulties of defining zones uniquely through boundary nodes. Alternately, consider the boundary elements. Each has node #2 associated with the line element, yet element 1 separates zones 1 and 2 uniquely. Similarly, element 2 separates zones 1 and 3 while the third element separates zones 2 and 3. Each boundary element has 4 integers associated with it: a) local boundary node #1, b) local boundary node #2, c) zone identifier to left of element, and d) zone identifier to right of element.

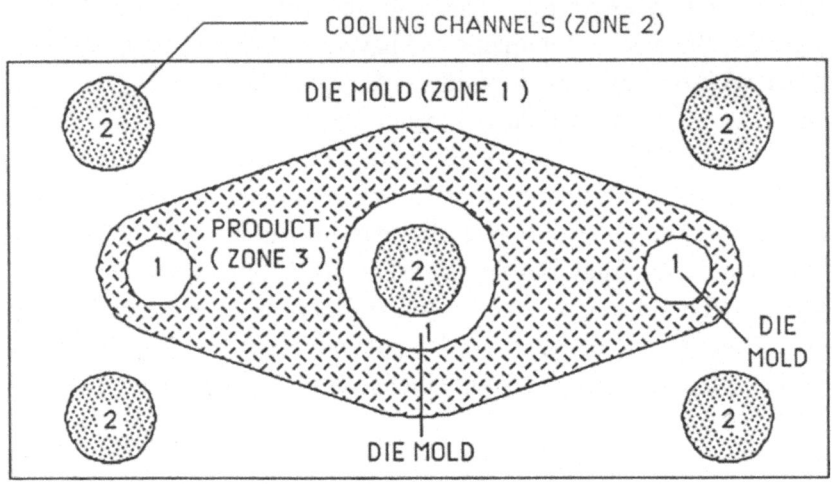

Figure 2 CAD Design with Unique Zone Identifiers

A CAD design of a mold is converted readily to this boundary element format, Fig. 2. The mesh generation routine reads the boundary node and boundary element files, determines the spatial range of the domain, number of unique zones, and average boundary element length. Nodes are uniformly placed over the entire domain range. The node density is a user specified variable with a default value equal to the average node separation on the boundary. Equilateral triangles provide the node connectivity. This numerical deployment of nodes is the fastest methodology of those sited in the literature review. Using the same boundary element format, regions of increased resolution are specified. All parent elements within the refinement zone(s) are subdivided at the midpoint of their sides. This operation (as opposed to a centroid node placement) maintains or improves the aspect ratio of all refined elements. The locally refined elements are knitted to the parent mesh such that continuity is maintained, Fig. 3. Following all user specified refinements, each domain node is classified via its zone location. This routine follows that of Lo

[3] wherein an odd number of boundary-line crossings identifies the zone uniquely. Elements are classified based on their node zones. Non unique node

Figure 3 Uniform Hexagonal Grid With Knitted Refinement Zones

classifications within an element identify the subset of elements that span physical boundaries. A simple intersection routine determines the closest element node to the physical boundary. That node is moved to the boundary and tagged. This classification process is followed for all elements spanning physical boundaries. A localized Laplician smoothing algorithm adjusts the

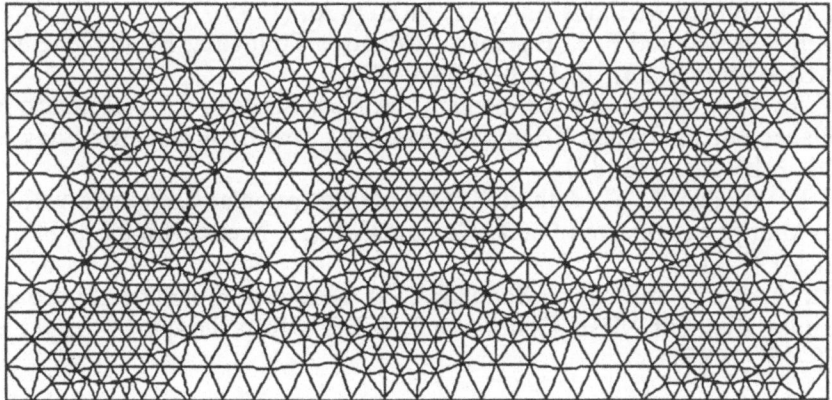

Figure 4 Final Triangulation Mesh Conforming to Physical Boundaries

elements to improve aspect ratios while restricting tagged nodes to tangential motion only along the boundary, Fig. 4. Note that the physical boundary elements and nodes are not part of the numerical mesh. These boundary parameters simply delineate the spatial zones of each material.

CONCLUSIONS: The objective of this example CAD design was to place cooling channels within the mold such that the part solidified rapidly while minimizing large thermal gradients on its surface in the tangential direction. These thermal specifications, interior and exterior boundary descriptions and a range or count of cooling channels (5 in this situation) were the design constraints. The initial CAD design was converted to a valid numerical mesh and the finite element solver determined the cooling time and thermal gradients. Afterwhich, the results were displayed with the option to interactively relocate and/or change the cooling channels. This iterative design option required a new or modified numerical mesh with each trial. Yet, the CPU consumption on a microVax II was less than 1 minute for the complete rediscretization process involving 1888 elements, Fig. 5. As a consequence of this rapid response time, an optimal design was determined without the need of sophisticated search algorithms in parameter space.

REFERENCES:
1.) J. Gabay, "CADKEY: A Streamlined PC-Based CADD Package", Engg. Tools, V 1, No.1, p56-62, 1988.
2.) J. Thompson (ed.), Numerical Grid Generation, North-Holland, Amsterdam, 1982.
3.) S. H. Lo, "A New Mesh Generation Scheme for Arbitrary Planar Domains", Int. J. Numer. Methods Eng., 21, 1403-1426, (1985).
4.) G. Erlebacher and P. R. Eiseman, "Adaptive Triangular Mesh Generation", AIAA Journal, 25, 10, 1356-1364, (1987).
5.) W. H. Frey, "Selective Refinement: A New Strategy for Automatic Node Placement in Graded Triangular Meshes", Int. J. Numer. Methods Eng., 24, 2183-2200, (1987).
6.) N. P. Weatherill, "A Method for Generating Irregular Computational Grids in Multiply Connected Planar Domains", Int. J. Numer. Methods Fluids, 8, 181-197, (1988).
7.) M. A. Yerry and M. S. Shephard, "A Modified Quadtree Approach to Finite Element Mesh Generation", IEEE Comp. Graphics Appl., 3, 1, 39-46, (1983).

Automatic Finite Element Mesh Generation in a Solid Modeling Environment: A General Framework and an Evaluation of Existing Algorithms

NICKOLAS SAPIDIS and RENATO PERUCCHIO[1]
Department of Mechanical Engineering
College of Engineering and Applied Science
University of Rochester, Rochester, New York, 14627

Abstract: In this paper we attempt a systematic analysis and evaluation of algorithms for automatic meshing from solid models. Given the limitations of space, we have included only three families of algorithms – element extraction, domain triangulation, and recursive spatial decomposition. Algorithms in these three families are particularly important because (1) they tend to satisfy more closely the conditions for genuine automatic operations, (2) they have been shown to be applicable to 3-D (solid) domains, and (3) they are well documented.

1 Introduction

The importance of Solid Modeling Systems (SMS) in modern CAD/CAM/CAE technology is well established. Current SMS provide the environment for creating and maintaining an informationally complete representation of a solid, either through Constructive Solid Geometry or Boundary Representation or both, and a set of high level operators for "editing" the representation, through regularized union, intersection, and difference with other solids [1]. Low level operators, which allow the classification of points, lines, and surfaces with respect to the solid, may also be available. To be usable for Finite Element (FE) analysis, the solid represented in the SMS must be transformed into a discretized model – the FE mesh – consisting of a collection of cells which must satisfy a number of geometrical and topological conditions dictated by the FE method. In general, even with the help of interactive computer graphics preprocessing, the operation of transforming a solid into a valid mesh is considerably labour intensive. Recently, a number of papers have been published describing algorithms that perform "automatically" FE decomposition using the SMS definition of a 2-D or 3-D object (see [2], [3] and references therein).

Automation of meshing from solid models implies that the user interaction must be limited to defining the solid in the SMS and specifying one or more mesh density parameters. Thus an automatic meshing algorithm should be regarded as a function G that relates a solid S (and a set of density parameters l) to a "valid" mesh M, such that $M = G(S, l)$. M is a *valid* mesh of S if: (1) M is a correct FE mesh in a general sense (that is, elements do not intersect each other, elements are geometrically correct, and so on), (2) no element of M is *totally* outside of S, (3) all nodes of M are either inside or on the boundary surface of S, and (4) all edges and vertices of S are represented in M. Note that the *validity* of M should not be a function of the density parameters

---

[1]Graduate Research Fellow and Assistant Professor, respectively.

l, that is, given S, G should always result in a *valid* mesh whatever the parameters in the l set. From the computational point of view, an automatic meshing procedure must exhibit the following behaviour: (1) the procedure must include a finite number of steps (subtasks) $G_1, G_2, ..., G_n$, (2) after completing step G_i, the next step G_{i+1} should be uniquely determined by G_i, and (3) each one of the steps $G_1, G_2, ..., G_n$ is guaranteed to terminate after a finite number of numerical operations.

Although unique regarding the meshing and geometric operations involved, each of the algorithms examined below follows a common two-stage strategy for constructing a *valid* FE mesh.

Stage 1: From the SMS description of the object, a *discrete* model is derived. This model provides the basis for the development of the FE mesh.

Stage 2: A FE mesh is obtained by transforming the elements of the discrete model into finite elements.

In the present study we focus on automation only. Other important aspects of meshing algorithms – such as overall efficiency, element quality, and applicability to self-adaptive procedures – will be examined in an extended version of this paper.

2   Element Extraction Algorithms

This family includes the algorithms introduced by Wördenweber [4] and Woo and Thomasma [5]. Since the algorithm in [4] is a generalization of the algorithm presented in [5], this section discusses Wördenweber's algorithm only.

Case A: Decomposition of Rectilinear Manifolds (Planar Polyhedra)
The decomposition algorithm for 3-D rectilinear manifolds includes 4 operators that extract tetrahedral elements from the solid: $OP0$ deals with the trivial case where the object is simply a tetrahedron. $OP1$ extracts a tetrahedron based on a convex vertex by introducing a single cut in the domain. $OP2$ operates on convex edges and introduces two planar cuts. $OP3$ embeds a new vertex inside the solid and then creates a tetrahedron by connecting the new vertex to an appropriate face of the solid (that is, $OP3$ extraction introduces three cutting planes). Operators $OP0, ..., OP3$ suffice for the volume triangulating any *simple* 3-D manifold. For the case of an arbitrary 3-D manifold, one needs two extra operators OPj, OPp to handle interior cavities and a nonzero genus, respectively. Finally, in order for the mesh to agree with the FE density specified by the user, refinement operators are used to further decompose tetrahedra into smaller tetrahedral finite elements.

Case B: Decomposition of Curvilinear Manifolds (Curved polyhedra)
In [4] Wördenweber proposes to decompose a curvilinear manifold S by first constructing a *planar equivalent manifold* PS of S and then apply the algorithm of case A to PS. Although this seems to be a reasonable approach, no robust algorithm is presented in [4] for constructing PS.

It is clear that Wördenweber's algorithm deals mainly with Stage 1 of the meshing process. When applied to planar polyhedra, it produces meshes that fulfill the *validity* requirements stated above. However, because it lacks a robust algorithm for constructing planar equivalent manifolds from curved polyhedra, this meshing procedure cannot be expected to operate correctly on general 3-D solids. As explained later, Wördenweber's operators play an important role in Stage 2 of meshing algorithms based on recursive spatial decomposition.

3 Domain Triangulation Algorithms

In the 1970's several authors presented 2D meshing algorithms that were based on the following scheme: (1)*Node Insertion*: nodes (points) are distributed within and on the boundary of the domain of interest in accordance with the FE density specified by the user, and (2) *Domain Triangulation*: the nodes are automatically triangulated so that a FE mesh of triangular elements is obtained. Step 2 is more or less automatic, but, step 1 always involves a considerable amount of user interaction. The first attempts to use domain triangulation for 3-D meshing are reviewed in [6]. Although these algorithms are labelled "automatic", Step 1 is still heavily based on user interaction. In the sequel we describe briefly the algorithm presented in [7], which seems to be the most successful approach towards 3-D domain triangulation:

(A) Calculate a bounding box for the solid and fill it with regular icosahedra. The radius of the icosahedra is determined by the prescribed FE density. Calculate center and surface nodes of the icosahedra. Add extra nodes for filling the voids among icosahedra. (B) Get *interior nodes*. These are nodes that lie inside or on the surface of the model. Calculate Delaunay volume triangulation for interior nodes. (C) Remove tetrahedra whose centroids lie outside the solid. Discard nodes forming *surface triangles* (i.e., element faces that belong to only *one* tetrahedron). (D) Place *boundary* nodes on the vertices, edges, and faces of the solid. The spacing of these nodes should be in accordance with the FE meshing density that the user has specified. (E) Derive Delaunay volume triangulation corresponding to boundary and interior nodes. Remove tetrahedra whose centroids lie outside the solid. (F) Identify and correct *very thin* tetrahedra.

Steps A, B, C, and E of this algorithm are completely automatic. Regarding step D, the authors assume [7, p.15] that the SMS provides an appropriate node-placement operator for performing this step. However, systematic node-placement is not one of the standard operators currently provided by a SMS. Indeed, reference [7] does not indicate a system that supports node-placement operations nor refer to an algorithm for performing systematic node-placement on a generic boundary surface. It is interesting to note that in [6] the same authors consider node placement as a meshing operation instead of a geometric operation (specifically, boundary node points are defined interactively by the user). Boundary nodes are introduced completely independently from interior notes. Thus, creation of badly shaped elements (*slivers*, according to [7]) is very probable. Step F deals exactly with this problem, but the procedures proposed for performing this step are not robust: it is possible that removing a sliver results into a new one [7, p.31]. In conclusion, Stage 1 of this algorithm – which includes steps A - D above – cannot be regarded as an automatic procedure.

4   Recursive Spatial Decomposition Algorithms

Recursive spatial decompositions approximate the solid with a union of non-inter-secting variably sized cells generated by recursively subdividing a spatial region enclos-ing the object. The recursive subdivision rule can be concisely described as follows: the solid is conveniently "boxed" and the box is decomposed into octants; octants are classified with respect to the solid: when an octant is totally inside or outside of the object, the decomposition ceases; when an octant cannot be so classified, it is further subdivided into octants; the process continues until some minimal resolution level is reached (in 2-D the decomposition proceeds by quadrants). Approximations produced by recursive spatial decompositions can be represented by logical trees whose nodes have eight sons (four in 2-D), hence the popular name "octree" ("quadtree" in 2-D). For the two algorithms presented below, stage 1 deals with the derivation of the oc-tree (quadtree) decomposition, and stage 2 with the transformation of the interior and boundary cells into valid finite elements.

4.1   The Shephard-Yerry (S-Y) Algorithm

In a sequence of papers (see [3], [8-9] and references therein) Shephard and Yerry present 2-D and 3-D versions of a mesh generator based on recursive spatial decom-positions. According to [3], this mesh generator has been designed so that it can be interfaced with the boundary representation provided in a SMS. Very little is known about the specific system (the RPI superquadrics modeler [8]) used by Yerry and Shep-hard. However, it appears that the limitations of this modeler have affected significantly the development of the S-Y mesher [8, p.35]. Weak points of the original S-Y algorithm and examples of unacceptable results are discussed in [9]. Here, we discuss the current version of the 2-D S-Y procedure as presented in [9]:

(A) *Initialization of the quadtree*: The quadtree is initialized to the required level without performing any classification of the cells. (B) *Discretization of boundary edges*: Edges are traversed, and intersections with quadrants are calculated. The discretized edge segments are entered into the tree, and the status of the corresponding quadrants is changed from *undefined* to *boundary terminal* or *vertex* quadrant (for cells that include edge segments and a vertex). (C) *Determination of interior octants*: Starting with a boundary cell one moves along the interior direction associated with the corresponding edge, changing to *interior* the status of each neighbouring *undefined* quadrant, until a boundary or interior cell is reached. (D) *Improvement of boundary quadrants*: Smooth-ing operations are applied to boundary cells in order to improve geometries that would lead to badly shaped finite elements. (E) *Triangular element mesh generation*: For each boundary cell, the node with the largest interior angle in the quad-loop is identified. One or two triangles are associated with this node, of which, the one having the best shape is selected to form an element (the shape of triangles is characterized by means of a numerical measure). The selected element is removed and the process is repeated until only three nodes are left.

Comments: (1) Cell classification is based on boundary traversing. The specific algorithm in [9] is ambiguous because it does not consider the case of an edge being totally inside of a quadrant (hence no intersections) or belonging to two quadrants (one totally inside and the other totally outside). Also, extending this algorithm to include traversing the faces and the edges of a 3-D object defined in a SMS environment is not a trivial task [11]. (2) The algorithms for performing some of the stage 2 meshing operations (e.g., determining triangles associated with the node with the largest interior angle – see step E) are not provided. (3) Most of the meshing operations used are closely related to 2-D geometry and their extension to 3-D is not obvious (e.g., see comment (1) above, and step E of the algorithm).

4.2   The Perucchio-Saxena-Kela (P-S-K) Algorithm

This algorithm was developed to provide the mesh generator for an integrated system for automatic meshing and self-adaptive analysis - see [2] for a description of the system and [10,11] for the 3-D algorithm. Stage 1 can be implemented in any SMS that supports the regularized intersection operation between solids and provides a boundary representation. Stage 2 requires a specialized modeling environment that supports element extraction operators. The following description refers to [11].

Stage 1: The octree is built according to the recursive subdivision rule given above. Each octant cell is intersected with the solid and is classified as IN – if the intersection is equal to the octant –, OUT – if the intersection is a null object –, or NIO – if the cell is neither IN nor OUT.

Stage 2: Only IN and NIO cells at resolution level are considered. IN cells are mapped onto hexahedral elements such that the vertices of the octants become nodes of the FE mesh. Each NIO cell C is intersected with the solid S and classified as *simple* or *complex* depending on the topology of $R = S \cap^* C$. NIO cells traversed by at most one boundary face of S are classified as *simple* NIO (SNIO), while all other NIO cells are labelled *complex* NIO (CNIO). SNIO cells are decomposed into finite elements by mapping an appropriate template mesh onto R. Seven templates cover all possible cases of SNIO cells. For CNIO cells, R is decomposed through the recursive application of Wördenweber's element extractors, and a special operator OP* [10] that extracts pyramid elements. Both template mapping and element extraction introduce finite element nodes either on the boundary of S or inside S. Stage 2 algorithms include provisions for constructing an all tetrahedral element mesh.

The P-S-K algorithm produces meshes that inherit three important properties from the underlying octree decomposition: *hierarchical structure*, *spatial addressability*, and *geometrical regularity*. These properties are exploited in the self-adaptive analysis system described in [2]. Stage 1 is robust and has been efficiently implemented in the PADL-2 solid modeler. Stage 2 is limited only by the fact that element extractors operate only on planar manifolds, while R might be curved and/or non-manifold [11]. As indicated in the discussion of Wördenweber's procedure, current algorithms for constructing the planar equivalent manifold of R are not sufficiently robust to insure the validity of the resulting mesh.

5 Conclusions

In the present paper we have briefly discussed the most promising approaches for the automatic construction of valid FE meshes from solid models. Because of the inherent complexity of the 3-D meshing problem and the stringent requirements posed by genuine automatic processes, all these approaches are algorithmically more complex – and thus computationally more demanding – than older procedures used for user-guided, mainly 2-D, interactive meshing. Although considerable progress has been made, no algorithm – in its present state of development – can be regarded as being fully automatic *and* applicable to curved polyhedra. We note, however, that all the limitations discussed in the previous sections are – at least in theory – removable. Thus, given the pressing need for automatic CAD/CAE tools in the production as well as in the research environment, we expect that robust algorithms for SMS-based automatic meshing will emerge in the near future.

References

[1] A. A. G. Requicha and H. B. Voelcker, "Solid modeling: Current status and research directions", *IEEE Comput. Graph. Appl.* vol. 3, no. 7, pp. 25-37, 1983.

[2] A. Kela, M. Saxena and R. Perucchio, "A hierarchical structure for automatic meshing and adaptive FEM analysis", *Engng. Comput.* vol. 4, no. 2, pp. 104-112, 1987.

[3] M. S. Shephard, P. L. Baehmann and K. R. Grice, "Automatic finite element modeling: geometry control for direct models", *Engng. Comput.* vol. 4, no. 2, pp. 119-125, 1987.

[4] B. Wördenweber, "*Automatic Mesh Generation of 2- and 3-Dimensional Curvilinear Manifolds*", PhD Diss. (Tech. Rep. No. 18), Cambridge University, 1981.

[5] T. C. Woo and T. Thomasma, "An algorithm for generating solid elements in objects with holes", *Computers & Structures*, vol. 18, no. 2, pp. 333-342, 1984.

[6] J.C. Cavendish, D.A. Field, and W.H. Frey "An Approach to Automatic Three-dimensional Finite Element Mesh Generation", *Inter. J. Num. Meth. Engng.* vol. 21, pp. 329-347, 1985.

[7] D.A. Field, and W.H. Frey, "Automation of Tetrahedral Mesh Generation", Research Publication GMR-4967, General Motors Research Labs., Warren, MI, 1985.

[8] M.A. Yerry and M.S. Shephard, "Automatic Mesh Generation for Three-dimensional Solids", *Computers & Structures* vol. 20, no. 1-3, pp. 31-39, 1985.

[9] P.L. Baehmann, S.L. Wittchen, M.S. Shephard, K.R. Grice and M.A. Yerry "Robust, Geometrically based, Automatic Two-dimensional Mesh Generation", *Inter. J. Num. Meth. Engng.* vol. 24, pp. 1043-1078, 1987.

[10] M. Saxena and R. Perucchio, "Geometrical and topological issues in octree based automatic meshing", *Proc. NAFEMS Int'l Conf. on Quality Assurance and Standards in Finite Element Analysis*, Brighton, UK, vol.I, paper no. 2.4, May, 1987.

[11] R. Perucchio, M. Saxena and A. Kela, "Automatic mesh generation from solid models based on recursive spatial decompositions", submitted to *Inter. J. Num. Meth. Engng.*, 1988.

A Vector/Solid Intersection Technique for Three-Axis NC Verification

James H. Oliver
International TechneGroup Incorporated
5303 DuPont Circle
Milford, Ohio USA 45150

Abstract

An efficient technique for calculating intersections between vectors and solid models is presented. This method is a small but crucial component of an algorithm for automatic verification of numerically controlled (NC) milling programs for sculptured surface parts. The verification algorithm produces graphical output depicting the desired part as shaded surfaces with out-of-tolerance areas highlighted.

Direct Dimensional NC Verification

A new technique for dimensional verification of NC milling programs is presented in [1] and [2]. With this approach, an *as designed* part model is represented by a non-uniform grid of surface points and corresponding outward-directed surface normal vectors. Each tool motion of the NC program is considered individually, in sequential order. A boundary representation (B-rep) solid model is created which represents the volume swept by a given motion of the mill tool. Tool motions are verified by calculating the directed distance along normal vectors from affected surface points to the swept volume models. If a calculated intersection distance (cut value) is less than the value previously stored for a given normal, the value is updated. Thus when all tool motions have been considered, only the closest or deepest excursion of the tool towards or into part is retained. Graphical output depicts the desired part as shaded surfaces with out-of-tolerance areas highlighted.

Three major modules form the basis of this NC verification technique. The first involves a method for discretizing the desired part model into a sufficiently dense grid of surface points and normals. The second module provides a means for extracting a subset of eligible points and normals to be considered for each tool motion (swept volume). The intersection of normal vectors with swept volume solid models constitutes the third module. Ongoing research is aimed at improved computational efficiency through alternative discretization and display techniques, as well as enhanced functionality for five-axis verification capability. This paper will focus on the third module; an efficient and robust technique for the calculation of vector/solid intersections for use in three-axis milling applications.

Vector/Solid Intersection

The intersection of a part surface normal vector with a mill tool swept volume is calculated efficiently by proceeding in hierarchical steps through a series of progressively more exact definitions of the shape of the tool swept volume. The calculation begins by modeling the swept volume as a parallelepiped. If necessary, the model is refined, but only in the region where the vector intersection could possibly occur. More precise bounding surfaces (cylindrical and spherical) are added to the model, in the region of the planar intersection, if the parallelepiped model yields an intersection distance which is less than a user prescribed maximum miss. Throughout the procedure, results of intermediate calculations are used to determine if further, more sophisticated swept volume intersection calculations are required. This structure ensures that redundant or superfluous intersection calculations are minimized.

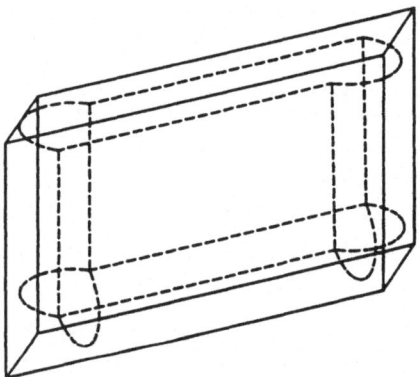

Figure 1. Parallelepiped surrounding a tool swept volume

Figure 1 depicts the smallest parallelepiped which surrounds a tool swept volume resulting from a three-axis mill motion. Three vectors are required to define the six planes which make up a given parallelepiped. The tool path vector **p** is calculated as the difference between successive tool center positions. The cross product of **p** with the mill axis **z** results in a vector **q** which defines the side planes of the parallelepiped. The end faces are defined with the vector generated from the cross product of **q** and **z**, and the cross product of **q** and **p** yields a vector perpendicular to the top and bottom faces of the parallelepiped.

Distances D_i are calculated from a surface point, along the normal vector, to each of the six infinite planes which define the parallelepiped. If none of the D_i values is less than the specified maximum miss L_m, then processing for the current surface point ends and the next one may be considered. Otherwise, the subset of intersection distances which are less than L_m is sorted from smallest to largest.

The sorted list of possible vector/solid intersections is processed to find the first plane to intersect the normal within the boundaries of the parallelepiped. This bounded plane check is accomplished with a modified version of the Roberts algorithm, which was originally applied to three-dimensional hidden line removal. [3] In this application, the algorithm is used to determine if a point on a plane lies within four bounding half-spaces (i.e., on a face). The parallelepiped is constructed so that the normal of each plane points toward the interior of the volume. Dot products are calculated between the vector representing the intersection point (in homogeneous coordinates) and vectors representing the coefficients of the bounding planes. If any of the four dot products yields a negative value, the intersection point is outside the face boundaries.

The planes which intersect the normal are processed in order, i.e., the ones with the smallest D_i values are considered first. If an intersection falls within a face, then the normal vector may intersect the actual swept volume, so the intersecting plane must be considered further; if not, the next plane in the list is checked, until the list of eligible planes is exhausted. If none of the plane intersection points falls within a face, then processing for the point is terminated and the next available one may be considered. Also, if the top face of the parallelepiped is found to be the closest intersecting plane, a warning is issued of possible tool interference with the workpiece, and processing for the point ends.

The first plane which yields an intersection within a face is processed to determine if a more accurate swept volume model is necessary. This is accomplished with further applications of the bounded plane check, this time with bounding planes moved inside the swept volume to represent the transitions from planar to cylindrical or from cylindrical to spherical surfaces.

For example, Figure 2 shows a side face of the parallelepiped surrounding a swept volume, divided into subregions labeled 1 through 8. The bounded plane check described above is performed with the intersection point and the planes which bound subregion 1. If the intersection point falls within this subregion, processing for this point ends, since the planar intersection is already exact. If it is not within subregion 1, the dot products are interpreted to determine which subregion should be examined next. For instance, if the plane separating subregion 1 and subregion 4 yields a negative value, then another application of the bounded plane check, with the planes which bound subregion 4, will determine if the intersection occurs within subregion 4, 5, or 6. At most, two bounded plane checks are sufficient to determine the subregion of the vector intersection for this (side) face of the parallelepiped.

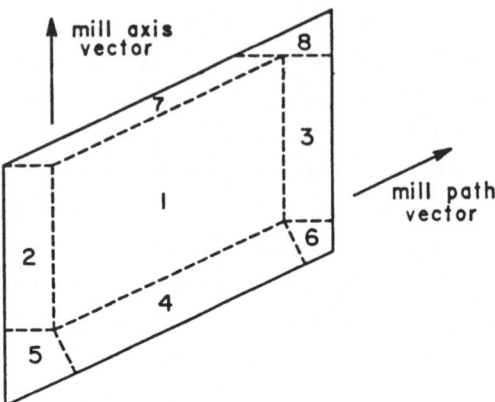

Figure 2. Subregions of a side face of a swept volume parallelepiped

Referring again to Figure 2, if subregion 2, 3, or 4 contain the intersection point, then the normal vector may intersect a cylindrical surface. If subregions 5 or 6 contain the point, a spherical surface must be considered. If the normal intersects the plane in subregion 7, the user is warned that the part may have been cut by the rearward facing top edge of the tool (if the mill is lifting), or forward facing top edge (if the mill is diving), and processing for the point is terminated. Similarly, if subregion 8 contains the point, the normal has missed the actual swept volume and processing terminates. Note that subregions 7 and 8 do not exist if the mill does not change its Z coordinate (height) during a motion. An analogous (although somewhat simpler) procedure is applied if the intersecting plane is an end face or the bottom face of the parallelepiped, except that only cylindrical or spherical subregions are possible.

Intersection of a vector with a sphere or cylinder requires a relatively straightforward application of vector algebra. The calculations are not computationally intensive, but are considerably more complex than the planar approximations used up to this point. Details of these vector/surface calculations are given in [2]. A normal vector can pierce the parallelepiped in a cylindrical or spherical subregion yet miss the actual swept volume completely. This condition can be determined with an intermediate result of the exact intersection calculation, so it is checked first to further reduce unnecessary computations.

Results

The vector/solid intersection algorithm outlined above has been implemented as part of a general verification program for three-axis NC milling applications. Figure 3 is an example of output from this program, depicting the results of an NC milling process for (one half of) a stamping die for an automobile hood.

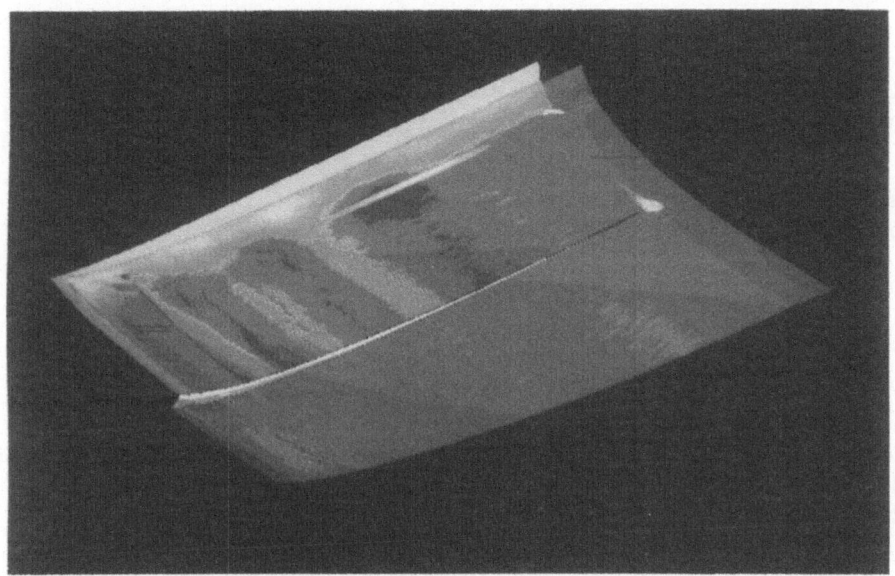

Figure 3. Direct dimensional NC verification of a die for an automobile hood

In this implementation, the output of direct dimensional NC verification is an image of the *as designed* part model in which color conveys information about the interaction between the mill tool and the part geometry. The color of each point on the image is made up of a hue and an intensity. Hue is specified by normal vector cut value. Areas of the part model cut within tolerance are assigned the hue green. Gouged areas are shaded via hue interpolation between red and yellow, where red represents "just beyond tolerance" and yellow represents "maximum gouge". Missed (under cut) areas are shaded via hue interpolation between blue and magenta, where blue represents "just beyond tolerance" and magenta represents "maximum miss". Independently of hue, the intensity of each point is based on the angle that the normal vector makes with a users-selectable light source, thus enhancing surface feature recognition.

Acknowledgement

The author is grateful to Chrysler Corporation for support of this work.

References

1. J.H. Oliver and E.D. Goodman, "Computational Verification of Numerical Control Programs for Sculptured Surface Parts," *Proc. Int'l. Conf. Computer Aided Design, Manufacturing and Operation in the Automotive Industries*, Geneva, Switzerland, March 1987.

2. J.H. Oliver, Graphical Verification of Numerically Controlled Milling Programs for Sculptured Surface Parts," Ph.D. Dissertation, Michigan State University, 1986.

3. D.F. Rogers, "Procedural Elements for Computer Graphics," McGraw-Hill Book Company, 1985.

Solid Modeling and Geometric Reasoning for Design and Process Planning

Dana Nau,[2] Nicholas Ide,[3] Raghu Karinthi,[4]
George Vanecek,[4] and Qiang Yang[4]

University of Maryland

Summary

This paper describes our work on the integration of techniques for solid modeling, geometric reasoning, and multi-goal planning, with application to computer-aided design and manufacturing. This work is being done with two long-term goals in mind: the development of a practical integrated system for designing metal parts and planning their manufacture, and the investigation of fundamental issues in representing and reasoning about three-dimensional objects. We believe this work will have utility not only for automated manufacturing, but also for other problems in design and multi-goal planning.

1. Introduction

One problem facing modern industry is the lack of a skilled labor force to produce machined parts as has been done in the past. In the near future, this problem may become acute for a number of manufacturing tasks. This has led to considerable interest in ways to automate various manufacturing tasks.

Our first work in this area was in the development of AI techniques for automated process selection. Since we believe that the rule-based approach used in most knowledge-based systems is not the most appropriate way to do process planning, we have developed a different approach, based on hierarchical abstraction. The implementation of this idea first resulted in SIPP, a process selection system written in Prolog, and later led to SIPS, a more sophisticated system written in Lisp. The evolution of SIPP and SIPS over the last several years have been described elsewhere [6,7,8,9,10,11,12], so SIPP and SIPS will not be described again here.

Recently we have increasingly become interested in integrating process planning with design and solid modeling, for two reasons. First, a good design system is essential to provide a decent interface to a process planning system. Second, there are process planning tasks which cannot be performed correctly without extensive interactions with a solid modeler. Our current work focuses on the following topics:

1. solid modeling techniques specifically suited for integration with automated reasoning systems such as process planning systems;

2. computer-aided design systems capable of reasoning about three-dimensional objects, both for use as a design aid and also for use in integrating design with process planning;

3. ways to reason about interacting features during design and planning.

These topics are discussed in Sections 2-4, respectively. Section 5 contains concluding remarks.

[1]This work has been supported in part by the following sources: an NSF Presidential Young Investigator Award to Dana Nau, NSF Grant NSFD CDR-85-00108 to the University of Maryland Systems Research Center, General Motors Research Laboratories, and Texas Instruments.
[2]Department of Computer Science, Institute for Advanced Computer Studies, and Systems Research Center.
[3]Current address: Century Computing, 1100 West St., Laurel, MD 20707.
[4]Department of Computer Science.

2. Solid Modeling

Most approaches to the integration of solid modeling with automated process planning have essentially involved using a geometric modeler as a front end to a process planning system. Two examples of this involve the use of SIPS as the process planning system: the interface produced at General Motors between SIPS and the MBF/X-Solid CAD system [12], and the interface being built at the National Bureau of Standards between SIPS and Unicad/Romulus [1]. Such interfaces make the process planning system more convenient to use, but in order to generate correct process plans for complex objects this approach is not sufficient. What processes can be used for some machinable feature—or whether the feature can even be made at all—may depend on geometric information not available solely from the descriptions of the features. To get this information will require the process planning system to interact extensively with the solid modeler during process planning.

For example, consider the task of drilling a hole in a flat surface. Although this is usually easy, it will be impossible if some other part of the object interferes with the tool trajectory. This condition can be recognized through the specification of geometric constraints and verification of these constraints through queries to a solid modeler. In more complex examples, the process planning system will need to make a large number of such queries.

Examples such as the one above can be handled by interfacing the process planning system to an existing solid modeler—and in fact, we have interfaced SIPS to the PADL-2 solid modeler for this purpose [4]. However, our experience at building this interface, as well as our experience with several other solid modelers, has led us to conclude that most existing solid modelers are not adequate for this purpose. One reason for this is that the primary focus guiding the development of most solid modelers has been the fact that they will be used by humans. Thus, much work has been done on efficient algorithms for operations such as rendering, but less attention has been paid to providing easy and efficient ways to answer queries, retrieve pieces of the objects being modeled, and make incremental changes.

The thorough integration of a solid modeler with a process planning system (or with various other automated systems) will require the ability to do several different kinds of solid modeling operations very quickly. Some of these operations include rotating or translating the solid, extracting bounding surfaces from it, and performing set operations such as union and intersection of solids. We believe that no existing approach to solid modeling can perform all of these tasks quickly and accurately enough.

In our opinion, the approach to solid modeling which comes the closest to fitting the above requirements is boundary representation. Using boundary representations, it is easy to do fast translations, rotations, and boundary extraction, but set operations are more time-consuming. Our approach is to enhance the capabilities of boundary representations, by developing fast algorithms for set operations.

When set operations are done on solid objects represented using boundary representation, the usual approach is to check each edge of one object against every edge of the other object. This results in a cost worse than $O(n^2)$. We have developed a faster algorithm based on the divide-and-conquer paradigm, in the form of non-regular decomposition of space [13]. This results in an average-case performance which has empirically been found to be $O(n \log n)$, where n is the total number of edges of both the input and the output. We have implemented a modeling system using non-regular decomposition for the representation and manipulation of two-dimensional polygons [14]. We are extending our algorithm to handle three-dimensional objects containing both flat and curved surfaces, and we are currently building a three-dimensional solid modeler using this approach.

3. Reasoning about Features

One of the greatest problems facing the manufacturing industry today is the differences in product description in various segments of the industry. Many tools created for aiding the design and the manufacturing processes seperately, but the problem is how to provide automatic integration of these tools.

CAD-generated objects can be defined in terms of the complete geometry of the part. The descriptions contain the faces, edges and vertices making up the part. For the purpose of manufacturing, the geometry and topology are the same, but the meaning associated with this geometric structure is different, and dictates a change in the description. An object which, to the designer, is a block minus a cylinder, is to the manufacturing engineer a block with a hole and certain tolerances.

One proposed way to handle this incompatibility is *automated feature extraction*, which consists of automating the task of determining the manufacturing features of a part from its geometry. This is an extremely difficult process, and the reader is referred to [3,5] for a discussion of the complexities involved. Some of the tougher problems include (1) inferring faces needed to describe the machining operations that do not appear in the CAD description, and (2) extracting a feature which intersects or otherwise interacts with other features, without disturbing those other features.

Another approach is *design by features*, in which the user builds a solid model of an object by specifying directly its "manufacturing features." For example, one might start with a model of a piece of metal stock, and modify it by adding holes, slots, pockets, and other machinable features. One problem with design by features is that it requires a significant change in the way a feature is designed. Traditionally, a designer designs a part for functionality, and a process engineer determines which are the manufacturable features are. However, design by features places the designer under the constraints of not merely having to design for functionality, but at the same time specify all of the manufacturable features as part of the geometry—a task which the designer is not normally qualified to do.

To overcome this problem, it would be desirable to allow the designer to use not manufacturing features, but instead "design features," which may not correspond directly to manufacturing operations, but which make sense to the designer. This would require the system to translate the design features into manufacturing features after the design of the part was completed. With an intelligently chosen set of available features and ways for combining them, this should be less complicated than extracting manufacturing features from an ordinary solid model.

Given a definition of a part as a combination of design features, there may be several possible ways to translate the part into a collection of machinable features. Different translations of the same object could result in very different process plans for that object, with different costs. For example, if a wide slot bisects a pocket, it may lead to a cheaper plan if the bisected pocket is considered to be two separate machinable features rather than just one. However, if the slot is narrow, it may be better to consider the pocket to be a single feature.

We intend to develop a system for feature-based design and analysis, with the ability to reason about interactions among the features in order to make good decisions about how to translate design features into machinable features. This system will make extensive use of the solid modeler described in Section 2, and will provide information about feature interactions for use by the process planning system (see Section 4).

4. Reasoning about Interacting Features

The SIPS process selection system works well when the plans for the various features are independent. However, the problem becomes much more complicated when one tries to handle interactions among features (for example, see [2,15]).

For example, consider an object containing two holes h1 and h2, both having the same diameter and the same machining tolerances. Suppose h1 can be created by either twist drilling or spade drilling. Then the least costly way to make h1 is twist drilling. If the depth of h2 is sufficiently large, h2 may require spade drilling rather than twist drilling. In this case, the cheapest way to make the entire object is to use spade drilling for both h1 and h2 in order to avoid a tool change—even though spade drilling would not be the cheapest way to make h1 if h1 were the only hole being made.

The problem described above can be characterized as a problem in multiple-goal planning, with the restriction that all interactions among the actions in the plans should be expressible in terms of partial ordering constraints, identity constraints, and the possibility of "merging" various actions [15]. In the case of process planning, each feature represents a separate goal, and merging corresponds to saving set-up or tool-change costs by performing two operations at the same time (such as the two twist-drilling actions mentioned above). In such problems, finding an overall plan to achieve all of the goals consists of selecting from among alternate plans for each of the goals and then merging certain of the actions.

As one might expect, the problem of finding an optimal overall plan is NP-hard, but it is possible to develop efficient approximation algorithms for this problem (i.e., algorithms which will produce results that are close to optimal, with reasonably fast average-case performance) [15]. We are developing such algorithms, and intend to develop them further. This will provide a way to produce process plans that take feature interactions into account.

5. Summary and Conclusions

This paper describes our work on the integration of techniques for design, geometric reasoning, and multi-goal planning, with application to computer-aided design and manufacturing. Our work focuses on the following tasks:

1. Knowledge representation and reasoning techniques for process planning. We believe that the rule-based approach normally used in knowledge-based systems is not the best approach to use in process planning. Instead, we have developed an approach based hierarchical abstraction, and implemented it in the SIPS process planning system.

2. Algorithms and data structures for solid modeling. We feel that existing solid modelers are inadequate for the kinds of interactions required for thorough integration with automated process planning systems, and we are addressing this issue by developing a new approach to solid modeling which we believe will satisfy the necessary requirements.

3. Ways to extract and reason about features and feature interactions. We believe that if a design-by-features system is to be made convenient to the designer, it is unrealistic to force the designer to design using manufacturing features. Thus, it will still be necessary to extract the manufacturing features from the model produced by the designer. However, we also believe that this task can be made less complicated than the task of extracting manufacturing features from an ordinary (non-feature-based) geometric model. We are developing techniques to handle this problem.

4. Ways to reason about feature interactions and their effects on the resulting plans. We have been developing fast algorithms to handle optimization in multi-goal planning problems, and intend to use these algorithms to handle feature interactions in process planning.

This work is being done with two long-term goals in mind: the development of a practical integrated system for designing metal parts and planning their manufacture, and the investigation of fundamental issues in representing and reasoning about three-dimensional objects. We believe this work will have utility not only for automated manufacturing, but also for other problems in design and multi-goal planning.

References

[1] P. Brown and S. Ray, "Research Issues in Process Planning at the National Bureau of Standards," *Proc. 19th CIRP International Seminar on Manufacturing Systems*, June 1987, pp. 111-119.

[2] C. Hayes, "Using Goal Interactions to Guide Planning," *Proc. AAAI-87*, 1987, 224-228.

[3] M. Henderson, "Extraction of Feature Information from Three Dimensional CAD Data," Ph.D. Dissertation, Purdue University, May 1984.

[4] N. Ide, *Integration of Process Planning and Solid Modeling through Design by Features*, Master's thesis, Computer Science Department, University of Maryland, College Park, 1987.

[5] L. Kyprianou, "Shape Classification in Computer-Aided Design," Ph.D. Dissertation, Cambridge University, July 1980.

[6] D. Nau and T. Chang, "A Knowledge-Based Approach to Generative Process Planning," *Production Engineering Conference at ASME Winter Annual Meeting*, Miami Beach, Nov. 1985, 65-71.

[7] D. Nau and T. Chang, "Hierarchical Representation of Problem-Solving Knowledge in a Frame-Based Process Planning System," *Jour. Intelligent Systems* 1:1, 1986, pp. 29-44.

[8] D. Nau and M. Gray, "SIPS: An Application of Hierarchical Knowledge Clustering to Process Planning," *Symposium on Integrated and Intelligent Manufacturing at ASME Winter Annual Meeting*, Anaheim, CA, Dec. 1986, pp. 219-225.

[9] D. Nau and M. Gray, "Hierarchical Knowledge Clustering: A Way to Represent and Use Problem-Solving Knowledge," *in* J. Hendler, *Expert Systems: The User Interface*, Ablex, 1987, 81-98.

[10] D. Nau and M. Luce, "Knowledge Representation and Reasoning Techniques for Process Planning: Extending SIPS to do Tool Selection," *Proc. 19th CIRP International Seminar on Manufacturing Systems*, June 1987, pp. 91-98.

[11] D. Nau, "Hierarchical Abstraction for Process Planning" *Second Internat. Conf. Applications of Artificial Intelligence in Engineering*, 1987.

[12] D. Nau, "Automated Process Planning Using Hierarchical Abstraction," Award winner, Texas Instruments 1987 Call for Papers on Industrial Automation, *Texas Instruments Technical Journal*, Winter 1987, 39-46.

[13] G. Vanecek and D. Nau, "Computing Geometric Boolean Operations by Input Directed Decomposition," Tech. Report, 1987.

[14] G. Vanecek, Jr. and D. Nau, "Non-Regular Decomposition: An Efficient Approach for Solving the Polygon Intersection Problem," *Symposium on Integrated and Intelligent Manufacturing at ASME Winter Annual Meeting*, 1987.

[15] Q. Yang, D. Nau, and J. Hendler, "Planning for Multiple Goals with Limited Interactions," submitted for publication, 1988.

Mechanism Design

Personal Computer Based CAD System for the Synthesis of Planar Four-Bar Mechanisms

Hyoung Jun Kim   and   Raj S. Sodhi

Department of Mechanical Engineering
New Jersey Institute of Technology
Newark, New Jersey 07102

Abstract

This paper presents a micro-computer based synthesis technique for designing planar four-bar mechanisms which generate the four user specified positions. The method provides a means for designer to design linkages free of order, branch, and Grashof type problems. A computer program is developed for use on an IBM compatible personal computer. This interactive computer method uses designer specified set of four arbitrary coupler positions and several other controlling parameters. An example is presented to illustrate the use of the proposed design technique.

Introduction

The technique for synthesizing a four-bar linkage to carry a lamina precisely through four given positions has been known for a very long time. It is based on the work of Burmester [1]. Since the basic Burmester synthesis technique yields spurious and undesirable solutions, many techniques have been developed including works by Filemon [2], Waldron [3-4], etc. to obtain defect free solutions. Waldron [3] describes three types of defective solutions which are branch problem, order problem, and Grashof problem. Branch problem occurs when a solution linkage does not traverse all the design positions without being disconnected and reassembled in a different configuration. Solution linkage with Grashof problem has neither crank capable of performing a complete rotation. Even though a crank capable of complete rotation is available, the order in which it moves the coupler through the prescribed positions may be different from the desired order, which is called order problem.

Geometric solutions of the branch and order problem for finitely seperated positions are available through the works of Filemon [2] and Waldron [3-4] but geometric solution of Grashof problem is not yet available. Chuang [5] and Wilhelm [6] developed an interactive linkage synthesis program which is just a computer implementation of Waldron and Strong's [7] geometric techniques. These programs do not guarantee solutions free of Grashof problem, and Chuang's application on a personal computer will be difficult because of considerable computation involved.

Since the interactive computer graphics on personal computer has provided new tools for performing engineering design, it's use is desirable to use for four position linkage synthesis on a personal computer, which is the objective of this paper.

Synthesis Equations

When the four positions of a planar lamina are given, the equation of the circle point curve becomes,

$$(x^2 + y^2)(Ax + By) + Cx^2 + Dy^2 + Exy + Fx + Gy + H = 0, \qquad (1)$$

where the coefficients can be found in a previous publication [8].

If we choose a circle point $D_1(x, y)$ on the curve, the corresponding center point, $D_0(x_0, y_0)$ can be determined. Since the circle point D_1 is a point in the moving plane at position 1, the points D_2, D_3, and D_4 are obtained by rotating $A_1 D_1$ about A_2, A_3, and A_4 respectively through angle θ_{1i} which is the angular position change of i th lamina position with respect to the first position of the lamina. Thus the position D_i becomes

$$D_i = [R_{\theta_{1i}}](D_1 - A_1) + A_i, \quad i = 2, 3, 4, \qquad (2)$$

where $[R_{\theta_{1i}}]$ is the plane rotation matrix, $A_i = (u_i, v_i)$, and $\theta_{1i} = \theta_i - \theta_1$. The constant length $D_0 D_i$ of the crank gives

$$(D_i - D_0)^T(D_i - D_0) = (D_1 - D_0)^T(D_1 - D_0), \quad i = 2, 3, 4. \tag{3}$$

Substituting equation (2) into equation (3) and solving equation (3) gives

$$x_0 = [(x^2 + y^2 - x_{D_2}^2 - y_{D_2}^2)(y - y_{D_3}) - (x^2 + y^2 - x_{D_3}^2 - y_{D_3}^2)(y - y_{D_2})]/\Delta, \tag{4}$$

$$y_0 = [(x^2 + y^2 - x_{D_3}^2 - y_{D_3}^2)(x - x_{D_2}) - (x^2 + y^2 - x_{D_2}^2 - y_{D_2}^2)(x - x_{D_3})]/\Delta, \tag{5}$$

where

$$x_{D_i} = (x - u_1)cos\theta_{1i} - (y - v_1)sin\theta_{1i} + u_i, \tag{6}$$

$$y_{D_i} = (x - u_1)sin\theta_{1i} + (y - v_1)cos\theta_{1i} + v_i, \tag{7}$$

$$\Delta = 2[(x - x_{D_2})(y - y_{D_3}) - (x - x_{D_3})(y - y_{D_2})]. \tag{8}$$

The crank link length $l_{D_0 D_1}$ becomes

$$l_{D_0 D_1} = [(x - x_0)^2 + (y - y_0)^2]^{\frac{1}{2}}. \tag{9}$$

As shown in Figure 1, which defines the angles θ, ϕ and ψ, the angular displacement ϕ_i of the crank relative to the base becomes

$$\phi_i = arg(\vec{D_0 D_i}), \quad 0 < \phi_i < 2\pi, \quad i = 1, 2, 3, 4. \tag{10}$$

The relationship between the angles θ, ϕ, and ψ at the i th position is

$$\psi_i = \phi_i - \theta_i - \pi, \quad 0 < \psi_i < 2\pi, \quad i = 1, 2, 3, 4. \tag{11}$$

The i th relative angular displacements ϕ_{1i} and ψ_{1i} with respect to position 1 are

$$\phi_{1i} = \phi_i - \phi_1, \quad -\pi < \phi_{1i} < \pi, \quad i = 1, 2, 3, 4, \tag{12}$$

$$\psi_{1i} = \psi_i - \psi_1, \quad -\pi < \psi_{1i} < \pi, \quad i = 1, 2, 3, 4. \tag{13}$$

And we define ϕ_{max} as the maximum value of ϕ_{1i}'s and ϕ_{min}, ψ_{max}, and ψ_{min} are defined in the same manner.

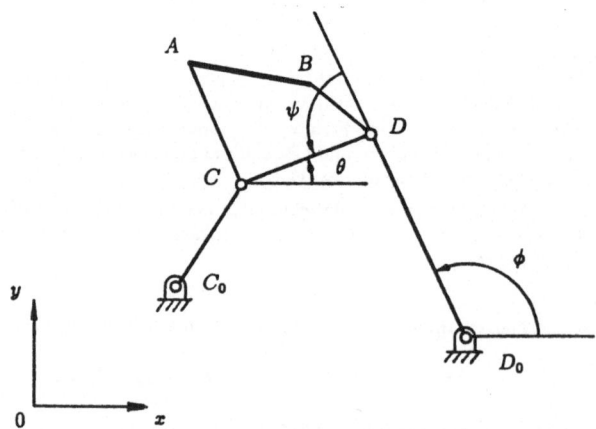

Figure 1.

Circle Point Curve Plotting

The cubic equation (1) of the circle point curve can be found in several papers. The coefficients of the equation are generated by the procedure used in reference [8]. In order to obtain a good distribution of points on the curve and to center the computation on the most useful portion of the curve, the four given positions of the lamina are located in the center of the view window, and the points on the curve are found by solving the cubic equation in both x and y with equal increments. After displaying the curve, the computer gives the designer options by which he can magnify, reduce, and/or translate the curve to produce the most useful portion of the curve.

After the circle point curve is generated, x_0, y_0, $l_{D_0 D_1}$, ϕ'_{1i}s, ψ'_{1i}s, ϕ_{max}, ϕ_{min}, ψ_{max}, and ψ_{min} are computed using the equations developed in the previous section for solving branch, order, and Grashof problems for each point on the curve.

Range of the Circle Point Curve for the Driven Crank

This range is generated by considering branch problem, order problem, and control of the crank link length. Waldron [3] has demonstrared a technique for locating the permissible region for driven cranks which subsequently assits in obtaining the Filemon's region for the location of the driving crank. The location is decided by choosing the points which have angular range of driven crank less than π. If we use the previously defined notations, the range of the circle point curve for the permissible driven cranks is the group of the points which satisfies the following equation:

$$\psi_{max} - \psi_{min} < \pi. \tag{14}$$

Since the ϕ_{1i}'s are in between $-\pi$ and π, we define the order of the crank as the order of these values. For example, if $\phi_{11} = 0, \phi_{12} = -0.8, \phi_{13} = 1.5$, and $\phi_{14} = -2.5$, then the order of the crank becomes 3124. In order to select a crank rocker mechanism, the order of the driven crank must be one of the orders shown in Table 1. If the order is 1423, the driven crank has two strokes during one rotation of the driving crank. Also for the drag-link mechanisms, the order of the driven crank must be one of the orders shown in Table 1, otherwise the crank links can not rotate continuously.

For the crank-rocker mechanisms, the maximum range of the angular displacement of the driven crank can not exceed π. Thus

$$\phi_{max} - \phi_{min} < \pi. \tag{15}$$

If the designer wants to control the link length of the linkage, the program asks for the maximum allowable link length l_{max}. The allowable range of the curve for the maximum link length is generated by the following equation:

$$l_{D_0 D_1} < l_{max}. \tag{16}$$

Each point on the curve is used to check the above equations. If all the equations are satisfied, the point is plotted for the range of the circle point curve for the driven crank. After the desired range of the curve is displayed, the cursor appears on the center of the screen. Using the cursor movement keys, the designer can select the circle point for the driven crank on this range of the circle point curve.

Range of the Circle Point Curve for the Driving Crank

This range will depend on the chosen circle point for the driven crank and is generated by considering branch, order, and Grashof problems. The range for the maximum link length of the cranks has already been checked in the previous section.

Fillemon [2] has studied the geometric solution to the branch problem. She has developed a construction which, for a given choice of the driven crank, locates all the driving cranks which, when combined with that crank, produce a linkage that can traverse the design positions without change of branch. Using her method, simple mathematical equations are developed for use in our computer technique.

As shown in Figure 2, the circle point and the center point of the driven crank are $D_1(x_{D_1}, y_{D_1})$ and $D_0(x_{D_0}, y_{D_0})$ repectively. The slopes of lines m and n become

$$m = tan(\phi_1 - \psi_{max}), \tag{17}$$

$$n = tan(\phi_1 - \psi_{min}), \tag{18}$$

The point $C_1(x_{C_1}, y_{C_1})$ is examined to check if it is located in the Filemon's region. The slope g of the line $D_1 C_1$ is

$$g = \frac{y_{C_1} - y_{D_1}}{x_{C_1} - x_{D_1}}. \tag{19}$$

The allowable point C_1 for the linkage free of branch change is obtained by checking the following conditions:

$$m < g < n \quad if \quad n > m, \tag{20}$$

$$g > m \quad or \quad g < n \quad if \quad n < m. \tag{21}$$

If the point C_1 satisfies above conditions, the resulting linkage is guaranteed to be free of the branch problem. Since the driving crank of crank- rocker or drag-link mechanism has to have a continuous rotation in clockwise or counterclockwise direction, the order of the driving crank must be one of the orders shown in Table 1.

For the crank-rocker mechanisms, the center point of the driving crank must be located in the unshaded area shown in Figure 3. Simple equations are developed as follows. The center pointer of driving crank is $C_0(x_{C_0}, y_{C_0})$. The slopes of lines m' and n' become:

$$m' = tan(\phi_1 + \phi_{min}), \tag{22}$$

$$n' = tan(\phi_1 + \phi_{max}), \tag{23}$$

The point $C_0(x_{C_0}, y_{C_0})$ is examined to check if it is located in the shaded area shown in Figure 3. The slope g' of line $D_0 C_0$ is

$$g' = \frac{y_{C_0} - y_{D_0}}{x_{C_0} - x_{D_0}}. \tag{24}$$

The allowable point C_0 for the crank-rocker mechanisms is obtained by checking the following conditions:

$$m' < g' < n' \quad if \quad n' > m', \tag{25}$$

$$g' > m' \quad or \quad g' < n' \quad if \quad n' < m'. \tag{26}$$

Since the driven crank has been chosen, the four lengths, $l_0, l_1, l_2,$ and l_3 can be decided for a circle point of the driving crank:

$$l_0 = [(x_{D_0} - x_{C_0})^2 + (y_{D_0} - y_{C_0})^2]^{\frac{1}{2}}, \tag{27}$$

$$l_2 = [(x_{D_1} - x_{C_1})^2 + (y_{D_1} - y_{C_1})^2]^{\frac{1}{2}}, \tag{29}$$

$$l_1 = L_{D_0 D_1}, \tag{28}$$

$$l_3 = L_{C_0 C_1}, \tag{30}$$

The designed linkage type is confirmed using Grashof's rule.

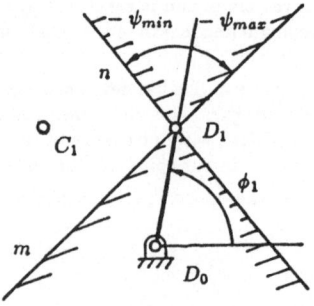

Figure 2.

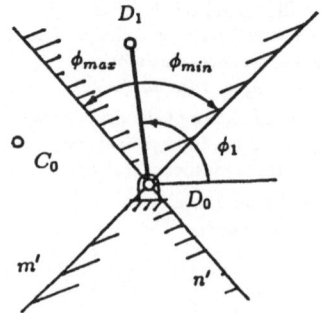

Figure 3.

Example Problem

A four-bar linkage of the crank-rocker type is to be designed having a coupler lamina which has to pass through the four specified positions as shown in Table 2.

1. The program asks for the input of the data for the four positions, type of four-bar linkage to be designed, and maximum link length allowed.
2. With the input data, the computer displays the four positions of the lamina and the circle point curve on the screen as shown in Figure 3. In order to locate the useful range of the curve in the center of the screen, we can magnify, reduce, and/or translate the curve by typing the controlling key shown in the dialogue area at the bottom of the screen.
3. After the circle point curve is established, the computer shows the range of the circle point curve for the driven crank and the cursor as shown in Figure 4. By moving the cursor, the designer selects the circle point for the driven crank. Then the program obtains the driven crank which is displayed in its first position on the screen. This driven crank can be retained by the designer or a new driven crank can be designed by typing the controlling key in the dialogue area. In this example, the circle point for the driven crank is chosen at (3.23, 0.81).
4. After the driven crank has been designed, The computer displays the range of the circle point curve for the driving crank and the designed driven crank as shown in Figure 5. The driving crank is selected in a similar fashion as before. In this example, The circle point for the driving crank is chosen at (-1.84, 7.07). The designed four-bar linkage is shown in Figure 6 along with its dimensions. This solution linkage can be animated by typing the controlling key to illustrate the movement of the lamina through the four desired positions.

Conclusions

This paper has presented a convinient micro-computer based technique for the four position synthesis of four-bar linkage mechanisms. The CAD method developed uses interactive computer graphics on a personal computer. The presented method permits the design of linkages that are free of branch problem, order problem, and Grashof problem. The progrom also controls the crank link lengths. The presented method is fast, efficient but simple to use.

References

1. Burmester, L., "Lehrbuch der Kinematik," Felix, Leipzig, 1988.

2. Filemon, E., "In Addition to the Burmester Theory," Proceedings of Third World Congress for Theory of Machines and Mechanisms, Kupari, Yugoslavia, Vol. D, 1971, pp. 63-78.

3. Waldron, K. J., "Elimination of the Branch Problem in Graphical Burmester Mechanism Synthesis for Four Finitely Seperated Positions," ASME, Journal of Engineering for Industry, Vol. 98, No. 1, 1976, pp. 176-182.

4. Waldron, K. J., "Graphical Solution of the Branch and Order Problems of Linkage Synthesis for Multiply Seperated Positions," Journal of Mechanism Design, Transactions of the ASME, August 1977, pp. 591-597.

5. Chuang, J. C., Strong, R. T., and Waldron, K. J., "Implementation of Solution Rectification Techniques in an Interactive Linkage Synthesis Program," Journal of Mechanism Design, Transactions of the ASME, July 1981, Vol. 103, pp. 657- 664.

6. Wilhelm, A. J., and Sodhi, R. S., "Burmester Synthesis for Four-Bar Linkages on a Microcomputer CAD System," Proceedings of the tenth OSU Applied Mechanisms Conferences, New Orleans, Dec 1987.

7. Strong, R. T., "Improvements to Mechanism Synthesis Methods," PH.D. Dissertation, University of Houston, 1978.

8. Kim, Hyoung Jun, and Sodhi, Raj S., "Derivation of Burmester Curve Equations Using Displacement Matrix and Matrix Threorem of Linear Equations," SBR-#119-000, NJIT, 1987.

Table 1.

Crank-rocker			Drag link	
Driving	Driven		Driving	Driven
1234	1234	3214	1234	1234
2341	1243	3241	2341	2341
3412	1423	3412	3412	3412
4123	1432	3421	4123	4123
4321	2134	4123	4321	4321
3214	2143	4132	3214	3214
2143	2314	4312	2143	2143
1432	2341	4321	1432	1432

Table 2.

Position	Coordinates		Angle
	x	y	
1	0.40	5.60	20
2	-5.40	3.45	92
3	-2.00	1.02	71
4	0.00	0.00	40

CIRCLE POINT CURVE AND FOUR POSITIONS

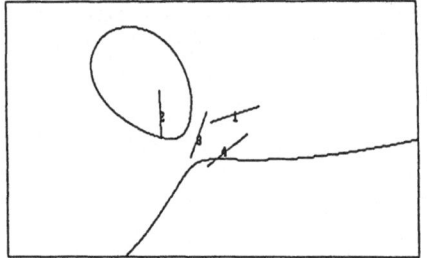

M : Magnify,   R : Reduce,   T : Translate

Figure 4.

RANGE OF DRIVEN CIRCLE POINT

Choose the driven circle point using cursor.

Figure 5.

RANGE OF DRIVING CIRCLE POINT

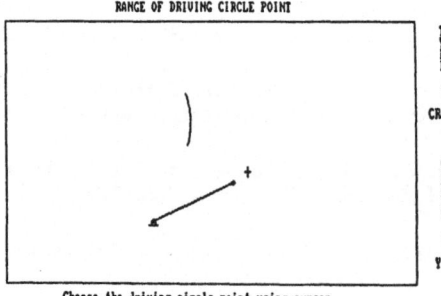

Choose the driving circle point using cursor.

Figure 6.

DATA OF DESIGNED FOUR-BAR LINKAGE

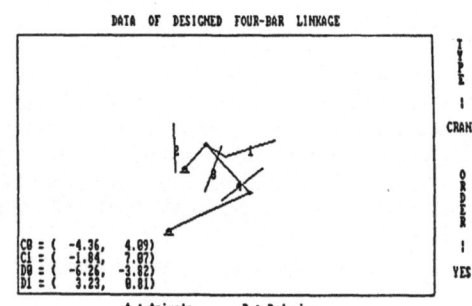

C0 = (-4.36, 4.89)
C1 = (-1.84, 7.07)
D0 = (-6.26, -3.82)
D1 = (3.23, 0.81)

A : Animate,   R : Redesign

Figure 7.

Simulation as a Design Tool

Parametric Study of an Electromagnetic Valve and the Fluid System

R. P. Sharma and P. K. Patel

Department of Mechanical Engineering
College of Engineering and Applied Sciences
Western Michigan University
Kalamazoo, Michigan 49008

Summary

In the last few years, the transportation industry has made significant improvements in the automobiles from a weight reduction, electronic controls and aerodynamic point of view. The electromagnetic valves have many applications in the engine and the vehicle subsystems because of increasing use of electronic controls and microprocessor. The electromagnetic valves are used as check valves in the fluid systems.

A mathematical computer model was developed to design and simulate the performance of an electromagnetic valve and its application in a fluid system. The mathematical model has been used to design and study the effects of various parameters on the performance of an electromagnetic valve and in turn on the dynamics of the fluid system. The different parameters were wire size, voltage, fluid pressure, flux density, valve travel, spring stiffness and other design variables of the electromagnetic valve and the system.

The valve travel, temperature, current and voltage have significant effect on the size and performance of the electromagnetic valve and in turn on the dynamics of the system.

Introduction

The electromagnetic valves have many applications in the engine and the vehicle subsystems. One of the simple applications of the electromagnetic valves is in the fuel injection system for any type of engine [1]. The solenoids are used to close and open the fuel supply to the fuel pump and indirectly control the fuel injection system's performance.

Fluid System Description

Figure 1 shows the generalized setup of a fluid flow system with electromagnetic valve for the fuel injection system application.

The fuel injection system consists of the fuel pump, injection line and injector. The fuel pump is a positive displacement pump. The fuel is supplied to the fuel pump at a pressure of 195 kPa. The fuel pump pressurizes the fuel and delivers it through the delivery valve to the injection line for onward supply to the injector.

In the electromagnetic pump, the timing and quantity of fuel injection is controlled by electromagnetic inlet-spill valves. The electromagnetic valves, one for each cylinder, are solenoid operated valves. These valves close the inlet-spill port for a different duration for required fuel injection into the combustion chamber.

Pump and Valve Operation

The face cam starts the upward stroke of the plunger from bottom dead center with an acceleration. The plunger moves with a constant velocity for a period of about 110° of pump shaft angle, after the acceleration period. During this period, the fuel displaced by the plunger can be pressurized and injected into the combustion chamber through the delivery valve, injection line and injector. The pressurization is achieved by controlling the closed duration of electromagnetic solenoid operated spill valve. Fuel injection commences when spill port is closed by energizing the electromagnetic (solenoid operated) spill valve of the particular plunger. Fuel injection ceases when the spill valve is de-energized and the fuel pressure in the plunger cavity forces the spill port to open. In this open spill port position the fuel displaced by the plungers is returned to the supply pump.

Valve Model

Figure 2 shows the lumped spring mass system model of the electromagnetic valve with electrical voltage and current to the solenoid as the driving force input. The fluid viscous damping was assumed as a constant value and calibrated using the fluid system available test data in literature.

Mechanical Equation

$$M\ddot{y} = \Sigma Fe = -B\dot{y} - Ky - Fo + F(t)$$

where $F(+) = 1/2 \ i \ \dfrac{2\partial L}{\partial y}$

$$L = \dfrac{\lambda}{L} = N^2 p = \dfrac{NC_1}{(L_1 - y)}$$

$$C_1 = \mu N^2 \pi r_1^2$$

Nomenclature

M = mass of moving armature
B = viscous damping (friction coefficient)
K = spring stiffness
R = resistance of coil
L = inductance of winding of coil
V = impressed voltage to coil

$$\frac{\partial L}{\partial y} - \frac{C_1}{(L_1-y)^2}$$

Electrical (Driving Force) Equation:

$$V - iR + L\frac{di}{dt} + i\frac{\partial L}{\partial y} \cdot \frac{dy}{dt}$$

Say: $Y_1 - i$, $Y_2 - Y$, $Y_3 - Y_2$

$$Y_3 - \frac{1}{M}[F(t) - Fo-KY-BY]$$

Y   - valve travel
P   - permeance
i   - current
F(t) - force function
        (electrical)
N   - number of turns
r_1 - radius
L_1 - valve travel limit
μ - permeability
Fo - fluid force (pressure)
λ - flux linkages

$$Y3 - - [F(t) - Fo-Ky_2-By_3]. . .(1a)$$

$$F(t) - \frac{}{2} Y1^2 \frac{C_1}{[L_1-y_2]^2} . . .(1b)$$

$$Y_3 - Y_2 \quad . . .(2)$$

$$V(t) + Y R+LY Y_1Y_3\frac{C_1}{[L_1-Y_2]^2} . . .(3)$$

For the given initial conditions these unknown variables, Y1, Y2, and Y3, are solved numerically for the fluid system for equations (1), (2), and (3). The model was solved and simulated using the RungeKutta Numerical Method (RKF - 45) on PDP-10 Computer. Once the model was calibrated then the effect of different design and operating variables on the delay time of the electromagnetic valve were studied. The simulated values were also compared with the available test data. There is good agreement between the test and simulated data from qualitative point of view. Effect of some of the variables is discussed here in the next part of the paper.

Parameters under simulation study:

1. Activation time versus supply voltage.
2. Activation time versus valve stroke
3. Activation time versus mass of the moving parts.
4. Activation time versus spring stiffness.
5. Activation time versus fluid force/pressure.

Voltage Effect

The delay (activation) time of the electromagnetic valve decreases linearly with the increase in voltage. Figure 3 shows this effect for three different valve travels (strokes).

Figure 4 shows the effect of supply voltage on activation time at three different valve travel settings.

Figure 5 shows the effect on activation time verses valve travel (stroke) at different fluid resistive pressures in the system.

Figure 6 shows the effect of increase in the moving mass on the delay (activation) time at different stroke settings.

There was no effect on the activation time of the valve because of the changes in the spring stiffness (Figure 7).

Conclusions

1. The mathematical model simulates the effect of parameters on the performance in a fluid system.
2. The activation time decreases linearly with increase in supply voltage.
3. There is no effect on system performance because of the changes in spring stiffness.
4. There is significant effect on the activation time because of the valve travel (settings).
5. There is significant effect on the activation time because of the changes in the moving mass of the system.

References
1. Sharma, R.P., "Parametric Study of the Fuel Injection System and Fuel Delivery Line," 1983 Annual ASME Conference, November, 1983, 83-WA/DSC-27.
2. Ledex Solenoid Manuals, Ledex, Ohio.

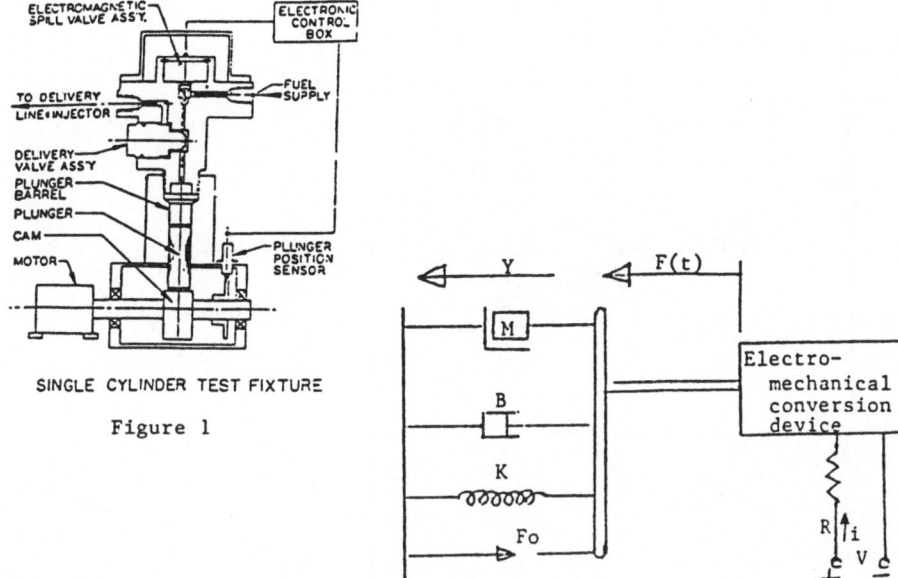

SINGLE CYLINDER TEST FIXTURE

Figure 1

Figure 2 - LUMPED SPRING MASS SYSTEM MODEL

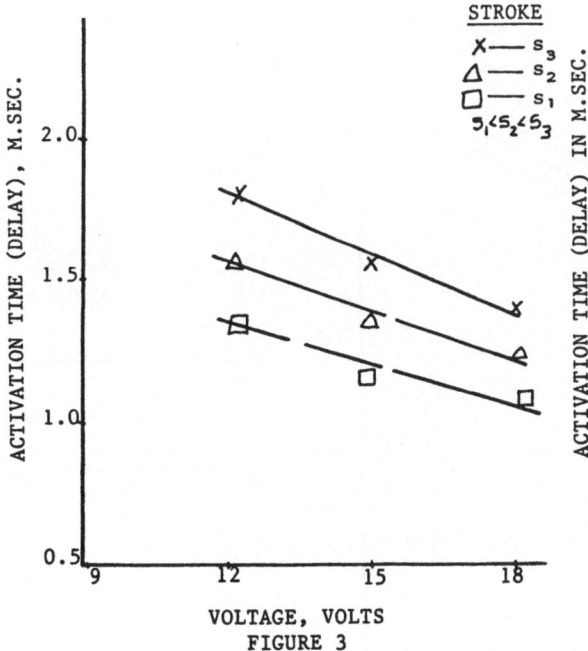

FIGURE 3

96

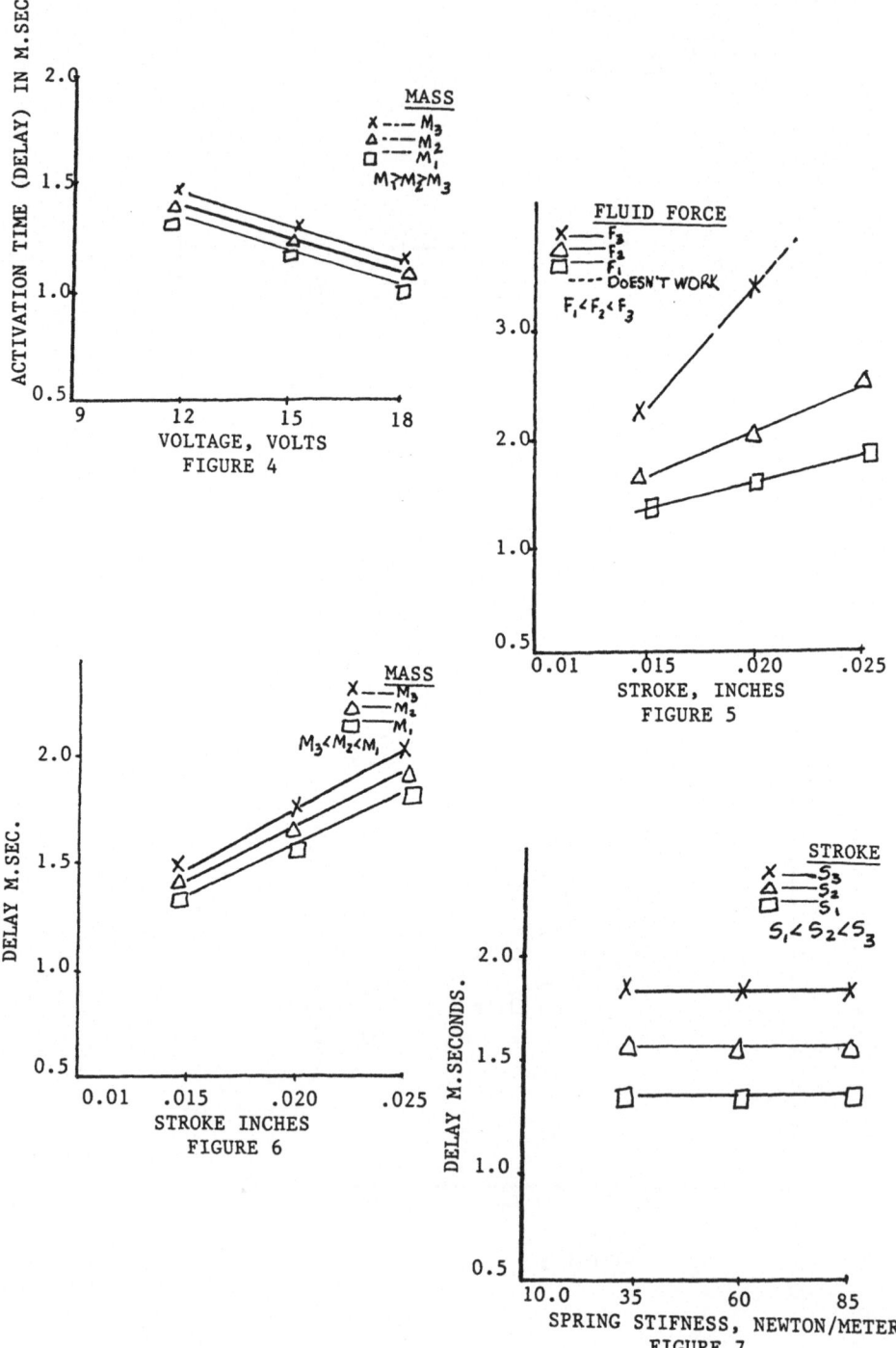

MASS
X ---- M₃
△ -·-· M₂
□ ---- M₁
M₁>M₂>M₃

ACTIVATION TIME (DELAY) IN M.SEC.

2.0

1.5

1.0

0.5

9   12   15   18
VOLTAGE, VOLTS
FIGURE 4

FLUID FORCE
X ---- F₃
△ ---- F₂
□ ---- F₁
---- DOESN'T WORK
F₁< F₂< F₃

3.0

2.0

1.0

0.5

0.01   .015   .020   .025
STROKE, INCHES
FIGURE 5

MASS
X ---- M₃
△ ---- M₂
□ ---- M₁
M₃< M₂< M₁

DELAY M.SEC.

2.0

1.5

1.0

0.5

0.01   .015   .020   .025
STROKE INCHES
FIGURE 6

STROKE
X ---- S₃
△ ---- S₂
□ ---- S₁
S₁< S₂< S₃

DELAY M.SECONDS.

2.0

1.5

1.0

0.5

10.0   35   60   85
SPRING STIFNESS, NEWTON/METER
FIGURE 7

Availability Analysis of a Repairable System Through Computer Simulation

K.C. KURIEN, G.S. SEKHON, O.P. CHAWLA

Indian Institute of Technology, Delhi (INDIA)

Summary

Availability analysis of repairable systems is of undeniable economic importance. The present study deals with an application of the discrete event simulation technique to the analysis of availability of a repairable system. A detailed theoretical study of the effect of the various system parameters has been carried out.

Introduction

Availability of repairable systems is of crucial practical and economic importance for any set-up equipped with such systems. Its analysis is however beset with several difficulties and complexities which include environmental factors, operating and service procedures, skills of operating and maintenance crews, age of the systems, management policies and so on. Published studies on availability of real-world repairable systems are quite scarce. However the concept of availability is quite extensively discussed in literature e.g. Barlow and Hunter(1), Sandler(2) and Myers et al (3). Also some studies on system availability analysis based upon the Markov modelling approach have been reported. Most of them e.g. Mann et al. (4) and Khan and Gupta (5) are limited to systems with two components and two or three transition states. The Markovian model proposed by Shetty et al. (5) analyses the availability of a system consisting of M subsystems with given mean failure and repair rates.

Proposed Model

A typical facility is equipped with a known number of repairable systems. During operations, each copy is subject to failure. On detection of a malfunction, the cause of failure is diagnosed and the system is either sent for repairs or sidelined pending procurement of spares. Afterwards the system may be reinspected to confirm its operation worthiness or otherwise. The system may have to wait in the queue for some time before it is pressed into service once again. A schematic representation of the typical activities involved in such a facility is given in Fig. 1.

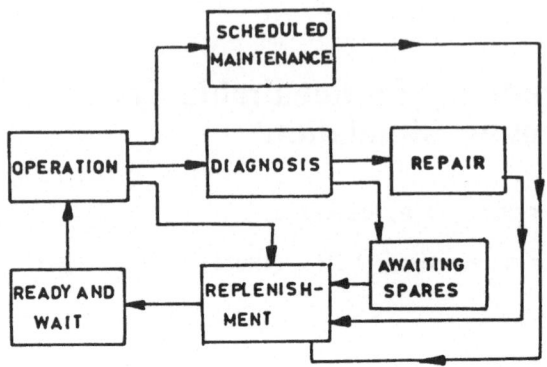

Fig.1. Activity Diagram

Computer simulation of a complex system as described above is based upon
the assumption that log-books or records of actual facility operations are
available for extraction of the following data. (i) Frequencies and time-to
-failure intervals for each subsystem, (ii) Probabilities of the spares
being available (when required) for repairing each subsystem, (iii)
Frequencies and time-to-repair intervals for each subsystem, (iv)
Frequencies and procurement-time intervals (hereafter referred to as AOG
clearance time intervals) in respect of the various subsystems, (v)
Scheduled maintenance intervals and the time required for their completion.

Application to Aircraft Training Facility

The different operations in an aircraft training facility can be divided
into six activities (Fig.2). At any given time, a particular aircraft of
the fleet can belong to only one of the six sets defined in the above
figure. The completion of an activity is marked as an event. On the

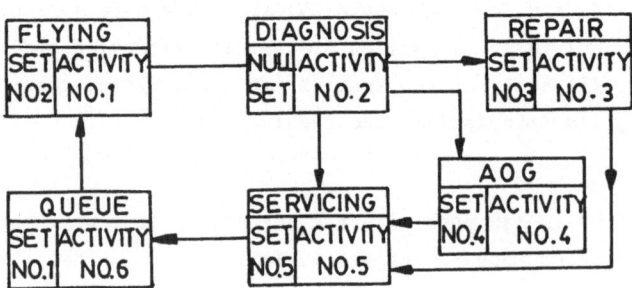

Fig. 2. Typical Activitives and Sets in an Aircraft Training Facility

occurrence of an event, an aircraft is shifted from one set to another. For example, all aircrafts after replenishment wait in queue in Set No.1. A specified minimum number of aircrafts are shifted to Set No.2. on satisfaction of the following conditions - (a) Shifting is possible only at fixed hours each day, (b) maximum number of sorties per day is prescribed, and (c) a minimum number of aircraft must be available in Set No.1 to launch a detail. At the beginning of simulation, all aircraft are assumed to be operable and held in Set No.1.

Computational Results

A Fortran program based upon the discrete-event simulation procedure was prepared for the model represented in Fig.2. Each aircraft was modelled as a system consisting of 13 repairable subsystems. A unit time was defined as a duration of 30 min., duration of a sortie was 2 time units, a working day constituted of 24 time units, flying activity was limited to 12 time units per day, 15 sorties in a day were aimed at and one month comprised of 25 working days. Typical computational results are presented in Table 1.

Table 1. Results describing state of the facility at the end of each month.

Month No.	NSORT	MISNSS	NRPRS	NAOGS	NSLST	AVLBTY	AVRPT	HRPRS
1	375	305	73	9	17	0.5082	7.4880	5.149
2	364	296	63	8	16	0.3888	8.4710	6.118
3	339	281	53	11	11	0.3744	9.9794	3.947
4	355	291	68	6	15	0.3309	10.4648	5.058
5	375	305	77	4	15	0.5960	10.9553	4.311
6	373	308	67	11	18	0.4411	7.4424	5.694
7	366	298	69	5	17	0.4524	8.4344	6.694
8	375	316	57	11	12	0.4095	9.0707	3.367
9	375	319	54	7	15	0.4981	9.6053	3.325
10	372	304	72	5	17	0.4240	10.4543	5.224

Legend.
    NSORT: number of sorties completed
    MISNSS: number of sorties successfully completed
    NRPRS: number of repairs
    NAOGS : number of times aircraft grounded pending procurement of spares
    NSLST: number of scheduled maintenance (second line services)
    AVLBTY: aircraft availability
    AVRPT: average repair time
    HRPRS: average maintenance time per flying hour

As part of simulation experiments, the failure frequencies corresponding to the smaller failure time intervals were reduced by 20, 40 and 60% to assess

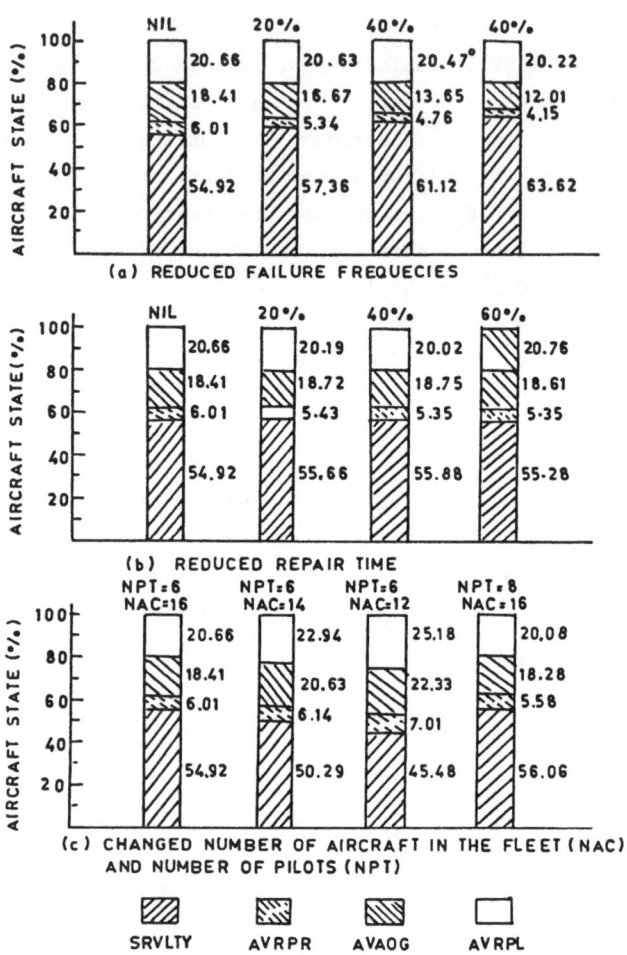

Legend

AVAOG: Monthly average number of aircraft on AOG

AVRPL: Monthly average number of aircraft on scheduled maintenance

AVRPR: Monthly average number of aircraft under repairs

SRVLTY: Monthly average number of serviceable (available) aircrafts

FIG. 3. Availability - related parameters
(results of simulated experiments)

their effect on system availability. Similar experiments by reducing repair times were also carried out. The effect of the above changes on the relative proportions of the number of aircraft in different sets is depicted in Fig. 3. It also shows the results of other experiments such as the number of aircraft in the fleet and the number of available pilots.

Discussion of Results

Computational results on subsystem performance show that subsystem 1 has a low availability and high average repair time. Its failures are fewer but the repairs take longer. The cause was found to be the large spares procurement time. Opposed to this, subsystem 12 fails more frequently but it also gets repaired faster. Monthly performance figures (Table 1) showed that the system availability varies from 0.3309 to 0.5960 even through the input resources are maintained at the same level. Without managerial intervention in the form of staggering of operating hours, rescheduling of maintenance or emergency procurement of spares etc., it is not possible to maintain a constant monthly average.

Conclusion

Availability of a complex repairable system in a given facility depends upon a number of operational and maintenance parameters. Due to inherent versatility of the simulation technique, it is however possible to include as many parameters in the analysis as desired. This flexibility and also the possibility of carrying out a systematic parameteric study of the effect of the various parameters on system availability makes computer simulation an attractive tool for optimizing a given facility.

References

1. Barlow, R; Hunter, L.C.: Optimum preventive maintenance policies, Op. Res.,8 (1960), 90-100.

2. Sandler, G.I.: System Reliability Engg., Prentice-Hall, Englewood Cliffs, (1963).

3. Myers, R.H.; Wong, K.L.; Gordy, H.M.: Reliability Engg. for electronics systems, John Wiley, N. York(1964).

4. Mann, N.R.; Schaffer, R.E.; Singpurwall, N.D.: Methods for statistical analysis of reliability and life data, John Wiley, N. York(1974).

5. Khan, N.M.; Gupta, A. : Availability analysis of 3-state systems, IEEE Trans. Rel., R-34(1985),86-87.

6. Shetty, B.N.; Sekhon, G.S.; Chawla, O.P.; Kurien, K.C.: Optimization of maintenance facility for a repairable system subject to availability constraint, Paper accepted for publication in the Microelectronics and Reliability J. (England)

Realtime Simulation of Manufacturing Systems

Sudhakar Paidy
Department of Industrial and Manufacturing Engineering
Rochester Institute of Technology
Rochester, New York 14623

S U M M A R Y

An increasingly important area of concern in implementing and
monitoring automated manufacturing facilities lies in avoiding
operational problems when expensive machine tools, robots and
conveyor systems are integrated into workcells. Simulations can
determine whether the plans for a system are accurate and
whether the chosen equipment meets the need. While analytical
and statistical simulations are very powerful and useful, they
lack the ability to portray on overall view of the system to
effectively integrate all elements of the system. Graphic
simulations can be used to supplement the traditional simulation
models. Further more, the "realtime" simulations closely
coupled with concurrent opeartion of the system will provide an
excellent analysis tool as well as realtime process monitoring.
This paper will present such realtime simulations developed in
the RIT's CIM laboratory.

I N T R O D U C T I O N

In the design and control of industrial facilities, designers
are faced with complex structural and operational aspects of
facility components as well as their integration into a unified
production system. Traditionally, in the academic environment,
the complexities have been analyzed using abstract, analytic
and/or numeric simulation models. The steadily increasing
capabilities of small scale computers (microprocessors to
microprocessors) coupled with their low cost, is resulting in an
added dimension to the field of simulation – realtime/dynamic
simulation.

This paper presents a study related to realtime simulation and
graphical animation of manufacturing situations using physical
simulators. The physical simulators are miniature prototype
models of real–world situations (e.g. an assembly line, an
automated storage and retrieval warehouse, a flexible
manufacturing center) with almost all of their essential
features and complexities. Under computer control, these models
mimic the operation of the desired system. Availability of
inexpensive and versatile modular building blocks for building
physical simulators is making them increasingly popular,
economical and convenient in both laboratory and real world
applications requiring realtime data acquisition and control.

The importance of visual simulations in the form of graphical
animation and/or development of physical simulators in studying
the manufacturing systems can not be over emphasized. The trend
towards dynamic simulations to supplement traditional
athematical and numerical simulations is also evident from the

emergence of the state-of-the-art grpahics oriented simulation software systems such as Cinema (Systems Modeling Corporation), TESS (Pritsker Associates), Autogram (Auto Simulations), among others. The realtime simulation has additional benefits in understanding the intricacies of the computer interfacing and integration of manufacturing elements, realtime data acquisition and control.

THE PHYSICAL SYSTEM

At Rochester Institute of Technology, within the Computer Integrated Manufactuirng Laboratory of the Industrial and Manufacturing Engineering Department, a computer controlled Automated Storage and Retrieval System (AS/RS) has been developed to serve a manufacturing cell and its receiving and shipping areas. The manufacturing cell consists of a milling station, a turning station, and a drilling station. The AS/RS is a single aisle-two bay warehouse. Conveyors are used to buffer the parts in the system while three mobile robots support the most of the material handling. One of the mobile robots is a stacker crane serving the warehouse. The second mobile robot moves the parts from the warehouse to the input conveyor of the manufacturing cell. The third robot routes the parts from the input conveyor, to and from the three machining stations, and to the output conveyor. The robot responsible for moving the parts from the warehouse conveyor to the cell's input conveyor also transports the parts from the cell's output conveyor to the warehouse conveyor. Figure 1 shows a schematic of the overall system. Figure 2 is a photographic view of the system.

The entire model/physical simulator is built with fisher-Tech model components. The model occupies approximately 4 ft. x 8 ft. area. The model resembles in its functionality the actual manufacturing cell in the departmental CAM laboratory. The current state of the model is an evolution over a period of about eight years and has been developed entirely by the undergraduate students in the realtime programming and related courses. The overall effort that shaped this model is in excess of 5 man years and perhaps can now be duplicated in about a year. Under the control of a mini-computer and a micro-computer the model can mimic the operations of the actual system. The model requires realtime data acquisition and control in its operation and provides an insight into the intricacies of real system. Although the physical simulators are expensive to build, the author has found these models to be very useful in teaching a variety of subjects including simulation, realtime data acquisition and control, computer integrated manufacturing, work sampling for almost a decade.

The scope of this paper allows the focus only into the development and control of the AS/RS simulator. As mentioned above, the AS/RS is a single aisle-two bay warehouse but can be easily expanded to additional bays. Each of the bays has storage space for pallets at four different levels vertically. Upto eight storage spaces have been used along the track of the stacker crane. The stacker crane is about 7 inches tall over a two inch high base platform with four wheels. It has four axes/degrees of freedom: (a) left/right movement along the track, (b) clockwise/counter-clockwise rotation, (c) up/down level adjustment for reaching one of the four vertical bins, and

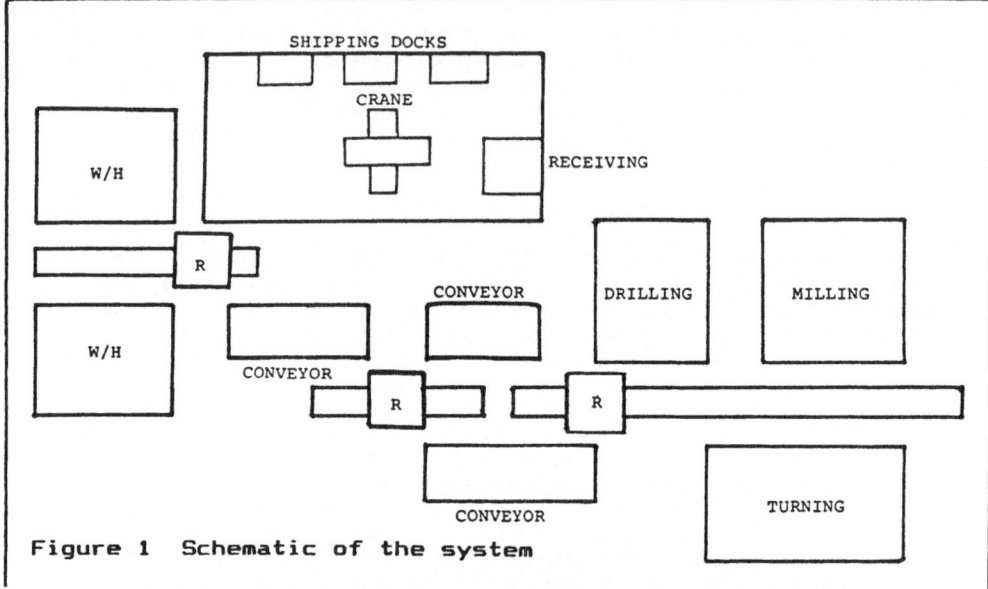

SHIPPING DOCKS
CRANE
RECEIVING
W/H
R
CONVEYOR
DRILLING
MILLING
W/H
CONVEYOR
CONVEYOR
R
R
CONVEYOR
TURNING

Figure 1   Schematic of the system

Figure 2   Physical simulator

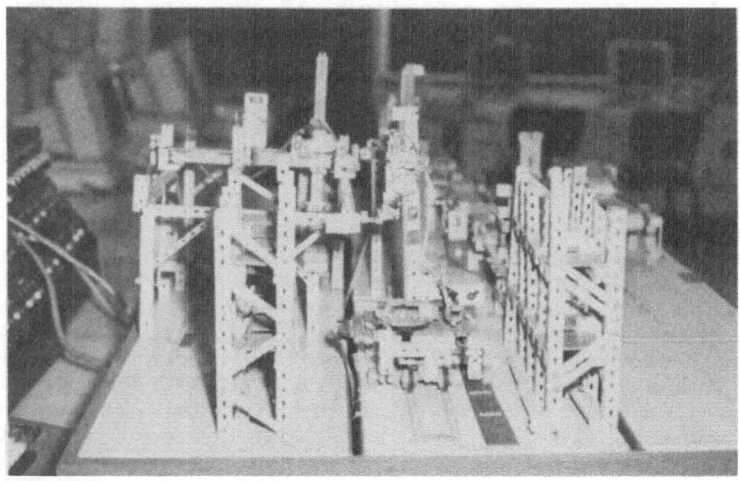

Figure 3   AS/RS simulator

(d) in/out fork movement. The forks are approximately 2 inches
long. Each of the axes has a bi-directional motor with limiting
switches at exterme positions. The intermediate positions on
the horizontal and vertical axes are determined using two light
sensors. Figure 3 is a photographic view of the model.

On demand, the stacker crane retrieves a pallet from the
warehouse and places it on the warehouse conveyor leading to the
manufacturing cell. When the pallets of the finished parts are
returned, the stacker crane will pick them up and store them in
the warehouse unitl requested by the shipping areas. Pallets of
the raw materials arrive from the receiving area. The pallets
used are about 2 in. x 2 in. x 3/16 in. with barcode labels on
the top. No real parts are carried on the pallets as such but
the barcode information is used to determine the type of the
part, intended process sequence, and the required material
handling. A fixed beam bar code scanner (skanamatic series S23)
is mounted on top of the conveyor connecting the warehouse to
the manufacturing segment. The code is read when
the pallet moves under the scanner. If a bad read is
registered, the conveyor is reversed (forward/backward) until
the code is read successfully. As the AS/RS warehouse is used
as a storage for raw material, work-in-process and finished
goods, the requests for the pallet storage can come from the
receiving dock as well as the manufacturing segment. The
retrieval requests come from three shipping docks and the
manufactuirng area. The strategies (FIFO, LIFO, random
selection, or nearest bin, etc.) for storage and retrieval for
each of the sources of requests are independent and can be
different and may change during the system's operation.

Simulation studies are conducted and software developed to
handle the four types of requests for the AS/RS. Various
priority schemes and typical storage and retrieval starategies
(FIFO,LIFO,nearest bin,Random location, etc.) are experimented
with model. The results at present are pedagogical. An
expansion is being designed for the actual CAM cell in which
these results will be used.

The realtime program design, computer hardware/software
interfaces, and data structures are presented in the following
sections.

R E A L T I M E   P R O G R A M   D E S I G N

Several definitions exist for "Realtime computing" [1]. In its
simplest form, a realtime program is one that performs all
system functions within the specified time constraints. The
realtime data collection and processing involves the
inter-connection of a system with a computer utilizing analog/
digital, digital/ analog, and/or generalized binary (digital)
input/ output data interfaces. The digital signals are required
to monitor the status of and control on/off type devices while
the analog input/ output signals are used to sense and control
real entities such as pressure, temperature, displacement, etc.
The program execution times as dictated by the program logic and
process time must be synchronized in order to monitor and
control the process in realtime. A common need of a process
control system is its ability to handle several events

concurrently. Thus, a hierarchy of programs, computers and/or computer based devices may be needed for a complex process control system.

The manufacturing cell is controlled by a Cocurrent Computer (model 3220 mini-computer) and the warehouse is controlled by an IBM PC-AT. The receiving and shipping areas are simulated using another PC. The minicomputer is capable of running multiple programs (multi-tasking) with inter-task communications and interrupt handling. Many choices are now available for a microcomputer operating system including PC/MS DOS, OS/2, PC MOS, CP/M 86, concurrent CP/M, VRTX, QNX, XENIX and many UNIX like systems. Details of multi-tasking featurs needed in realtime applications and how several micro-computer operating systems provide these are documented elsewhere[2].

DOS is perhaps the most popular operating system with its largest microcomputer user base and software availability. In this study, the PC is run under DOS which is strictly single task oriented (but multi-tasking can be and is simulated is this study as discussed later) and is connected to the mini-computer using its RS232 serial port.

MULTIPLE TASKS
The software for the AS/RS interfacing and manufacturing system simulation is implemented as eight different tasks on the minicomputer. Tasks 1-3 control the three maching stations. Tasks 4-5 control the two robots. Task 6 provides the priority scheme and a que for all storage and retrieval requests to the AS/RS. Task 7 is a slave task directing all digital I/O requests from all other tasks to a microprocessor controlled device. Task 8 is the supervisory task providing initial conditions, inter-task coordination, and system performance statistics.

The control and animation software for the AS/RS is implemented as four modules on the PC. These three modules are setup to run independently and are called sequentially from the main program to perform their intended operations. The first module controls the motions of the stacker crane. The second module controls the barcode reader and the conveyor motions. The third module is responsible to interface with the supervisory task on the mini-computer. The fourth module is responsible to display and animate the AS/RS system graphically on the PC screen.

CONCLUSION
A realtime simulation study to enhance the traditional simulation strudies is discribed. The significance of the study is developing a completely computer controlled model to simulate the operations of a planned system. In addition to the usual benefits of simulation studies, this provided a learning tool for computer interfacing, integration and multi-tasking concepts.

REFERENCES
1. Mellichamp, D.C., ed., "Real-Time Computing", Van Nostrand, New York, 1983
2. Paidy, S., Wolff, S. and Stewart, B., "Realtime Operating Systems for microcomputers", under review

A Micro-Processor Controlled Carburetor and Fuel Injector Test Stand

R.P.Sharma and W.H. Kirk

Western Michigan University   Triad Engineering Inc.
   Kalamazoo, Michigan USA   Berkley, Michigan USA

SUMMARY

It has become apparent that a commercial requirement for a fully automated small engine carburetor test stand exists today, with test stand accuracies capable of handling present and future micro-processor based engines. It was recognised that high test accuracies could be achieved only on stands with adaptive controls using modern computer technology and that all electromechanical devices associated with such controls would have to meet exceptional requirements.

Modern internal combustion engine technology has placed exceptional demands on test equipment used in this area. Although automotive technology has improved tremendously in the last few years, as related to emissions, power per unit displacement and fuel economy, emissions standards have not been set in small engine design. In a matter of time, motorcycle, outboard and small industrial engines will need to meet more stringent environmental conditions and will inevitably become microprocessor based.

The advantages of computer control are obvious: a greater expectation of reliable results, reduction of the human variable, an ability to integrate into the CAM environment, improved accuracies. Disadvantages are increased maintenance complexity, the inability to qualify data easily, the general inflexibility of the programmed test environment. This paper will describe a fully automated small carburetor test stand, built with the above considerations in mind.

GENERAL DESCRIPTION

A three station test stand was designed to perform production qualification of a variety of small side draft carburetors. The operator would load and unload each station in succession ,start the test sequence and monitor a set of GO-NO GO lamps. The test stand would automatically set throttle position, downstream vacuum and air flow select functions, collect all relevant data , perform an ACCEPT-REJECT function and print out collected data. The tests performed were:

1. Idle flow test. The carburetor throttle plate was set to an idle stop, a specified downstream vacuum set, air flow nozzles selected, and tests of air and fuel flow conducted. An idle fuel screw (if a part of the carburetor) would be adjusted to a specified air/fuel ratio. An ACCEPT/REJECT light would be illuminated for operator guidance.

2. Similar tests were run at roughly 1/4, 1/2 and WOT conditions.In these tests, air flow was not measured, but downstream vacuum and fuel flow were logged. The fuel flow/vacuum relationship was compared to design conditions and again, ACCEPT/REJECT lights selected.

3. At end of the test sequence, an acceptable carburetor would be
    automatically unclamped for operator unload. The operator would
    be required to manually unclamp a rejected unit.

4. A column printer was incorporated in the test stand to log
    actual data obtained for the GO/NOGO decisions reported by the
    front panel lights.

Each station of three incorporated in the stand was identical and
self contained, such that three tests could be performed simultaneously.

AIR FLOW CONTROL

Precision air flow control and resulting downstream vacuum stability
were recognised as the most critical stand functions. Particular attention had to be placed on vacuum control equipment. Since the general
form of the air flow equation involves a square root ($Q=KA\sqrt{P\Delta P}$) the
air flow control valves would be called upon for extremely close control
of "A" while at the same time covering a wide range of "A" values, if
precision ΔP values were to be maintained. In particular, vacuum
stability at the small throttle opening at idle would demand extremely
close control of air flow. A diametrically opposed condition exists at
large throttle openings where very large control valve area changes have
little influence on ΔP. Added to these problems was the range of carburetor throat sizes (from 15mm dia. to 30mm dia.). Two valves were
specified for air control, a relatively large unit with a smaller
"trimmer" in parallel. The dual design afforded some protection against
the unknown factors that would present themselves during stand qualification tests. See Fig. 1 for graph relating two orifices in series.

A factor also considered was control element sensitivity. Insuffi-
cient sensitivity results in conditions of hunting or instability of
control. In this case the finest adjustment the control can produce
causes the feedback signal to step in larger increments than can be
tolerated by set point requirements. The change in feedback signal per
step of control adjustment is larger than the system deadband. Insuffi-
cient sensitivity will prevent the control system from settling within
the deadband. Excessive sensitivity will prevent the system from stabi-
lizing within the deadband (and quite often outside as well) and will
have excessively slow response. Since mechanical and electrical filtering as well as software averaging can often control excessive sensitivity, this condition is preferable to low sensitivity, which is difficult to correct without re-design.

A survey of existing control valves produced nothing that appeared
capable of the repeatability, size and sensitivity deemed necessary for
this application. Specially designed valving was constructed for this
purpose. The design chosen was a poppet valve style with long outlet
diffuser for the following reasons:

1. Angularity of the diffuser could easily be altered to
    change the valve characteristics if needed.

2. The design with a long actuating stem, lent itself to
    linear ball bearing support of an actuating cylinder.

3. The valves could easily be actuated by a frictionless diaphragm actuator.

4. A relatively heavy return spring could be used for good repeatability and freedom of hysteresis. A piston ring style Teflon stem seal was also used for this reason.

5. The design lent itself to system space requirements.

The valves were designed with strokes of approximately 3" and driven by Analog Devices eight bit D/A converters operating thru Fairchild E/P converters. Valve opening resolution was 3.00"/256 or approximately .012" per step. Area change for the small valve was .0197 sq.cm. per step. A check of valve movement verified that the system could resolve .012 +- .003" using the above approach. To take advantage of this inherent sensitivity, precision vacuum transducers were used as the feedback elements. An algorithm was developed that measured the characteristics of the valve at the same time that control was being exercised. If a set point of 1.8"Hg below atmospheric was desired, and the vacuum reading before control was 3.2"Hg., a movement of the valve of ten steps (as an example), or .120" would be commanded and vacuum read at the end of valve movement, as example 2.4"Hg. The valve characteristic would then be calculated as 2.4"Hg./.120" or 20"Hg./inch. The difference between set point of 1.8" and 2.4"Hg. (or .6"Hg.) would then be divided by 20 to obtain the next valve motion step series, or .6/20=.030, roughly 2 steps. Of course the algorithm also had to take into account approaching set point from either direction as well as special conditions relating to the operation of fine and course control valves in tandem. The valve geometries initially selected were workable over the the full range of conditions for which the test stand was designed.

AIR FLOW MEASUREMENT

Air flow measurement capability was provided at idle test conditions only. Solenoid valves were incorporated in the air flow stream to by-pass low flow air thru two ASME configured nozzles. The nozzles in turn were individually valved such that either the larger or smaller or both could be used to measure air flow over a wider range of air flows without sacrificing useable P conditions. Nozzles were chosen for this project due to NBS traceability and long term calibration stability, in spite of the admittedly narrow rangeability typical of nozzles.

FUEL FLOW MEASUREMENT

Fuel was measured with piston type displacement meters. These meters are the accepted standard in the industry for low flow measurement. In the case of some carburetors tested, a "surge" condition existed whereby the carburetor float valve would not stabilize. This was generally the case for designs where the float valve admitted fuel on top of the float resulting in an oscillatory balance between weight of fuel admit-

ted on the float top and normal compensating buoyancy of the float.
Computer programs were devised to nullify this effect. The test fluid
was STANISOL, a close equivalent to gasoline in density. During testing,
liquid Stanisol precipitated in the air passages would be dumped thru
an automatic solenoid valve.

MECHANICAL ADJUSTMENTS

Certain carburetor models used needle screw adjustment for idle
fuel and/or idle air flow. The test stand incorporated stepper motor
driven screwdrivers that performed the needle valve adjustment and
throttle positioning under computer control. The screwdriver bits
themselves were advanced using an air loaded diaphragm operator. Methods
of reading air/fuel ratio versus needle valve setting were developed
such that no operator intervention was required to completely adjust
and calibrate any of the carburetors tested.

COMPUTER

The custom computer used was built from a selection of STD BUS card
products. This bus structure is popular for small industrial level
applications. A main processor board, a memory board, digital I/O,
analog I/O and counter/timer I/O boards were used to make up the com-
puter/controller system for this project. The main processor was programmed in Basic. Additional electronic interfaces were custom designed by
Triad. The Basic language has been found to have adequated speed for real
time controls of this type. Basic literacy is the most common and is a
factor when designing equipment for customers having weak or non-existent
support for computer systems. In this case, the customer's personnel
could handle limited programming tasks in Basic and were able to maintain
these systems as well as experiment with the software.

TEST STAND PERFORMANCE

Initial qualification tests of the stands uncovered a number of conditions that dictated a well reasoned approach toward stand development,
repeatability, stability and accuracy. As with any highly sensitive
equipment, precautions had to be taken to assure that the feedback control signals were of highest quality. It was found on idle tests of certain carburetors that idle fuel would collect around the throttle plate
periphery effectively sealing off air flow. As a result, vacuum would
increase until sufficient to pull the collected fuel thru the throttle
plate/bore clearance and again increasing air flow. As a result, The vacuum transducers were outputting a signal with broad band noise, inter-
fering with stand stability. Identification of this problem led to
improvements in the carburetor design with corresponding improvement in
end user engine idle stability. Another problem on certain carburetors,
was throttle plate jamming in the throttle bore at idle, also contributing to stand instability and poor engine idle to off-idle transition.

Due to these and other effects of the carburetors on test stand
performance, qualification was performed with a series of orifice plates.
By the nature of their constant area, they provided the stability needed
to qualify the stands to their close tolerance vacuum and air flow set
points.

CONCLUSIONS

1. The design and fabrication of a fully automatic, computer-controlled small carburetor test stand is feasible. It has been demonstrated that a computer can be programmed to perform the delicate tasks of setting test variables to close limits.

2. It is doubtful that the project could have succeeded without some form of adaptive control. The controls must continually reference themselves from highly accurate measurement of the controlled variable.

3. Definite consideration must be given to custom components where 0.1% (10 bit) precision is demanded. It is probably not possible to develop a high sensitivity machine with off-the-shelf components only.

4. Considerable ingenuity must be exercized in the software constructs. A computer programmer without intimate knowledge of electro-mechical device and instrumentation limitations could not write effective software. The ability to comprehend and implement strict cause and effect experimentation is a mandatory requirement.

5. In some cases, stand sensitivity to carburetor operating characteristics were so extreme, that test stability was compromised. It was found that end user carburetion difficulties could under certain circumstances, be linked to the degree of difficulty in performing a stable test.

6. The end user/customer must be made well aware of all the ramifications of the high accuracy design, particularly when the equipment will be used in a production environment as opposed to research or engineering test application. It is essential that customer personnel cooperate closely with the equipment designer when establishing and modifying guideline specifications. The customer must understand and be prepared for the essentially experimental nature of test equipment design so that unrealistic expectations are not developed.

In summary, the design approach used for these stands was successful and demonstrated that totally automatic high-accuracy testing is feasible on small carburetors.

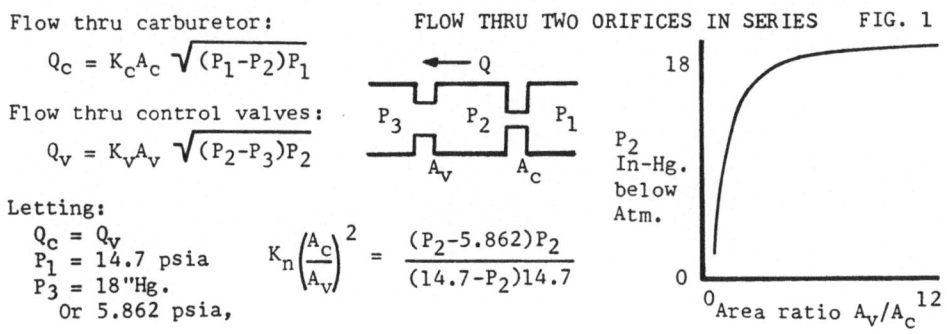

Flow thru carburetor:
$$Q_c = K_c A_c \sqrt{(P_1-P_2)P_1}$$

Flow thru control valves:
$$Q_v = K_v A_v \sqrt{(P_2-P_3)P_2}$$

Letting:
$Q_c = Q_v$
$P_1 = 14.7$ psia
$P_3 = 18$"Hg.
Or 5.862 psia,

$$K_n\left(\frac{A_c}{A_v}\right)^2 = \frac{(P_2-5.862)P_2}{(14.7-P_2)14.7}$$

FLOW THRU TWO ORIFICES IN SERIES   FIG. 1

Design for Manufacture

Virtual Manufacturing: A New Concept in Automated Design Aids

John J. Mills, W. Furth, Y. Sekine., E. Wysocki, and A. Burzio
Martin Marietta Laboratories

Summary

Designers are increasingly using computer based tools such as solid models
to provide feedback on how their devices will perform in operation. VIRTUAL
MANUFACTURING is a concept in which computer based tools help the designer
analyze and visualize how his design will be built and tested on the shop
floor or in flexible robotic assembly cells. The thrust of this approach is
to capture and use information generated (and usually discarded) during the
design phase to provide -- among other things -- analysis of components,
assembly sequence and task plans, and animation of the results of these
plans where appropriate. Such feedback will, by informing the designers of
potential difficulties in manufacturing, facilitate rapid prototyping of
hardware.

Introduction

Defense contractors in the 1990's and beyond will be required to design,
build, and test devices whose precision exceeds anything being manufactured
today, with the result that humans will not be able to assemble them.
Moreover, the anticipated clean room requirements will be incompatible with
non-encased humans: flexible robotic assembly will be the only option[1].
However, for robotic assembly to be cost effective, the devices must be much
more producible than at present. This means that the current
"manufacturing" process must be much improved. Such a major change can be
brought about only by a "systems" approach in which each step in this
process is integrated, as much as possible, with that preceding and
succeeding it. In particular, it is vital that designers receive feedback
from production, quality, and test at the conceptual stage [2].

Approach

In one current approach to these problems — concurrent engineering — designs are produced by teams consisting of engineers from each of the requisite disciplines [3]. Teams have their own problems, however, and are only as good as the individual members. Our concept is called "Virtual Manufacturing" (VM), in which designers have access to a set of software tools that allow them to "virtually" manufacture a design on a high-performance graphics workstation.

After investigating many approaches to realizing this concept, we consider one critical element of any such system to be the capability to show the designer the automatic assembly of a proposed device via solid model animation. This simulation would make immediately apparent such problems as complex motions during assembly, lack of accessibility, and interference. Solid modellers already provide much of this visualization but are currently used only for static visual inspection. Our VM system would automatically plan assembly sequences, program robot tasks and paths, and then use these plans to drive the solid modeller's animation capabilities. Not only would the designer receive visual feedback, but the time between concept creation and actual assembly would be considerably reduced since the plans generated for the simulation could also be used for a real assembly cell. The result would be a true Integrated Concurrent Engineering (ICE) capability.

In addition, the animated robot task plan would reveal errors in the task planning, permitting correction of the plan before it is executed in an actual workcell and minimizing the need for sophisticated real-time exception handling algorithms in the cell controller. An automatically generated robot task plan also eliminates one of the barriers to applying robotics to low-volume assembly: the cost of developing code.

Our concept is based on certain beliefs, as follows: 1) solid modelling for design will be widespread [4]; 2) since designers usually produce bad designs out of ignorance, providing them with the right information at the right time will allow them to correct their mistakes and improve their concept; 3) much of the information generated during the design process is lost, only to be regenerated during the preparation for production — capturing this information, much of which is geometric in nature, will thereby facilitate the planning process; and, 4) most of the technology required for our VM system is available today or is under development.

Issues and Challenges

There are a number of issues and challenges facing us in the development of
the VM system, all of which are currently being investigated. One concerns
how the system will really be used. Should designers be able to use it on
their own, or should production engineers use it to feed back appropriate
information to the designer? Another issue is the efficient transmission,
representation, and manipulation of design information, features, and
restraints. We need not only geometric information to provide constraints
for assembly sequence planning, but also logical relationships among
components. Can we use the data structures in solid models as our source of
this kind of information [5] or will the proposed Product Definition
Exchange Specification (PDES) provide it?

In the planning arena, how do we best develop an assembly sequence plan,
given that the number of feasible ways may be astronomically huge? Various
task planners are described in the literature [6 . How viable are they for
our purpose?

An important feature envisioned for the VM/ICE system is the ability to
determine the complexity of a design; however, the determination of a means
of calculating a complexity index is in itself a complex task. Possible
factors being evaluated include the number of components in a design, the
number of surfaces created by machining operations, and the intended use of
the final assembly.

Status

The architecture of our VM/ICE system has been developed to a sufficient
degree to permit the acquisition of segments of the software and hardware.
We identified some of the basic requirements for a solid modeller,
including: use of exact mathematical expressions for representing surfaces
and primitives instead of facets, ability to perform interference checking,
and the capability to simulate assembly and robotic motions. Based on these
and other factors, we selected the I-DEAS software from SDRC, which
incorporates the GEOMOD solid modeller.

We also acquired the HP 350 SRX workstation [7] because of its speed of execution and graphics display, which has enabled us to develop a series of animations representing robot motions and simple assemblies. The UNIX-based operating system and availability of the new standard X-Windows were also factors. We are presently exploring GEOMOD's facility for internal representations of objects and the interfacing capabilities of the HP with other systems.

Considerable attention in the project has been devoted to the planning problem, both at the assembly sequence and task planning levels. Work on the assembly sequence planning has been initially restricted to literature searches and the evaluation of some planning techniques on simple assemblies. The largest portion of the planning effort has been devoted to the task planning problem, where we have focused on evaluating the ITA (Intelligent Task Automation) system developed by the Martin Marietta Astronautics Group [6]. Our goal is to understand the basic operation of the ITA system so that it can be adapted for use in the VM system. Modifications will be required to interface it to the GEOMOD solid modeller, and to allow it to work with other types of robots and perform different types of tasks than those for which it was originally developed.

References
1. Nevins, J.L. and Whitney, D.E.: What progressive companies are doing to raise productivity. In Rethinking DoD Manufacturing Improvement Strategies. Defense Manufacturing Forum, Institute for Defense Analysis, October 1986.

2. Product design for manufacture and assembly: The five fundamentals. In 2nd International Conference on Product Design for Manufacture and Assembly. G. Boothroyd and B. Huthwaite, eds. Troy Conferences, 134 W. University Dr., April 1987.

3. Vasilash, G.S.: Simultaneous engineering. Production 99(7) (July 1987) 36-41.

4. Johnson, R.H.: The acceptance of solid modelling in industry. In AUTOFACT'87. SME, Dearborn, MI, 1-1 to 1-14, November 1987.

5. Kohl, D.L.: M.S. thesis. Rensselear Polytechnic Institute. December 1987.

6. Becker, J.M. and Garrett, F.L.: An architecture for intelligent task automation. In AAAI 87, 2. M. Kauffmann Publishers, 95 First St., Los Altos, CA, 672-676, July 1987.

7. Burgoon, D.: Pipelined graphics engine speeds 3-D image control. Electronic Design July 23, 1987.

An Innovative Control System for Resilient Manufacture

Anthony W. Smith

CIM Centre, Kingston Polytechnic, Millennium House, 21 Eden Walk, Kingston upon Thames, Surrey KT1 1BL, England

Summary

This paper describes a new approach to the control of manufacture being investigated at Kingston. A new kind of controller has been designed which derives from the characteristics of the brain. This controller, in certain circumstances, allows production in a manufacturing system to continue despite machine breakdown.

Introduction

This paper summarises research currently underway at Kingston into the application of 'connectionist networks' to real-time planning and control in a manufacturing system. First, conventional manufacturing control is summarised to make clear its inherent weakness. A consideration of the properties of connectionist networks has led to the development of a new 'style' of manufacture which avoids the problem affecting conventional control. The main concern of the paper is to describe this new style. The connectionist controller which exploits the new style is also briefly described.

The problem with conventional manufacturing control

The conventional approach to the control of manufacture consists of two distinct stages: 'planning', which takes place before production begins, then 'control' as parts are manufactured.

During the planning stage the 'process planner' must identify and sequence the machining operations required to transform raw material into end-product. Thus the 'process plan' (at its simplest) is an ordered list of machining operations. Each of these operations must then be assigned to an appropriate shop-floor machine. The final form of the process plan is shown in Figure 1, where each letter identifies a particular machine tool able to perform the required operation.

```
-->    A
-->    C
-->    D
```

Figure 1: A simple process plan identifying the machine tool to perform each operation

The control activity is invoked once production of the part type begins, and is responsible for executing the process plan to manufacture the required end-product. Control is thus a relatively straight-forward task, involving the movement of parts along predetermined paths between machines.

From this description, it is clear that the conventional style of manufacture can be characterised as 'serial', since parts follow one another along a predefined path and are successively transformed until they emerge as finished products.

Conventional planning and control is perfectly adequate where the shop-floor conditions assumed during process planning can be guaranteed during production. However, this is not often the case in practice. There are several ways that production may be disrupted, the cause of most interest to the current research being machine breakdown. It is clear that conventional serial manufacture requires that all machines specified in a process plan must remain available for the duration of the production run. If just one of the machines in a particular path breaks-down then production of the entire path must halt until action has been taken to restart production.

The simplest response to machine breakdown is to wait for the faulty machine to be repaired and returned to service so that production can continue with the same plan. If this is not satisfactory for any reason then production must be replanned to avoid the broken-down machine. However, it is found in practice that 'panic' replanning is often not done effectively. In Ranky's experience [3], when production is disrupted the manufacturing system eventually becomes "partly controlled by (usually wrong) decisions taken under heavy pressure".

The ultimate effects of machine breakdown and subsequent panic replanning are to decrease productivity and increase manufacturing costs and lead times. Costs accrue while machines stand idle during replanning, and delays accumulate on the completion time of parts. Added to these are the costs of running the manufacturing system to an inefficient process plan once production is restarted.

The parallel style of manufacture

A new approach to manufacture has arisen at Kingston as a result of a consideration of connectionist networks for real-time control. Connectionism [1, 2, 4] is an approach to the modelling of intelligence which is currently attracting a great deal of interest in the fields of Artificial Intelligence and Cognitive Science. A connectionist network is based upon the information processing properties of neurons in the brain and, like the brain, consists of very many, interconnected, simple processing elements which operate upon simple signals in parallel. A 'parallel' style of manufacture has been developed to take advantage of the massively-parallel nature of connectionist networks.

Parallel manufacture relies upon the nature of the modern production facility, in which there are often several machines able to perform any particular type of operation. If *several* alternative machines can be specified for each step of a process plan, this potentially offers a way to deal with the problem of machine breakdown. When one machine in a process plan step breaks-down, parts can simply be routed to an alternative machine and production can continue without interruption. This is the basic underlying principle of parallel manufacture, although the idea is considerably extended to prove practical in a manufacturing context.

A process plan with alternative machines per step is called a 'production program' to distinguish it from conventional process plans, making a loose analogy between computer programs executed by a computer and production programs executed by a production facility. An example production program is shown in Figure 2. (The Figure is read as 'A or K or L, followed by C or R or Z or S', and so on.)

```
-->     A  |  K  |  L
-->     C  |  R  |  Z  |  S
-->     D  |  F
```

Figure 2: An example production program

The naive approach to the execution of a production program is to allow only a single machine from each step to take part in production while the alternative machines stand idle waiting to take over when a breakdown occurs. *It must be stressed that this is not what happens in parallel manufacture.* Instead, any number of machines in each step may be operating simultaneously. Thus when a part completes at one machine, it may then be machined by *any* of the machines in the production program step following. In this way the 'breadth' of the production program is exploited and no machine is obliged to stand idle, waiting for another to fail. When a machine in a production program does break-down, parts are simply routed to the alternative machines and production continues without the need for replanning.

Production planning for parallel manufacture remains similar to conventional production planning, requiring only that several machines be identified for each machining operations, rather than one.

However, unlike serial manufacture, parallel production places great demands upon the method used to control the manufacturing system. Many parts at different stages of completion must be routed and scheduled amongst the many machines of the production program, based upon the current availabilities of the shop-floor machines. Thus the controller has very many, interdependent decisions to make, and in real-time so that parts do not have to wait while their next destinations are determined.

The connectionist controller

A connectionist network has been designed to control parallel manufacture, and is described briefly below.

First the connectionist network must be built to represent the manufacturing system it is to control. This is easily done, with groups of network 'units' used to represent machines on the shop-floor and connections between groups of units representing paths which parts may follow between machines.

When a production program is to be executed, it must first be 'loaded into' the connectionist controller. The parallel execution of the computational processes in the network units establishes 'dynamic connections' between units, effectively identifying all the possible paths that parts may follow to visit the production program machines in the correct order. This set of paths is thus a subset of the total paths in the manufacturing system.

During production, when a part completes an operation one of the subset of paths available to it is chosen by propagating signals through the network from the appropriate unit. Signals propagate in parallel through the network, so the next paths for any number of parts can be decided simultaneously. Conflicts between parts for the same machine are resolved by a competition mechanism built-in to the propagation process. If there are no paths available to a part, propagation continues until a path becomes available. A broken-down machine is unavailable to perform a machining operation, and is effectively ignored by the controller.

A Pascal program has been written to verify the controller. This allows the user to define a constrained manufacturing system, and then to observe the system's behaviour as it manufactures parts under connectionist control. The implementation has confirmed that the connectionist controller can make dynamic routing and scheduling decisions, in real-time, for any number of parts, determined by the availabilities of the shop-floor machines.

The simulation program has also been used to make performance comparisons between serial and parallel manufacture. The comparison is necessarily artificial because the two approaches are so different, but encouraging results have been obtained. The general conclusion from experiments over a small number of manufacturing situations is that, when no breakdown occurs, conventional serial manufacture is more productive. However, when breakdown occurs, the ability of the connectionist controller to 'work around' the broken-down machine means that parallel production is often more productive.

Conclusions

In closing, it is appropriate to emphasise the limitations of the new approach. First of all, the potential for parallel manufacture is a feature of a given manufacturing system. If there is little or no overlap in capability between machines then there is no opportunity for anything other than conventional, serial manufacture. Also, parallel manufacture does not seem appropriate for one-off and small batch production. Dedicating a large number of machines to the production of a small number of parts will inevitably result in low machine utilisations.

Given that parallel manufacture may be possible and appropriate, the connectionist controller currently also has several limitations. However, it is thought that these can be overcome with further work in two main areas:

• The connectionist controller must be extended to represent and control a more realistically varied range of manufacturing situations.

• The controller currently makes no attempt to optimise its routing and scheduling decisions, it merely generates feasible solutions. It is thought that the quality of decision might be improved either by using more complex, symbolic signals between network units, or by incorporating a higher level of control to oversee and improve the basic network functioning.

In conclusion, a new approach to manufacture seems to offer an advantage over conventional methods in dealing with machine breakdown. This is potentially significant since the advantage translates into improved productivity and profitability. This initial work is to be developed further towards an exploitable product for use in industry.

References

1. Feldman, J. A. and D. H. Ballard. Connectionist models and their properties. Cognitive Science, 6, (1982) 205-254.
2. McClelland, J. L. and D. E. Rumelhart (Eds.). Parallel Distributed Processing, Vol. 2, Cambridge, Mass.: MIT Press 1986.
3. Ranky, P. G. Computer Integrated Manufacturing, London: Prentice-Hall International 1986 p. 363.
4. Rumelhart, D. E. and J. L. McClelland (Eds.). Parallel Distributed Processing, Vol. 1, Cambridge, Mass.: MIT Press 1986.

CHAPTER II:
Integrated Engineering Analysis

Introduction

The papers in Chapter II have a substantial interest in the theory
or application of engineering analysis. It is clear from the papers
in Section II.1 that the finite element method is both a mature and
growing field and from Section II.2 that the boundary element method
continues to show promise as an alternative method for certain classes
of problems. In both sections the more theoretical papers are given
first and followed by those dealing with more specific applications.
As Section II.3 illustrates we must not overlook other numerical
methods for various problems. The Chapter concludes with several
papers on the dynamics of mechanical systems.

The development of computer-based analysis tools is one of the major
success stories of the computer revolution. They have matured largely
as stand alone tools, but in recent years have been integrated with
other applications. Now the long process of integration with computer-based design and manufacturing tools must be speeded up. For those
looking for integration, this chapter is merely an introduction to an
advanced chapter on optimization included in Volume II proceeding,
which shows that significant progress in integrated Engineering Analysis and Design Automation has been made.

Finite Element Method

Hierarchic Models for Plates and Shells

B. A. Szabó
Center for Computational Mechanics
Washington University
St. Louis, MO 63130

Summary

Some recent developments in computer aided design technology are surveyed with emphasis on methods for controlling the errors of idealization and discretization. It is shown that the technological foundations for computer based design certification are now in a mature stage of development.

Introduction

One of the focal topics of this conference is computer aided design, that is the use of computers in the engineering decision making process. So far computers have been used mainly for obtaining insight and ranking alternative designs in accordance with criteria based on strength or durability considerations. Design certification, however, is generally left to testing. The often quoted proof of validity of finite element analysis results is that they agree with experimental data. Agreement with experimental data may be reassuring, but can be very misleading also: Often displacement and strain measurements are correlated. Close agreement in displacements does not guarantee close agreement in strains or stresses and close agreement in strain values at locations where the solution is smooth does not guarantee that computed maximum strain values at fasteners, notches, weldments and similar critical regions are accurate. Also, testing is expensive and time consuming. In general, testing cannot cover more than a few representative cases. Of necessity, extrapolation of experimental data is by mathematical models. Thus, estimation and control of errors in mathematical models is an essential requirement.

Errors of discretization

A large number of papers have been written on error estimation in finite element analysis. In the case of elliptic partial differential equations, the class of mathematical problems which includes elastostatics, and therefore problems of mechanical design, a complete understanding of the relationship between finite element discretization and smoothness measures of the exact solution on one hand, and error measured in energy norm on the other, exist. A recent survey paper gives a good overview of this subject [1]. The use of error estimation in engineering practice is discussed in [2-5].

Measurement and estimation of error in energy norm is not in itself of great practical importance, although it is closely related to the root-mean-square error in stresses. When coupled with extraction procedures, however, it is possible to compute any engineering data with about the same efficiency as the strain energy and an energy based error bound is available. A brief outline in the general framework of elastostatics follows: We denote the displacement function by $u = (u_r, u_z)$, the exact displacement function by u_{EX} and the finite element solution by u_{FE}. We denote an arbitrary admissible virtual displacement function by $v = (v_r, v_z)$ and the virtual work of internal stress corresponding to the displacement function u, when the elastic body is subjected to virtual displacement v, by $B(u, v)$. To compute a particular functional $\psi(u)$ we select a virtual displacement function w such that

$$\psi(u) = B(u, w) + Q(w) \tag{1}$$

where $Q(w)$ is a functional which depends only on the input data. If w is Green's function for $\psi(u)$ then $B(u, w) = 0$ and $\psi(u)$ can be computed from the input data. It has been shown in [6] that

$$|\psi(u_{EX}) - \psi(u_{FE})| \leq \|u_{EX} - u_{FE}\|_E \|w - w_{FE}\|_E \tag{2}$$

where $\|u\|_E \overset{\text{def}}{=} \sqrt{B(u, u)}$ is the energy norm of u, w_{FE} is the finite element approximation to the extraction function w. Since in general $\|w - w_{FE}\|_E \to 0$ at roughly the same rate as $\|u_{EX} - u_{FE}\|_E \to 0$, the error in the computed functional decreases at about the same rate as the error in energy norm squared. That is, 10% error in energy norm is roughly 1% error in the computed value of $\psi(u)$.

Errors of idealization

Control of the error of discretization is not sufficient if we wish to perform certification of engineering design by computer. It is also important to control the errors of idealization. Any particular idealization involves a number of simplifying assumptions, the validity of which can be tested once a reasonably accurate approximate solution is available. Because of space limitations, we will confine our attention to the control of errors of idealization in linearly elastic plate and shell structures.

Hierarchic models for plates and shells

Plate and shell theories are nothing more than restrictions imposed on the space of admissible displacement functions in some generalized formulation of the theory of elasticity. By systematically removing the restrictions, a hierarchic sequence of plate and shell theories can be created starting with a theory similar to the Reissner-Mindlin theory and ending with the fully three dimensional representation. Passing from one theory to another does not require modification of the mesh. In this way the error of idealization can be efficiently controlled by adaptive selection of the appropriate plate or shell model.

The details of the formulation and computer implementation of hierarchic theories for plates and shells are described in [7]. One interesting aspect is that the traditional use of curvilinear coordinate systems is abandoned and a completely general family of mappings, based on the blending function method, is employed. Another important aspect is that a hierarchic system of basis functions is used which is compatible with the hierarchic basis functions used in fully three-dimensional models. Thus joining plate and shell elements with stiffeners and solids is simple and straightforward.

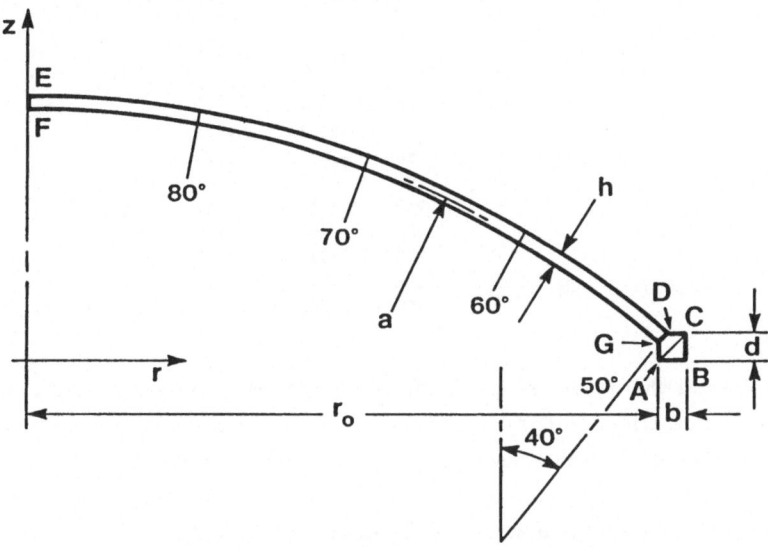

Fig. 1. Spherical shell with footring. (Not to scale.)
$a = 919.2$ in, $h = 2.36$ in, $b = 23.64$ in, $d = 19.68$ in.

Example: Girkmann's problem

For our example we briefly discuss a spherical shell with a footring. This axisymmetric problem, shown in Fig. 1, was discussed by Girkmann [8] and Timoshenko and Woinowsky-Krieger [9]. The shell is loaded by its own weight (41.0 lbf/ft²). We assume that the reinforcing ring rests on rigid frictionless supports and we are interested in computing the bending moment between the ring and the support. By definition, the moment per unit length is:

$$M \overset{\text{def}}{=} \frac{1}{r_m} \int_{r_0}^{r_0+b} \sigma_z^{(u)}(r - r_m)\, r\, dr \tag{3}$$

where $r_m = r_0 + b/2$. The moment can be computed directly from the finite element solution or by extraction, that is by selecting $w = (w_r, w_z)$ such that

$$M = \frac{1}{r_m} \oint \left(\begin{bmatrix} \sigma_r^{(u_{FE})} & \tau_{rz}^{(u_{FE})} \\ \tau_{rz}^{(u_{FE})} & \sigma_z^{(u_{FE})} \end{bmatrix} \begin{Bmatrix} n_r \\ n_z \end{Bmatrix} \right)^T \begin{Bmatrix} w_r \\ w_z \end{Bmatrix} r\, ds = \frac{1}{r_m} B(u_{FE}, w) \tag{4}$$

132

where $n = (n_r, n_z)$ are the unit normals to the generating section of the shell. Specifically, we let

$$w_z = (r - r_z) \text{ along AB}, \quad w_r = w_z = 0 \text{ along DE, EF, FG}.$$

Elsewhere w is not restricted. The results of computation for a 6-element mesh are given in Table 1. in the convention that the moment is positive counterclockwise. p is the polynomial degree, N the number of degrees of freedom. The moment and the estimated absolute error are given in in lbf/in units.

Table 1. Extracted moment. 6-element mesh.

p	N	M_{AB} in lbf/in	Est.'d Error
1	22	−2188.8	±4824
2	56	−1571.8	±1616
3	92	−765.7	±697
4	140	−523.8	±339
5	200	−454.6	±128
6	272	−444.8	±55
7	356	−443.3	±36
8	452	−442.4	±25

In the case of the 6-element mesh the estimated error in energy norm, computed by the method described in [2,4] is 1.73% at p=8 but the error in the extraction function is 28.3% at p=8. For this reason the error estimate computed from (1) is very conservative. If we solve the same problem with a finer mesh, using 22-elements, then the estimated error in energy norm is 0.03% and the estimated error in the extraction function it is 2.89% at p=8. The estimated absolute error bounds on the bending moment are shown in Table 2.

Table 2. Extracted moment. 22-element mesh.

p	N	M_{AB} in lbf/in	Est.'d Error
1	56	−913.3	±2376
2	156	−480.3	±279
3	274	−450.0	±55.8
4	436	−445.3	±16.4
5	642	−440.5	±5.6
6	892	−439.4	±1.1
7	1186	−439.3	±0.2
8	1524	−439.4	±0.04

Acknowledgement

The work reported herein is being supported by the National Science Foundation under grant No. DMC-8606533.

References

[1] Babuška, I., "The p- and hp-Versions of The Finite Element Method. The State of the Art", *Finite Elements: Theory and Applications*, edited by D. L. Dwoyer, M. Y. Hussaini and R. G. Voigt, Springer-Verlag New York, Inc. (1988).

[2] Szabó, B., "Estimation and Control of Error Based on P-Convergence" in: I. Babuška, J. Gago, E. R. de A. Oliveira and O. C. Zienkiewicz, editors, *Accuracy Estimates and Adaptive Refinements in Finite Element Computations*, John Wiley & Sons Ltd., pp. 61-78 (1986).

[3] Schiermeier, J. E. and Szabó, B. A., "Interactive Design Based on the p-Version of the Finite Element Method", *Finite Elements in Analysis and Design*, Vol. 3, pp. 93-107 (1987).

[4] Szabó, B. A., "Mesh Design for the p-Version of the Finite Element Method", *Computer Methods in Applied Mechanics and Engineering*, Vol. 55, pp. 181-197 (1986).

[5] Szabó, B., "On the Errors of Idealization and Discretization in Finite Element Analysis", *Advances in Computer Methods for Partial Differential Equations - VI*, R. Vichnevetsky and R. S. Stepleman, Editors, Int.'l Association for Mathematics and Computers in Simulation, pp. 70-74 (1987).

[6] Babuška, I. and Miller, A., "The Post-Processing Approach in the Finite Element Method - Part 1: Calculation of Displacements, Stresses and Other Higher Derivatives of the Displacements", Int. J. num. Meth. Engng., Vol. 20, pp. 1085-1109 (1984).

[7] Szabó, B. and Sahrmann, G. J., "Hierarchic Plate and Shell Models Based on p-Extension", Report WU/CCM-87/5, Center for Computational Mechanics, Washington University in St. Louis (1987). To appear in the *Int. J. num. Meth. Engng.*

[8] Girkmann, K., *Flächentragwerke*, Fourth Edition, Springer-Verlag, Wien, pp. 442-447 (1956).

[9] Timoshenko, S. and Woinowsky-Krieger, S., *Theory of Plates and Shells*, Second Edition, McGraw-Hill, pp. 555-558 (1959).

Order 2p Derivatives from p-Differentiable Finite Element Solutions by a Spectral Method

Eric R. Johnson and David L. Bonanni

Aerospace and Ocean Engineering Department
Virginia Polytechnic Institute and State University
Blacksburg, Virginia 24061-0203

Summary

The displacements and their first p-1 derivatives are continuous across element boundaries for conforming elements in the finite element representation of a C^{p-1} variational problem. That is, a variational problem in which the highest derivative of the state variable in the functional is order p. For plate bending problems based on Kirchhoff theory, the state variable is the out-of-plane displacement and p = 2. This paper discusses the problem of computing derivatives of order 2p for a laminated composite plate from the assembly of elements in which the element displacements have p-differentiablity within the element. Order 2p derivatives of the out-of-plane displacement are necessary to compute interlaminar stresses.

Introduction

Delamination is a common failure mode in laminated composite structures. However, most finite element analyses for built-up structural components are based on Kirchhoff theory which assumes a state of plane stress and hence neglects interlaminar stresses. The engineering approach to compute the interlaminar stresses from the Kirchhoff theory is to integrate the three-dimensional equilibrium equations for the out-of-plane stress components using the linear distribution in the thickness coordinate of the in-plane stress components from the Kirchhoff theory [1]. This is difficult to implement in finite element solutions because such a proceedure implies that fourth order derivatives of the out-of-plane displacement are required when the finite element formulation only requires continuity of the displacement and its first derivatives between elements. Thus taking fourth order derivatives of the finite element representation of the displacements within an element is meaningless. The method presented in this paper uses the Discrete Fourier Transform of the finite element displacement data over the whole domain of the plate to determine a finite number of Fourier Series coefficients for the displacement. The truncated Fourier Series is differentiated to obtain the higher order derivatives. This approach to compute derivatives from discrete data was used by Tielking and Schapery [2] in a shell contact problem.

Spectral Method

The spectral method is used in this paper as an efficient computational tool to determine Fourier Series coefficients from discrete displacement data. The method is reviewed in the context of representing the out-of-plane displacement of a plate. Designate this displacement as $f(x,y)$, $0 \leq x \leq a$, and $0 \leq y \leq b$, where a and b denote the length and width , respectively, of a rectangular plate. It is assumed the plate is modelled in the finite element analysis such that there are M equally spaced intervals between nodes in the x-direction, and N equally spaced intervals between nodes in the y-direction, where M and N are even integers. The extension of the function $f(x,y)$ outside the domain of the plate, or the protracted function, is defined by periodicity in x with period a, and periodicity in y with period b. The protracted function and its first derivatives are assumed to be continuous; i.e., $f(x,y)$ has C^1 continuity.

The complex Fourier Series representation of $f(x,y)$ is

$$f(x,y) = \sum_{m=-\infty}^{\infty} \sum_{n=-\infty}^{\infty} c(m,n) \exp[2\pi i(mx/a + ny/b)], \quad i = \sqrt{-1}. \tag{1}$$

Evaluating $f(x,y)$ in eqn. (1) at the nodes $(x_j, y_k) = (ja/M, kb/N)$, where $j = 0,1,2,\ldots,M-1$, and $k = 0,1,2,\ldots,N-1$, we obtain

$$f(j,k) = \sum_{m=-\infty}^{\infty} \sum_{n=-\infty}^{\infty} c(m,n) \, W_M^{jm} \, W_N^{kn} \tag{2}$$

in which $f(j,k) = f(x_j, y_k)$, and $W_M = \exp(2\pi i/M)$ and $W_N = \exp(2\pi i/N)$ are the weighting kernels. The doubly infinite sum in eqn. (2) may be restructured in the form

$$f(j,k) = \sum_{m=0}^{M-1} \sum_{n=0}^{N-1} c_p(m,n) \, W_M^{jm} \, W_N^{kn} \tag{3}$$

where

$$c_p(m,n) = \sum_{r=-\infty}^{\infty} \sum_{s=-\infty}^{\infty} c(m + Mr, n + Ns); \quad \begin{array}{l} m = 0,1,\ldots,M-1 \\ n = 0,1,\ldots,N-1 \end{array} \tag{4}$$

Eqn. (3) defines the Inverse Discrete Fourier Transform, and shows that the sequences $f(j,k)$ and $c_p(m,n)$ are Discrete Fourier Transform (DFT) pairs.

Thus, the sequence $c_p(m,n)$ is determined from the DFT of the sequence $f(j,k)$ by the formula

$$c_p(m,n) = 1/(MN) \cdot \sum_{j=0}^{M-1} \sum_{k=0}^{N-1} f(j,k) \; W_M^{-mj} \; W_N^{-nk}; \qquad \begin{array}{l} m = 0,1,\ldots,M-1 \\ n = 0,1,\ldots,N-1 \end{array} \qquad (5)$$

The Fast Fourier Transform algorithm can be used to compute the DFT of the sequence $f(j,k)$ in eqn. (5).

For large values of M and N in eqn. (4), the terms $c_p(m,n)$ are good approximations to the Fourier Series coefficients. The dominate Fourier coefficients in the infinite sums on the right-hand-side of eqn. (4) occur for values of r and s equal to either -1 or 0. Thus

$$c(m,n) \approx c_p\left[m+M \cdot H(-m), \; n+N \cdot H(-n)\right], \; |m| < M/2 \text{ and } |n| < N/2 \qquad (6)$$

in which $H(\;) = 1$ if the argument is > 0, and $H(\;) = 0$ if the argument is ≤ 0. If $m = \pm M/2$ or $n = \pm N/2$, then eqn. (6) is valid if c_p is divided by two. If both $m = \pm M/2$ and $n = \pm N/2$, then eqn. (6) is valid if c_p is divided by four. With the Fourier coefficients approximated by eqn. (6), the truncated Fourier Series representation of $f(x,y)$ is

$$f(x,y) \approx \sum_{m=-M/2}^{M/2} \sum_{n=-N/2}^{N/2} c(m,n) \; \exp\left[2\pi i(mx/a + ny/b)\right] \qquad (7)$$

Orthotropic Plate Example

Although the objective is to use the spectral method to compute derivatives of finite element data, the methodology is tested here by sampling an analytic function, computing its derivatives by the spectral method, and comparing derivatives computed by the spectral method to exact values. The analytic function chosen for this purpose is an approximate buckling mode for a rectangular plate subject to uniform compression at $x = 0$ and $x = a$. The plate is clamped along edges $x = 0$, $x = a$, and $y = 0$, and is free along the edge $y = b$. It is laminated from AS4/3502 graphite-epoxy tape with material properties $E_1 = 18.5$ msi, $E_2 = 1.64$ msi, $\nu_{12} = 0.30$, and $G_{12} = 0.87$ msi. The laminate consists of sixteen plies with a stacking sequence $[\pm45/0/90]_{2s}$. The plate's dimensions are $a = 2.5$ in., $b = 1.0$ in., and the thickness is 0.080 in. The approximate analysis neglects the twist curvature-bending moment coupling of anisotropic plate theory, and uses Kantrovich's method in Trefftz's criterion to determine the buckling mode.

The buckling mode is normalized such that the maximum displacement at x = a/2 and y = b is equal to the thickness of the plate. The result is

$$w(x,y) = \left[1 - \cos(2\pi x/a)\right] \left[K_1 \exp(\alpha y/b) + K_2 \exp(\alpha y/b) + K_3 \sin(\beta y/b) + K_4 \cos(\beta y/b)\right] \tag{8}$$

where $K_1 = 5.6196 \times 10^{-4}$ in., $K_2 = 1.4571 \times 10^{-2}$ in., $K_3 = 2.4399 \times 10^{-2}$ in., $K_4 = -1.5133 \times 10^{-2}$ in., $\alpha = 3.0919$, and $\beta = 1.7753$.

The protracted function obtained from $w(x,y)$ in eqn. (8), and its first partial derivative in y (denoted $w,_y$), are discontinuous at integer multiples of y = b. The truncated Fourier Series representation of $w(x,y)$ would exhibit Gibbs phenomena near y = b, because of the nonuniform convergence at the discontinuity. To avoid the Gibbs phenomena, a polynomial function $w_p(x,y)$ is defined such that the difference function

$$f(x,y) = w(x,y) - w_p(x,y) \tag{9}$$

has the C^1 continuity properties discussed previously. The polynomial is selected to match the essential boundary conditions, i.e., w and its derivative normal to the edge, at the nodal points along the edge. For this example, the polynomial selected is

$$w_p(x,y) = x^2 (x-a)^2 y^2 \left[f(x) + (y-b)g(x)\right] \tag{10}$$

in which $f(x)$ and $g(x)$ are Mth order interpolation polynomials. Polynomials $f(x)$ and $g(x)$ are defined by the M+1 nodal values of $w(x_j,b)$ and $w,_y(x_j,b)$, respectively, where $x_j = ja/M$, $j = 0,1,\ldots,M$.

Denoting the approximation to the function $w(x,y)$ as $\tilde{w}(x,y)$, then $\tilde{w}(x,y)$ is the sum of $w_p(x,y)$ and the approximation of $f(x,y)$ as given by eqn. (7). The norms of $w(x_j,y_k)$ and $\tilde{w}(x_j,y_k)$, and their first four derivatives, are compared in Table 1 for M = 20 and N = 10. The norm of a discrete function is defined as the square root of the sum of the squares of the discrete function values over all nodal points. The approximate function norms in Table 1 are determined from smoothed data values; i.e., for an interior node, the value of the function at that point is replaced by the average value of the function at the node plus the function values at the eight adjacent nodes. The smoothed data gives slightly better results than raw data. It is clear from Table 1 that for derivatives of the same order

138

errors increase as the number of differentiations in x increase with respect
to those in y. Also, errors increase with increasing order of the
derivative. It is likely that the high order interpolation polynomials f(x)
and g(x) cause severe oscillations in the higher order derivative data,
especially for derivatives in x. A least squares fit of lower order
polynomials for f(x) and g(x) may decrease this oscillation in higher order
derivatives with respect to x. However, continuity of the protracted
function at the node points along y = b is sacrificed if lower order
polynomials for f(x) and g(x) are used. The loss of continuity leads to
Gibbs phenomena, which can also result in severe oscillations.

Acknowledgement

The research for this paper sponsored by NASA Grant NAG1-537.

References

1. Bonanni, D. L., Johnson, E. R., and Starnes, J. H., Jr., "Local Buckling
and Crippling of Composite Stiffener Sections," Center for Composite
Materials and Structures Report CCMS-88-08, Virginia Polytechnic Institute
and State University, Blacksburg, VA 24061, pp. 126-136.

2. Tielking, J. T., and Shapery, R. A., "A Method for Shell Contact
Analysis," **Computer Methods in Applied Mechanics and Engineering**, Vol. 26,
1981, pp. 181-195.

Table 1 Norms of the exact and approximate displacements and there
first four derivatives on a 21 by 11 rectangular grid

Derivative	Norms Exact	Approximate	Percent Error
w	0.322498	0.312629	3.06
w,x	0.467957	0.446934	4.49
w,y	0.694032	0.681892	1.75
w,xx	1.17610	1.02820	12.58
w,xy	1.00707	0.974863	3.20
w,yy	1.45716	1.43758	1.34
w,xxx	2.95587	18.1876	-515.30
w,xxy	2.53104	2.23268	11.79
w,xyy	2.11439	2.05507	2.81
w,yyy	4.84802	4.39844	9.27
w,xxxx	7.42892	991.130	-13,241.
w,xxxy	6.36120	43.8807	-589.82
w,xxyy	5.31405	4.76466	10.34
w,xyyy	7.03466	6.28657	10.63
w,yyyy	14.6630	41.6433	-184.00

An Analysis of an Incompressible, Fiber-Reinforced Material Subjected to Finite Strains in the Direction of Reinforcement

D. L. Cox

Naval Underwater Systems Center
New London, CT   06320

INTRODUCTION

The purpose of this work is the development of a general stress analysis method for application to problems involving fiber-reinforced polymeric materials. More specifically, the problem of interest is one where finite elastic deformations are allowed in the direction of reinforcement.

In the approach taken here, the material is modeled as an incompressible, fiber-reinforced continuum with no constraint of inextensibility. The problem is formulated for finite element implementation following much of the modern numerical work in finite elasticity for isotropic materials.

Incremental Form of the Continuum Formulation

The application of the finite element method to problems in nonlinear solid mechanics requires a linearized, incremental statement of the total Lagrangian form of the equilibrium equation. This equation is given as

$$\int_{^0V} C_{IJRS} \, \Delta E_{RS} \, \delta(\Delta E_{IJ})^0 dV + \int_{^0V} {}^N S_{IJ} \, \delta(\Delta \eta_{IJ})^0 dV$$

$$= {}^{N+1}R - \int_{^0V} {}^N S_{IJ} \, \delta(\Delta e_{IJ})^0 dV \qquad (1)$$

where the superscript N refers to an increment

C_{IJRS} is an incremental material matrix

e_{RS} is the infinitesimal strain tensor

S_{IJ} is the 2nd Piola-Kirchoff stress tensor

n_{IJ} is the nonlinear term of the Green-Lagrange strain tensor

and R^{N+1} represents externally applied loads.

For a hyperelastic material C_{IJRS} is given by the following expression:

$$C_{IJRS} = \frac{\partial S_{IJ}}{\partial E_{RS}} \qquad (2)$$

where E_{RS} is the Green-Lagrange strain tensor.

Constitutive Equation for a Fiber-Reinforced Material - One Family of Fibers

The general constitutive equation for a hyperelastic material relating the 2nd Piola-Kirchoff stress tensor, S_{KL}, to the right Cauchy deformation tensor, C_{KL}, is

$$S_{KL} = 2 \frac{\partial W}{\partial C_{KL}} \qquad . \qquad (3)$$

The determination of a specific constitutive equation therefore requires a specific expression for the strain energy function W. The general form of the strain energy function for a fiber-reinforced material can be derived following the principles of modern continuum mechanics; that is, by applying the principle of material frame-indifference and invoking the appropriate symmetry arguments. Once the functional dependence has been determined, it is then possible to develop a specific form for W by first expressing it as a power series in terms of its arguments, and then either through geometrical argument or experimental investigation reduce the resulting expression to a simpler form, truncating the series at the desired order. In contrast with the objective approach followed in developing the appropriate functional dependence, the development of the specific form of the function requires geometrical argument, judgment, and experimentation, or a combination of all three.

Form of W for a Fiber-Reinforced Material

It is assumed at the outset that there exists a strain energy function W per unit volume which is a function of the deformation gradients F. Following Spencer,[1] the effect of the fibers is introduced by assuming the strain energy function W to also depend on the direction cosines, A, of the reinforcement. For a single family of fibers this may be written as

$$W = W(F,A) \ . \tag{4}$$

Invoking the principle of material frame indifference, W is forced to be form invariant for a rigid rotation of the deformed body, with the result that

$$W = W(C,A) \tag{5}$$

At this point, arguements of material symmetry are invoked that further constrain the allowable form of W.

Following Spencer's work it is assumed that "the only anistropic properties of the material are those that arise from the presence of the fibers." This implies that W is unchanged if we choose a new reference configuration obtained by a rigid body rotation of the undeformed material and the fibers, with the result that

$$W = W(I_1, \ I_2, \ I_3, \ I_4, \ I_5) \ \ . \tag{6}$$

where

$$I_1 = \text{tr } C,$$

$$I_2 = 1/2[\text{tr } C^2 - \text{tr } C^2] \tag{7}$$

$$I_3 = \det C$$

$$I_4 = \text{tr}(AA)C$$

and $\quad I_5 = \text{tr}(AA)C^2.$

As a function of these five invariants, the strain energy function derived for the case of one family of fiber reinforcement is equivalent to that of a transversely isotropic material.

A Specific Expression for W

In 1975, Sun and Ueng[2] proposed a specific form of the strain energy function for an incompressible, transversely isotropic material subject to finite deformation. They developed their expression following the work of Adkins,[3] Smith and Rivlin,[4] and Mooney.[5] Their resulting expression for W,

$$W = C_{1000} (I_1 - 3) + C_{0100} (I_2 - 3) + C_{0010} I_4$$

$$+ C_{0001} I_5 + C_{0020} I_4{}^2 + C_{1020} (I_1 - 3) I_4{}^2 \tag{8}$$

$$+ C_{1001} (I_1 - 3) I_5$$

where

$$I_1 = \lambda_1^2 + \lambda_2^2 + \lambda_3^2$$

$$I_2 = \frac{1}{\lambda_1^2} + \frac{1}{\lambda_2^2} + \frac{1}{\lambda_3^2}$$

$$I_3 = \lambda_1^2 \lambda_2^2 \lambda_3^2 = 1 \tag{9}$$

$$I_4 = \frac{1}{2}(\lambda_3^2 - 1)$$

$$I_5 = \frac{1}{4}(\lambda_3^2 - 1)^2$$

turns out to be less general than the one desired in this work in that at the outset they assume symmetry with respect to a particular axis. Replacing the expressions for I_1, I_2, I_3 and I_4 in Equations (9) with those given by (7) gives the more general expression for W where the axis of transverse isotropy depends on the direction of fiber reinforcement. Hence,

$$\bar{W} = C_1 \, (I_1 - 3) + C_2 \, (I_2 - 3) + C_3 \, (I_4 - 1) + C_4 \, (I_5 - 1)$$

$$+ \, C_5 \, (I_4 - 1)^2 + C_6 \, (I_1 - 3) \, (I_4 - 1)^2 + C_7 \, (I_1 - 3) \, (I_5 - 1) \quad . \tag{10}$$

The constitutive equation and incremental material matrix are then obtained by substituting (10) into expressions (3) and (2).

Experimental Investigation

In order to provide a quantitative basis for evaluating this approach to the finite deformation analysis of fiber-reinforced polymers, a series of tests were performed on a thin (0.09 inch thick) sheet of neoprene reinforced with nylon cords. This work involved collecting finite strain data from a fiber-reinforced elastomer, deducing material coefficients from the appropriate data, using these coefficients to make finite element predictions, and finally comparing these predictions with the experimental data.

As shown in the figure below, the finite element and experimental data are in close agreement.

144

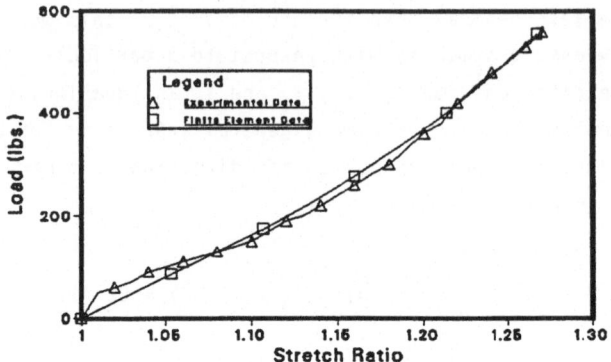

Figure 1. Comparison of Finite Element with Experimental Data.

References

1. Spencer, A. J. M., _Deformations of Fiber-Reinforced Materials_, Oxford University Press, New York, 1972.

2. Sun, Y. S. and Ueng, C. E. S., "Non-Linear Stress-Strain Equations for Incompressible Transversely Isotropic Elastic Materials," Int. Journ. Non-Linear Mechanics, Vol. 10, pp. 129-135, 1975.

3. Adkins, J. E., "Finite Deformation of Materials Exhibiting Curvilinear Aeolotropy," Proc. R. Soc., Vol. A 229, 119, 1955.

4. Smith, G. F. and Rivlin, R. S., "The Strain Energy Function for Anisotropic Elastic Materials," Trans. Am. Math. Soc., Vol. 88, pp. 175-193, 1958.

5. Mooney, M., "A Theory of Large Elastic Deformation," Journ. of Appl. Physics, Vol. 11, pp. 582-592, 1940.

On Analytical Modeling of Thick Composite Laminates

J. N. Reddy and E. J. Barbero

Department of Engineering Science and Mechanics
Virginia Polytechnic Institute and State University
Blacksburg, VA 24061 USA

J. L. Teply
ALCOA
Alcoa Technical Center
Alcoa Center, PA 15069

Summary

A review of the layer-wise continuous expansion of the displacement field
suggested by the senior author to accurately model the stress field in
thick composite laminates is presented. Analytical solutions of the
(linear) theory are compared with the 3-D elasticity solution and the
classical laminate theory solution to bring out the accuracy of the
theory.

Introduction

All laminate plate and shell theories derived from the Basset type
expansion [1] assume that the displacements vary through the thickness of
the laminate according to a single expression, not allowing for possible
discontinuities in the slopes of the deflections at the interfaces of two
individual lamina. For example, a composite laminate made up of layers of
stiff material adjacent to layers of flexible material inevitably
experiences a discontinuous membrane strain field at the interface of the
two layers. None of the existing theories, including the refined theories
(see [2,3]), accommodate such kinematic behavior. Recently, Reddy [4]
proposed a laminate plate theory that allows layer-wise representation of
displacements through the thickness of a laminated plate. The theory is
much simpler than those proposed by Epstein and Glockner [5] and Murakami
[6]. The generalized laminate theory reduces the equations of a three-
dimensional layered medium to equations governing several functions of
position in the reference surface. A review of the theory for plates is

presented here and its analytical solutions are compared with the 3D-elasticity and classical laminate theory solutions.

A Review of the Theory

Consider a laminated plate composed of N orthotropic lamina whose material axes are oriented arbitrarily with respect to the laminate (x,y) coordinates, which are taken to be in the midplane of the laminate. The displacements (u_1,u_2,u_3) at a point (x,y,z) in the laminate are assumed to be of the form

$$u_1(x,y,z) = u(x,y) + U(x,y,z)$$
$$u_2(x,y,z) = v(x,y) + V(x,y,z)$$
$$u_3(x,y,z) = w(x,y), \tag{1}$$

where (u,v,w) are the displacements of a point $(x,y,0)$ on the reference plane of the laminate, and U and V are functions which vanish on the reference plane:

$$U(x,y,0) = V(x,y,0) = 0. \tag{2}$$

In order to reduce the three-dimensional theory to a two-dimensional one, it is assumed that U and V are approximated as

$$U(x,y,z) = \sum_{j=1}^{n} U^j(x,y)\phi^j(z)$$

$$V(x,y,z) = \sum_{j=1}^{n} V^j(x,y)\phi^j(z), \tag{3}$$

where n is an integer depending on the number of layers, U^j and V^j are undetermined coefficients, and ϕ^j are any continuous functions that satisfy the condition

$$\phi^j(0) = 0 \text{ for all } j = 1,2,\ldots,n. \tag{4}$$

The approximation in Eq. (3) can also be viewed as the global semi-discrete finite-element approximations of U and V through thickness. In that case ϕ^j denote the global interpolation functions, and U^j and V^j are the global nodal values of U and V (and possibly their derivatives) at the nodes through the thickness of the laminate.

The equilibrium equations of the theory can be derived using the principle of virtual displacements (see [4]):

$$N_{x,x} + N_{xy,y} = 0 \; , \; N_{xy,x} + N_{y,y} = 0,$$

$$Q_{x,x} + Q_{y,y} + q = 0,$$

$$N^j_{x,x} + N^j_{xy,y} - Q^j_x = 0, \quad N^j_{xy,x} + N^j_{y,y} - Q^j_y = 0, \tag{5}$$

for $j = 1,2,\ldots,n$. Thus there are $2n + 3$ differential equations in $(2n + 3)$ variables (u, v, w, U^j, V^j) when Lagrange interpolation through thickness is used. The laminate constitutive equations are given in [4].

Analytical Solution

To asses the quality of the theory we consider a three-ply (0/90/0) square laminate, simply supported, and subjected to sinusoidal transverse load. This problem has the 3-D elasticity solution [7,8] and the CPT solution. The accuracy of the solutions obtained with this theory can be fully appreciated considering the variation of the inplane displacement u and stress distributions $(\sigma_{xx},\sigma_{yz})$ through the thickness (see Figs. 1-3). The material properties used for each ply are:

$$E_1 = 25.0 \times 10^6 \text{ psi}$$

$$E_2 = 1.0 \times 10^6 \text{ psi}$$

$$G_{12} = G_{13} = 0.5 \times 10^6 \text{ psi}$$

$$G_{23} = 0.2 \times 10^6 \text{ psi}$$

$$\nu_{12} = \nu_{23} = 0.25. \tag{9}$$

All stresses are nondimensionalized with respect to the applied load. The deflection $w(x,y)$ coincides with the exact 3-D solution and is not shown here. In all cases the present solution for stresses are in excellent agreement with the 3-D elasticity solution, whereas the CPT solution is in considerable error.

Conclusions

The generalized laminate theory yields very accurate results both for
displacements and stresses in thin and thick composite laminates.
Compared to other refined plate theories proposed in the literature, the
present theory is simple both in concept and formulation, and it is
suitable for finite-element analysis of composite laminates. The theory
can be extended to include geometric imperfections, delamination between
layers, and other manufacturing defects. The theory allows for transverse
shear deformation and requires no shear correction factors. The finite
element based on this theory is developed and preliminary results were
presented at a recent meeting [9].

References

1. Basset, A. B.: On the extension and flexure of cylindrical and
   spherical thin elastic shells. Phil. Trans. Royal Soc., (London)
   Ser. A, Vol. 181, No. 6, pp. 433-480, 1890.

2. Mindlin, R. D.: Influence of rotatory inertia and shear on flexural
   motions of isotropic, elastic plates. J. Appl. Mech., Vol. 18,
   pp. 31-38, 1951.

3. Reddy, J. N.: A simple higher-order theory for laminated composite
   plates. J. Appl. Mech., Vol. 51, pp. 745-752, 1984.

4. Reddy, J. N.: A generalization of two-dimensional theories of
   laminated composite plates. Commun. Applied Numer. Methods, Vol. 3,
   pp. 113-180, 1987.

5. Epstein, M.; Glockner, P. G.: "Nonlinear analysis of multilayered
   shells. Int. J. Solids Structures, Vol. 13, pp. 1081-1089, 1977.

6. Murakami, H.: Laminated composite plate theories with improved
   inplane responses. ASME Pressure Vessels and Piping Conference,
   pp. 257-263, 1984.

7. Pagano, N. J.: Exact solutions for composite laminates in
   cylindrical bending. J. of Composite Materials, Vol. 3, pp. 398-411,
   1969.

8. Pagano, N. J.: Exact solutions for rectangular bidirectional
   composites and sandwich plates. Journal of Composite Materials,
   Vol. 4, pp. 20-35, 1970.

9. Reddy, J. N.; Barbero, E. J.; Teply, J. L.: A plate bending element
   based on a generalized laminate theory. Proc. 29th SDM Conference,
   Part 2, Paper No. 88-2322, pp. 937-940, April 18-20, 1988, Williamsburg,
   Virginia.

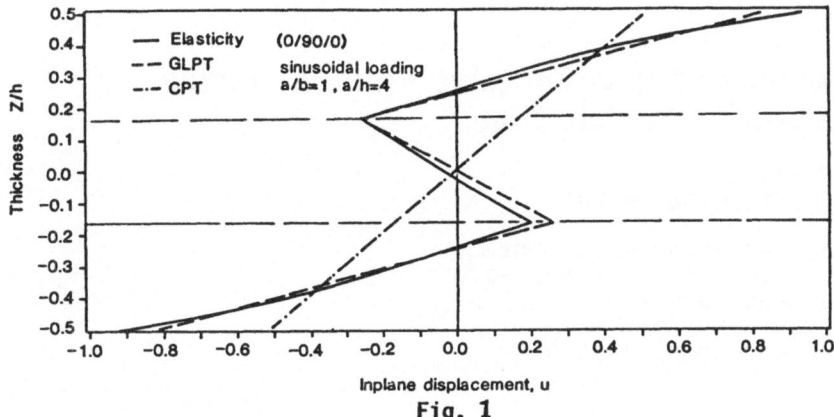

Fig. 1

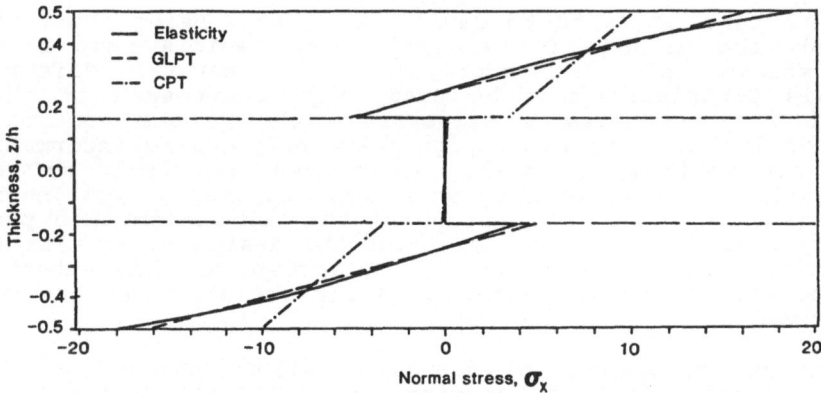

Fig. 2

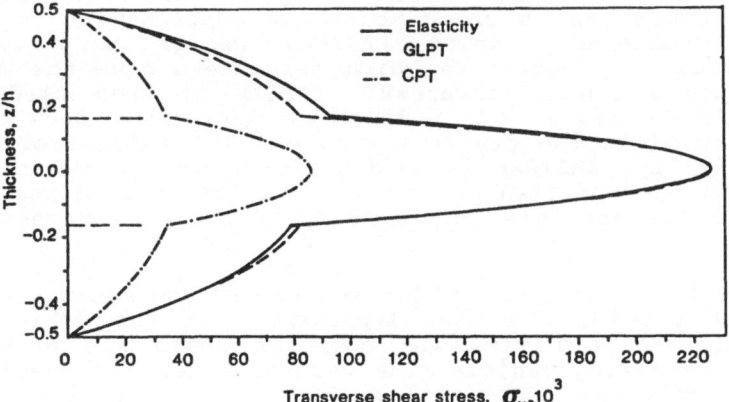

Fig. 3

A Systematic Approach Using Finite Elements for Improving Vehicle Ride

Brian William Deutschel

Chevrolet-Pontiac-Canada Group
C-P-C Engineering Center L-105
General Motors Corporation
30003 Van Dyke Avenue
Warren, Michigan 48090-9060

ABSTRACT

With the time it takes to develop a vehicle being constantly reduced, the methods for analyzing the vehicle must become more streamlined. To accomplish this several different analysis techniques must be used, exploiting each of their strengths to optimize a structure while in the conception phase of design. In this design phase only coarse information about the vehicle is usually known such as, front or rear wheel drive, unibody or body on frame, suspension type, number of passengers, etc. The design of the vehicle in this phase is very fluid and analysis can lead the design if results can be obtained in a timely manner. To accomplish this attention must be focused on those parts of the vehicle that are most sensitive.

The process described in this paper allows the analysis to lead the design without becoming computationally intensive or costly. The programs used in the process are NASTRAN, MODAL and FASTAR. NASTRAN is used to determine the baseline and create a database for the other programs to use. MODAL provides direction as to which modes are contributing to the undesirable response. Next, NASTRAN Design Sensitivity Analysis is used to determine which parameters have the most effect on the modes of interest. FASTAR is then used to evaluate the effects of modifying the most sensitive parameters found in the previous step on the response of the vehicle. Finally, NASTRAN is used to reanalyze the vehicle's response with the modified parameters. If the desired results are achieved the analysis is complete, if not the process is repeated.

Using this systematic approach focuses effort and resources on changes that provide the most improvement in the vehicle's ride. Also, analysis time and computer costs are greatly reduced for improving vehicle ride because after the baseline NASTRAN run only one additional NASTRAN run is required per iteration.

DISCUSSION

The constant pressure to reduce the time required to bring a new vehicle to market is making computer simulation an integral component of new vehicle programs. An important part of the computer simulation is the vehicle system model, which is a mathematical model of the vehicle based on finite element methods. The vehicle system model is used to predict interaction between the vehicle's subsystems and overall vehicle behavior. Evaluation of proposals and ideas without the cost of hardware construction can be quickly and efficiently accomplished with such models. To be effective, the vehicle system model must be able to provide design direction in the early phases of a vehicle program. Currently, no one analysis code exists that can completely interpret the behavior predicted with the vehicle system model. Therefore, several different analysis codes must be used, with the strengths of each exploited without repeating previous steps. When analyzed in a systematic approach the vehicle system model can yield great amounts of insight into a vehicle's behavior and characteristics.

Since the vehicle system model is most effective in the early phases of a vehicle program, early system models generally have the body structure represented by large plate and beam elements. Due to the large size of the plate elements, second order TRIA6 and QUAD8 NASTRAN elements are typically used. The chassis is represented by rigid elements for chassis components and springs for elastic mounts. The total number of elements in an early system model is usually below 2000. These models aid in the design of vehicle structure, suspension geometry and chassis mounting.

One use of such vehicle system models is for improving vehicle ride where proposals are evaluated by comparing seat track acceleration levels for a forcing function of a given road profile. The goal generally is either to reach a predetermined acceleration target or reduce seat track acceleration levels with respect to the baseline system model. Most often the second type of evaluation, reduce seat track acceleration, is done on proposals. Generally, these early system models are only used for ride improvement analyses up to 30 Hertz.

An approach to systematically analyze vehicle system models

for improving vehicle ride will be presented in this paper. This approach utilizes several analysis codes consisting of MSC NASTRAN solution sequences (i.e SOL63, SOL71, SOL53) and two General Motors(GM) internal programs FASTAR (1)[*] and MODAL (2). The answers these analysis codes generate when used systematically provide the most effective solutions for improving vehicle ride in the least amount of time.

The systematic approach starts with a solution 63 NASTRAN normal modes analysis of the finite element vehicle system model. This analysis yields the natural frequencies of the vehicle's body, suspension, drivetrain and other components. The mode descriptions of each mode are documented for later reference in the analysis sequence.

A solution 71 NASTRAN forced response analysis is then performed on the vehicle system model using a measured road profile to determine the response at the drivers seat. The analysis is a restart utilizing the NASTRAN DB01 database which was generated in the previous normal modes analysis. By utilizing the restart capability, several forced response analyses using different road profiles can be done without incurring the substantial cost of dynamic reduction each time. Thus, the seat response levels can be compared to targets set for the vehicle program.

Once the baseline forced response is determined, a modal participation analysis of the response at the seat for the fore/aft, cross car and vertical directions is done. The participation of each mode is calculated utilizing the program MODAL developed at GM which uses an output file created by the solution 71 NASTRAN forced response analysis as input. The output generated by MODAL is a tabular listing of percent mode participation versus frequency. Also, three dimensional plots of mode participation as a function of mode and response frequency are generated by MODAL. Using the tables and plots, it can be determined which modes are the major need to be modified. The MODAL program focuses resources on

[*]Numbers in parentheses designate references at end of paper.

contributors to the response at the seat and thus which modes only those modes that effect the response.

To determine how to modify the modes indicated by the MODAL program, a solution 53 NASTRAN design sensitivity analysis is done to determine the sensitivities of the problem modes to the vehicle system model properties. Sensitivities with respect to any property card field such as beam area, moments of inertia, torsional constant, plate thickness or spring rate can be calculated. The NASTRAN design sensitivity analysis is efficient because it uses the DB01 data base as input, thus the cost of dynamic reduction is again not incurred. The computer time required is typically a fraction of the original solution 63 NASTRAN normal modes run that created the database.

Using the results from the solution 53 NASTRAN design sensitivity analysis and MODAL, an effective set of modifications for reducing seat track response in the system model can be determined in FASTAR. The FASTAR program, developed at Chevrolet-Pontiac-Canada(C-P-C) Group allows interactive modification of the system model. The input required for FASTAR is generated from a NASTRAN output file created in the solution 71 NASTRAN forced response analysis done previously. FASTAR has several types of elements consisting of springs, dampers, bars, rods and masses which can be added to the vehicle system model. Using FASTAR, the vehicle system model is modified and the effects to mode shapes, frequency and forced response evaluated. Since interactive modifications to the vehicle system model can be done, several independent sets of modifications can be developed which meet the ride targets. To quantify the effects of the structural modifications, FASTAR can compute the modal assurance criteria(MAC)(3) on the modified and baseline models. The MAC listing saves time since all the mode shapes need not be viewed after the modifications have been added to the system model.

The last step in this systematic approach, is a confirmation analysis using the original NASTRAN vehicle system model and

154

the best set of modifications as determined in FASTAR. Generally, a solution 71 NASTRAN forced response analysis is used for confirmation since the target for customer satisfaction is the acceleration response at the seat. If the targets are not achieved the first time using this systematic approach, the process is iterated until they are.

CONCLUSIONS

The systematic approach discussed in this paper minimizes the computer resources and engineering time required to analyze vehicle system models for improving vehicle ride. There are two key elements in this approach. The first consists of all the intermediate steps, which are NASTRAN post processors or restarts, which use information generated in the baseline NASTRAN normal modes analysis as input. Thus, the computer time required is minimal when compared to the baseline NASTRAN normal modes analysis. The second key element is the capability to evaluate many proposals in FASTAR, independently of NASTRAN thus reducing the time and cost requirements. Also, the approach directs engineering time and resources to the most effective modifications for improving vehicle ride.

ACKNOWLEDGEMENTS

The author wishes to thank Richard Katnik from C-P-C Engineering for his assistance and suggestions during the course of this work. Also, the author would like to thank General Motors Corporation for ermission to publish this report.

REFERENCES
1. Katnik, Yu and Wolf; "Interactive Modal Animation and Structural Modification"; 6th International Modal Analysis Conference Proceedings; Kissimmee, FL; February 1988.

2. "Vehicle Systems Analysis Procedures - Structural Analysis", Finite Element Center of Expertise (FESCOE) General Motors, Report No. 4 June 1987.

3. Allemang and Brown; "A Correlation Coefficient for Modal Vector Analysis"; 1st International Modal Analysis Conference Proceedings; Orlando, FL; November 1982.

Advanced Dynamic Analysis of Vehicle Structures

R. KANNAN, S. SUTHARSHANA, and D. LAM

Engineering Mechanics Research Corporation,
1707 West Big Beaver Road
Troy, Michigan 48084

Summary

A methodology for evaluating ride comfort and failure probability of vehicle structures using advanced dynamic analysis methods is presented. This methodology is demonstrated on a large scale model of an automobile.

Introduction

Dynamic analysis is an integral part of the design of ground vehicles for evaluating ride comfort and structural strength. Often a 'stick' finite element (FE) model of the structure is used for dynamic analysis. Such a crude model may be used to uncouple the beaming and torsional modes of vibration in the vehicle and for ride evaluation. The latter is achieved by deriving the statistics of the response from stochastic descriptions of standard road surfaces via a random vibration analysis. Although such analyses are efficient, they provide little information regarding stress levels in various components. A full FE model consisting of continuum and structural elements gives more accurate estimates of stress levels and the associated probabilities of exceedence, enabling a rigorous structural reliability analysis.

This paper presents a methodology for evaluating the structural performance and failure reliability of vehicles using advanced dynamic analysis procedures on more accurate large-scale models. The methodology is demonstrated on a detailed model of an automobile using the general purpose FE program NISA II. Two algorithms of eigenvalue extraction are evaluated. Results from the eigenvalue analysis are then used for random vibration analyses. The RMS values and probabilities of exceedence from the random vibration analysis are verified using Monte Carlo (MC) simulations.

Eigenvalue Analysis

All commercially available finite element analysis packages usually provide several methods for eigenvalue extraction, since there is no method that is efficient and reliable for all situations. These algorithms can be classified into two categories; tracking methods and transformation methods. Here, two algorithms available in NISA [1], the accelerated subspace iteration (a tracking method), and the Lanczos algorithm (a transformation method), are examined as to their applicability in eigenvalue analysis of a large scale model of an automobile.

The model (shown in Fig. 1) used for the study was that of a two passenger coupe and has 4,824 elements and 26,450 free degrees of freedom (DOF) which is typical for a full model of an automobile. Consistent (coupled) mass formulation was employed to achieve a higher degree of accuracy. The total number of mass (dynamic) DOF was 26,125 with a maximum wavefront of 614. Feature line plot for the first torsional mode (Mode No. 10) with a frequency of 12.99 HZ as obtained from the eigenvalue analysis is shown in Fig. 2.

Several runs were made to study the efficiency of the two algorithms in terms of the CPU time for eigenvalue extraction. For problems of this size, the Lanczos algorithm is generally faster and requires less core memory than the subspace iteration for the number of eigenvalues typically sought, while disk space is nearly the same. For example, to evaluate 20 eigenvalues, Lanczos required 3.30 CPU hours and 222kB core memory on ELXSI System 6400 whereas the subspace iteration needed 4.92 CPU hours and 639kB in core. The disk usage in both cases was around 158 MB. However, to extract more eigenvalues the Lanczos algorithm required higher core space although the requirement of subspace iteration was almost independent of the number sought.

Random Vibration Analysis

The dynamic loading on a vehicle in reality is not known uniquely and at best can only be predicted. One way of representing a possible dynamic load is by a family or ensemble of histories. The structure may be analyzed for each of the load histories to derive the response statistics. Alternatively, the same results can be obtained more efficiently by introducing the

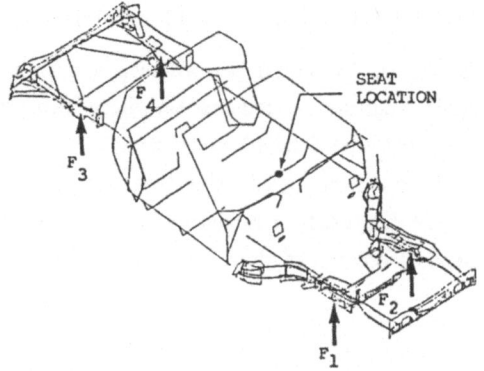

Fig. 1 Boundary Line Plot of the Model
of an Automobile

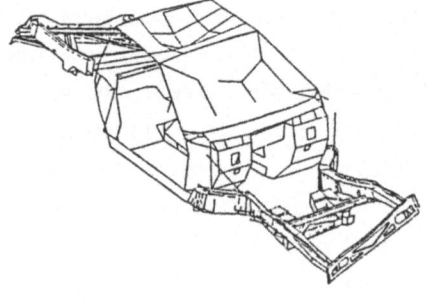

Fig. 2 First Torsional Mode
Frequency = 12.99 HZ

statistics of the load process in the beginning via a power spectral density (PSD) function and by performing a single random vibration (RV) analysis.

When a structural system is subjected to more than one load process, in addition to the description of each load process given by their PSDs, a measure of correlation between the processes has to be provided. In NISA a cross-PSD may be specified to characterize the correlation between two processes. The loading due to terrain roughness at the four wheels of an automobile (shown in Fig. 1), may be given by a 4x4 Hermitian PSD matrix. The loading at the rear wheel lags behind that at the corresponding front wheel by a delay in time (L/V) where L is the distance between the front and the rear wheels and V, the velocity of the automobile. This results in the following elements for the PSD matrix of the two load processes $F_1(t)$ and $F_3(t)$.

$$S_{31} = S_{11}\, e^{i\Omega L/V} \qquad\qquad S_{13} = S_{11}\, e^{-i\Omega L/V} \qquad\qquad S_{22} = S_{11}$$

where S_{11} is the auto PSD function for the loading on the front wheel. In the absence of data supporting any other assumption, full correlation can be assumed between the loadings from the right and the left track resulting in a PSD matrix for all the processes given by:

$$[S] = \begin{bmatrix} S_{11} & S_{11} & S_{11}\,e^{-i\Omega L/V} & S_{11}\,e^{-i\Omega L/V} \\ S_{11} & S_{11} & S_{11}\,e^{-i\Omega L/V} & S_{11}\,e^{-i\Omega L/V} \\ S_{11}\,e^{i\Omega L/V} & S_{11}\,e^{i\Omega L/V} & S_{11} & S_{11} \\ S_{11}\,e^{i\Omega L/V} & S_{11}\,e^{i\Omega L/V} & S_{11} & S_{11} \end{bmatrix}$$

The function S_{11} can be derived by processing time histories obtained from terrain roughness records using a spectral analysis program like DYSPAN [3].

A flat (white noise) acceleration PSD function S_{11} of intensity 0.03125 (g^2/HZ) was used. This vertical acceleration was applied at the support points by using the 'large mass' concept. The vertical acceleration response at a seat location was chosen as the parameter of interest. Different values of delay corresponding to various velocities of the automobile including the zero delay case were used. The PSD, RMS values or standard deviations, zero mean crossing rates or apparent frequencies, and the expected rate of peaks were obtained from RV analysis. Figure 3 shows seat acceleration response PSD at 55 MPH.

To verify the accuracy of the results obtained from the random vibration analysis, MC simulations were done as follows. Random time history samples were generated for the given PSD using DYSPAN, and simulations were performed using transient dynamic analysis of the model for each sample. A sample seat acceleration response time history is shown in Fig. 4. The resulting response time histories were processed to derive statistics of the response and the values were compared to the RV analysis estimates (see Table 1).

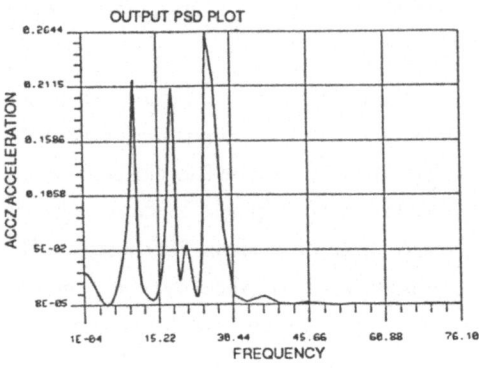
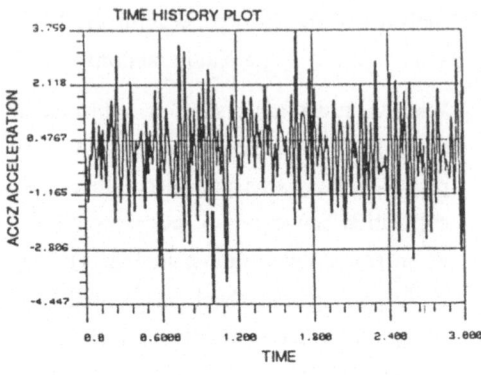

Fig. 3 PSD Plot for Seat Acceleration   Fig. 4 Time History Plot for Seat Acceleration

Reliability Estimation

The ultimate objective in performing a random vibration analysis is to enable estimation of the probability of failure of a design. This may be achieved crudely by employing the standard deviation (i.e.,RMS value) of the response to obtain bounds or safety factors. Alternatively, additional descriptors of the response process such as zero mean crossing and expected peak rates obtained from a random vibration analysis can be employed to accurately estimate the probability of failure [2].

To evaluate the probability of failure of a design, it is useful to estimate *level crossing rates* and the *distribution of the peaks* for a process. If Y(t) is assumed to be a Gaussian process, the positive crossing rate at a level y = a may be expressed as:

$$\eta_a^+ = \eta_0^+ \exp\left(\frac{-a^2}{2\sigma_y^2}\right)$$

where η_0^+ is the zero mean crossing rate and σ_y the RMS value of the response, both of which are available from a NISA analysis. The level crossing rates for normalized thresholds of upto five standard deviations is shown along with the MC simulation estimates in Fig. 5.

Table 1 Comparison of RMS Values and Zero Mean Crossing Rates of Seat Acceleration

Speed (MPH)	Delay (Sec)	RMS Value (g)		Mean Crossing Rate (/sec)	
		RV Analysis	MC Simulation	RV Analysis	MC Simulation
—	0.0	1.9462	1.80	22.283	23.33
55	0.109	1.4446	1.44	22.454	22.72
10	0.598	1.5963	1.40	21.722	22.87

The *probability density function* (PDF) p(a) of the response peaks may be estimated based on σ_y and an irregularity factor α of the process. The expression for p(a) is given in Ref. 2. The irregularity factor is defined as $\alpha = \eta_0/m_0$ and its range is $0 < \alpha < 1$. The parameter m_0 is the expected rate of the peaks. The factor $\alpha = 1.0$ for a narrow band process, and p(a) reduces to a Rayleigh distribution. On the other extreme, when $\alpha = 0$, p(a) becomes a Gaussian PDF. The use of α, which accounts for the nature of the response process provides better estimates of p(a). Figure 6 shows the cumulative peak distribution which is compared to the MC simulations. The irregularity factor here was $\alpha = 0.77$.

In addition to the acceleration, statistics of other response parameters such as stress, strain, and displacement may be used to evaluate a design. Two types of design failure probabilities may be estimated using η_a^+ and p(a). They occur *either* when the response Y(t) reaches a certain failure level y = a for the first time (e.g., yield failure), *or* when the excursions of Y(t) do a small but definite amount of damage depending on the amplitude y = a, and these damage increments reach a fixed total (e.g., fatigue failure). The software package ENDURE [4] employs these statistical parameters derived from a NISA RV analysis to estimate crack initiation and propagation lives for a structure.

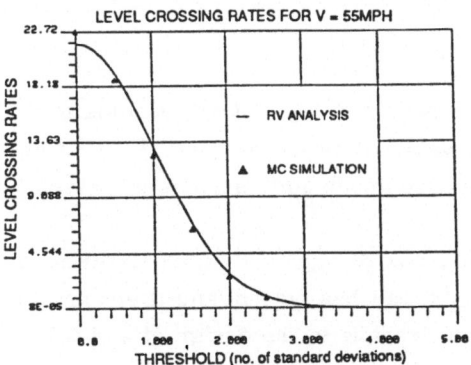

Fig. 5 Level Crossing Rates for Seat Acceleration; V = 55 MPH

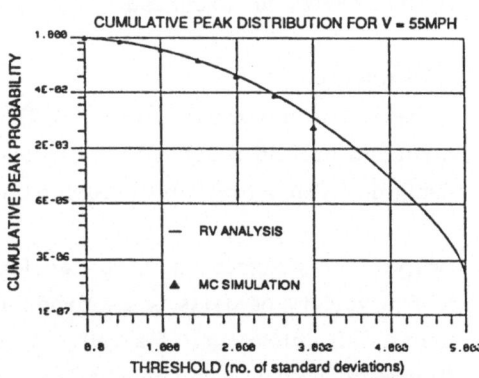

Fig. 6 Cummulative Probability Distribution for Peak Seat Acceleration; V = 55 MPH

References

1. NISA II User's Manual, Engineering Mechanics Research Corporation, Troy, MI

2. Lin, Y.K., *Probabilistic Theory of Structural Dynamics*, McGraw-Hill Inc., 1967.

3. DYSPAN User's Manual, Engineering Mechanics Research Corporation, Troy, MI.

4. ENDURE User's Manual, Engineering Mechanics Research Corporation, Troy, MI.

Design of Flexible Fixtures for the Broaching Process Using the Finite Element Method

KOFI NYAMEKYE and LAEJOON YI

Department of Industrial Engineering
Auburn University, Alabama

Summary

The fundamental philosophy of a flexible manufacturing cell is to produce a family of parts using the same collection of machines. For the cell to be truly flexible, the fixtures should accommodate parts of various geometrical configurations within a parts family. In this paper, an approach in the design analysis of flexible fixtures for the broaching process using the finite element method (FEM) is discussed.

Introduction

A flexible manufacturing cell (FMC) consists of a group of CNC machine tools arranged around a robot [1]. In recent years attention has been focused on designing FMCs for manufacturing low-volume goods and varieties of parts.

Despite the progress in FMC development, the design of flexible fixtures to hold a variety of parts to be produced by a cell has not advanced enough to permit full automation of a cell. A typical example is the design of a flexible fixture to hold a parts family for a broaching process [2].

Several researchers have used the FEM to model various machining processes, but none of those researchers analyzed the design of the fixture that held the parts during machining [3,4,5].

Finite Element Model of Flexible Fixture and Workpiece

In this article, a V-block work-holding device is designed to hold a parts family for a broaching process. Furthermore, the elastic deflection of the workpieces, due to the clamping forces, is analyzed to determine whether the final size of the keyway is within the specified tolerance.

A finite element model was used to predict elastic deformation and stresses in the workpiece-fixture structure. The modeling was based on the general-purpose finite element program, SAP-V2. To model the workpiece-fixture several assumptions were made:

1. The workpiece and fixture shapes will not be changed along the axis of a workpiece's centerline. Therefore the workpiece-fixture can be modeled as two-dimensional plane stress elements.
2. The contact region between the workpiece and the fixture is a line contact in a two-dimensional analysis.
3. In modeling the workpieces, the left edges of the workpieces, which are axes of symmetry of the workpiece-fixture structure, were restrained horizontally.
4. The V-block was considered to be subjected to repeated mechanical stresses clamping after clamping due to high clamping pressure. These stresses could cause fatigue failure.

Design of Flexible Fixtures

Initially, the main cutting forces in the broaching process were estimated using empirical relationships recommended by Kaczmarek [6]. From the known cutting force values, the clamping forces were determined. Table 1 shows the properties of the fixture and workpiece materials as well as the values of the cutting and clamping forces for various keyway sizes. The table also shows the fatigue strength of fixture material [7].

Next, the V-block size was chosen arbitrarily based on the diameter of the largest workpiece. An algorithm for the optimal design technique for flexible fixtures was developed as follows:

1. Design the fixture with the initial dimensions (a = 6 in., b = 6 in., c = 3 in.) (Fig. 1).
2. Check the deformation and stresses due to the clamping force.
3. If the stresses are less than the working fatigue stress, reduce the dimensions of the fixture.
4. If the stresses are greater than the working fatigue stress, increase the dimensions until satisfactory dimensions are achieved.

From the above procedures, the optimal dimensions of the fixture were obtained. Figure 2 shows the optimal design (a = 3.375 in., b = 6 in., c = 2.625 in.). The maximum normal stress around the region in the fixture where the force is applied is 17,459 psi which is less than the working stress. Therefore, this design was considered to be the most satisfactory. Any further analysis will produce an oversized design of the flexible fixture.

Table 1

Workpiece material type (AISI)	1146	1146	1146
Part number	2011	2022	2033
Yield strength (kpsi)	85	85	85
Outside diameter (in.)	5.5	4.5	3.4
Inside diameter (in.)	1.25	1.0	0.75
Keyway size (in.)	.25x.25	.25x.25	.19x.19
Cutting force (lb.)	1376	1376	812
Clamping force (lb.)	48,000	48,000	28,200
Max. keyway deflect. (in.)	8.4×10^{-5}	1.1×10^{-4}	5.4×10^{-5}
Tolerance (in.)	2×10^{-4}	2×10^{-4}	2×10^{-4}
Fixture material type (AISI)	3150	3150	3150
Tensile strength (kpsi)	200	200	200
Working fatigue strength (kpsi)	19.8	19.8	19.8

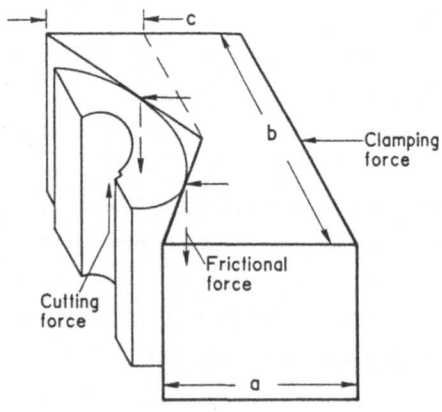

Fig.1. Workpiece-fixture structure

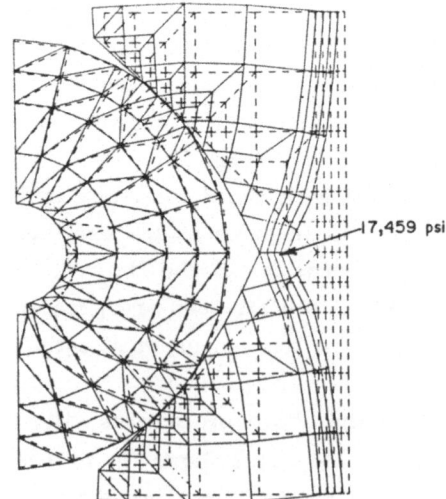

Fig.2. Modeling of optimal design of the fixture

To permit the fixture to hold all parts of the family, the FEM was extended to design a 0.663-inch-thick adapter plate (Fig. 3).

Analysis of Workpieces

The workpiece-fixture structure was modeled by constructing 214 nodal points and 248 quadrilateral elements. From these 248 elements, 136 were used to model the workpiece. This number was considered large enough for predicting the stresses and deformations near the workpieces' contact points and keyways. Figure 4 shows the elastic deformation around the keyway of a typical workpiece. Table 1 also shows the maximum elastic deflection around all of the workpieces' keyways. Notice that none of the deflections exceeds the tolerances of the workpieces; therefore the workpieces are considered to be under the tolerance after the clamping forces are released.

Conclusions

A finite element model for the design of flexible fixtures for the broaching process is presented in this article. Clamping forces are input into this structural model to simulate the stresses and deformations of workpieces and fixtures under different clamping conditions (different cutting conditions). Emphasis is given to the development of a finite element model for workpiece-fixture structures and to the use of the simulation methodology to study the broaching process.

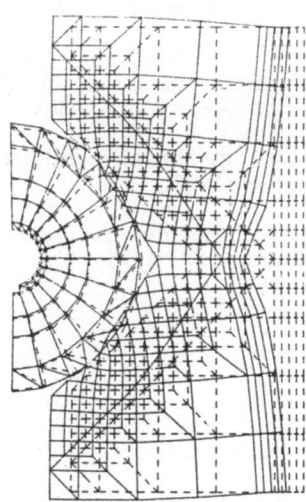

Fig.3. Modeling of workpiece, adapter and fixture

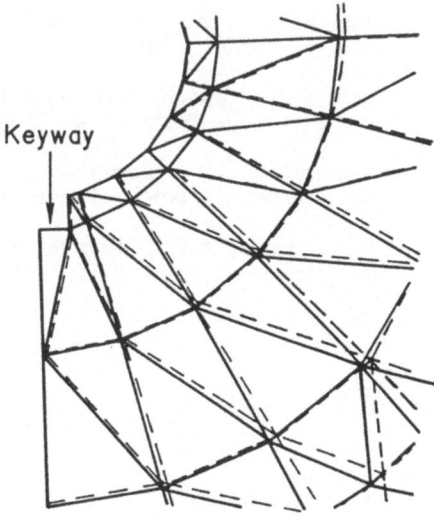

Keyway

Fig.4. Elastic deformation around a keyway

Given the dimensional specifications of the workpieces such as the finished part tolerances (0.0002 inch) and permissible stresses (19,800 psi for the fixture) as constraints, an algorithm has been constructed to obtain a satisfactory fixture design.

The largest deflection of the machined keyway was 0.00011 inch. This deflection is under the specified tolerance for the keyway. The length of the V-block was chosen as 3.375 inches. The maximum normal stress of the fixture was 17,459 psi which is less than the working fatigue strength.

Acknowledgment

The authors thank the Advanced Manufacturing Technology Center for its support and A.H. Honnell and P.S. Flick of the AMTC Information Resources Laboratory for production of the manuscript.

References

1. Nyamekye, K.; Black, J.T.: Rational approach in the design analysis of flexible fixtures for an unmanned cell. 15th North Am. Manuf. Res. Conf. Proc. (1987) 600-607.

2. AMTC unmanned robotic cell for flex-hub manufacturing. Advanced Manufacturing Technology Center. Auburn University, Alabama.

3. Vijayaraghavan, L.: Evolution of stress and displacement of tool and workpiece on broaching. Int. J. Tool Des. Res. 121 (1981) 263-270.

4. Jablonowski, J.: New ways to build machine structures. Am. Mach. Autom. Manuf. 131 (1987) 87-94.

5. Lee, S.J.; Kapoor, S.G.: Cutting process dynamics simulation for machine tool structure design. J. Eng. Ind., Trans. ASME. 108 (1986) 68-74.

6. Kaczmarek, J.: Principles of machining by cutting, abrasion and erosion. Stevenage, England: Peter Peregrinus Limited 1976.

7. Deutschman, A.D.; Michels, W.J.; Wilson, C.E.: Machine design--theory and practice. New York: Macmillan Publishing Co. 1975.

Boundary Element Method

The Solution of Elastic Contact Problems Using a Boundary Element Flexibility Approach

S. TAKAHASHI*

Material Research Lab., Central Engineering Labs,
Nissan Motor Co. Ltd., 1, Natsushima-cho,
Yokosuka 237, JAPAN

C.A. BREBBIA

Computational Mechanics Institute, Ashurst Lodge, Ashurst, Southampton, UK.

Summary

A flexibility approach using boundary elements for elastic contact problems
with and without friction is proposed. A system of equations is required
only for the elements which are expected to come into contact, and hence
is usually small which results in efficient computations. In order to
demonstrate the accuracy and validity of the approach a Hertzian contact
problem is examined in detail. The results are in good agreement with
analytical solutions.

1. Introduction

Boundary elements have developed rapidly and can now be used to solve
efficiently many industrial problems. They have several advantages over
other numerical techniques, i.e., i) only surface discretization is required, ii) traction and displacement unknowns in stress analysis are computed with the same degree of accuracy.

2. Theory

The approach followed in this paper has important advantages, namely that
it is very efficient and easy to implement. The technique consists of
forming the flexibility matrix for each elastic body using boundary elements
and once this is done the contact analysis can progress using always the
same matrices. Furthermore, the only elements of the flexibility matrix
required are those corresponding to the nodes which are coming into contact.
Since the system of equations required is usually small, efficient computations can be achieved. The flexibility approach presented in this paper
has been implemented using constant boundary elements and an incremental
load procedure because of the non-linearity of the problems. This method
comprises two main steps as follows,

> STEP 1: Boundary element analysis including formulation of the
> flexibility matrices of different components

---

* Presently at: Computational Mechanics Institute, Ashurst Lodge, Ashurst,
  Southampton, UK.

STEP 2:   An analysis to follow the process when the deformation progress and the nodes come in contact.

(i) Boundary element analysis

The most important step of boundary element analysis is the formation of
the flexibility matrix. The elements of this matrix express the displacements resulting when a unit traction in the normal or the tangential direction to the contact angle (θ_s) is applied on the element under consideration. The contact angle is defined as the angle of the contact surface
and given by the following equation,

$$\theta_s = \theta_c - (\theta_c - \theta_f) \cdot \frac{E_f}{E_c + E_f} \tag{1}$$

where θ_c and θ_f are the angles of the contact elements on the contactor
and the foundation, E_c and E_f are Young's moduli, respectively. The flexibility coefficients for a particular body can be defined as $f_{k\ell}^{ij}$ where i
represents the node number at which the displacements occur, j is the node
number at which a unit traction is applied, k is the direction of the displacement components and ℓ is the direction of the applied tractions. The
flexibility matrix F can now be written as,

$$\underset{\sim}{F} = [f_{k\ell}^{ij}] \tag{2}$$

This matrix is used in the contact analysis to calculate the normal and
the tangential tractions at the nodes coming into contact.

(ii) Contact Analysis

Once the flexibility matrix has been formed, the contact analysis can comm-
ence. If nodes come into contact, the displacements and the tractions
conditions need to be satisfied in the contact region and the boundary
tractions are computed. Then, the analysis can proceed with the next load
increment until the total load is reached. The displacements of the contactor ($\underset{\sim}{u}_c$) are written as follows,

$$\underset{\sim}{F}_c \cdot \underset{\sim}{t}_c = \underset{\sim}{u}_c \tag{3}$$

where $\underset{\sim}{F}_c$ is the flexibility matrix of the contactor and $\underset{\sim}{t}_c$ are the tractions
on the contact region. For the foundation similar relationship can be ob-
tained, i.e.,

$$\underset{\sim}{F}_f \cdot \underset{\sim}{t}_f = \underset{\sim}{u}_f \tag{4}$$

The contact analysis is then performed by using equations (3) and (4) in conjunction with contact conditions such as stick and slide conditions. If pairs of nodes come into contact, one applies stick conditions first. Equilibrium of tractions is expressed as follows,

$$t_{cn} + t_{fn} = 0 \tag{5}$$

$$t_{ct} + t_{ft} = 0 \tag{6}$$

where subscripts n and t are the normal and the tangential directions of the tractions. The following displacement conditions are then introduced, i.e., i) the distance between the pair of nodes, when the contact occurs, is maintained. ii) the line which passes through the pair of nodes has the same angle as the contact angle θ_s. These conditions are introduced into (3) and (4) and then the tractions in the contact region are computed. The calculated tractions are added to the tractions which are accumulated during the previous load increment. Sliding conditions should be introduced if the accumulated tractions (T) violate the following conditions.

$$\left| \frac{T_{ct}}{T_{cn}} \right| \leq \mu \tag{7}$$

$$\left| \frac{T_{ft}}{T_{fn}} \right| \leq \mu \tag{8}$$

where μ is Coulomb friction coefficient. The relationship between the normal and the tangential tractions in sliding conditions is represented by the following equations.

$$t_{cn} + t_{fn} = 0 \tag{9}$$

$$t_{ct} = \pm \mu \, t_{cn} \tag{10}$$

$$t_{ft} = \pm \mu \, t_{fn} \tag{11}$$

For the displacement conditions, the line which passes through the pair of nodes also has the same angle as the contact angle θ_s.

In certain cases tensile tractions may occur in the normal direction when considering perfect contact between the two surfaces, but they may disappear when sliding conditions are introduced. Because of this, if these tensile tractions occur one needs to examine if they are still valid for the more realistic friction conditions before one assumes that the nodes are really separated.

3. Numerical Applications

In order to test the present method, a classical Hertzian contact problem [3] has been studied, i.e., an elastic cylinder with radius R = 50 mm pressed on an elastic rectangular punch with 190 mm × 190 mm as shown in Figure 1(a). Figure 1(b) shows the boundary element mesh discretization; the cylinder and the foundation have 52 and 57 constant elements, respectively. The problem is assumed to be plane strain and non-friction condition is employed.

Two cases are studied on this problem. The first is the one in which different values of applied forces are considered. The material properties and applied force (P) are selected as follows,

$$\text{Force } P = 398 \quad 1,178 \quad 1,962 \quad 3,926 \text{ N}$$
$$\text{Young's Modulus} \quad E = 10,800 \text{ N} /\text{mm}^2$$
$$\text{Poisson's ratio} \quad \nu = 0.35$$

Figure 2 gives the normal tractions against distance from the centre of the model and shows the results corresponding to the Hertz solutions.

The second case consists in testing a series of combinations of Young's modulus of the cylinder and E_f is Young's modulus of the foundation (Table 1), i.e.

Table 1   Ratio of E_c to E_f

E_c (N /mm^2)	E_f (N /mm^2)	E_c/E_f
10,800	54,000	0.2
10,800	27,000	0.4
10,800	10,800	1.0

The Poisson's ratio is 0.35 and the force equal to 3,926 N . Figure 3 shows the contact pressure versus the length. The results are in good agreement with the Hertz solution for different Young's modulus ratio.

Conclusions

This paper studies two-dimensional contact problems between elastic bodies using the boundary elements flexibility approach. The method has been ex-

tended to solve frictional as well as non-frictional problems. The calculated results are in good agreement with analytical Hertz solutions. It
can be concluded that the flexibility approach can give accurate results
for the problems studied here.

References
1. Andersson, T., Fredriksson, B. and Persson, B.G.A.,   The Boundary
    Element Method Applied to Two-dimensional Contact Problems, Proc.
    2nd Int. Seminar on Recent Advances in Boundary Element Methods,
    Southampton,   C.A. Brebbia (Ed.) CML Publications, March 1980.

2. Kuich, G. Applications of the Boundary Element Method to Contact Prob-
    lems,   BETECH/86, MIT, C.A. Brebbia and J.J. Connor (Ed), Computa-
    tional Mechanics Publications, pp.499-519, June 1986.

3. Timoshenko, S.P. and Goodier, J.N. The Theory of Elasticity, McGraw-
    Hill, New York, pp.409-420, 1970.

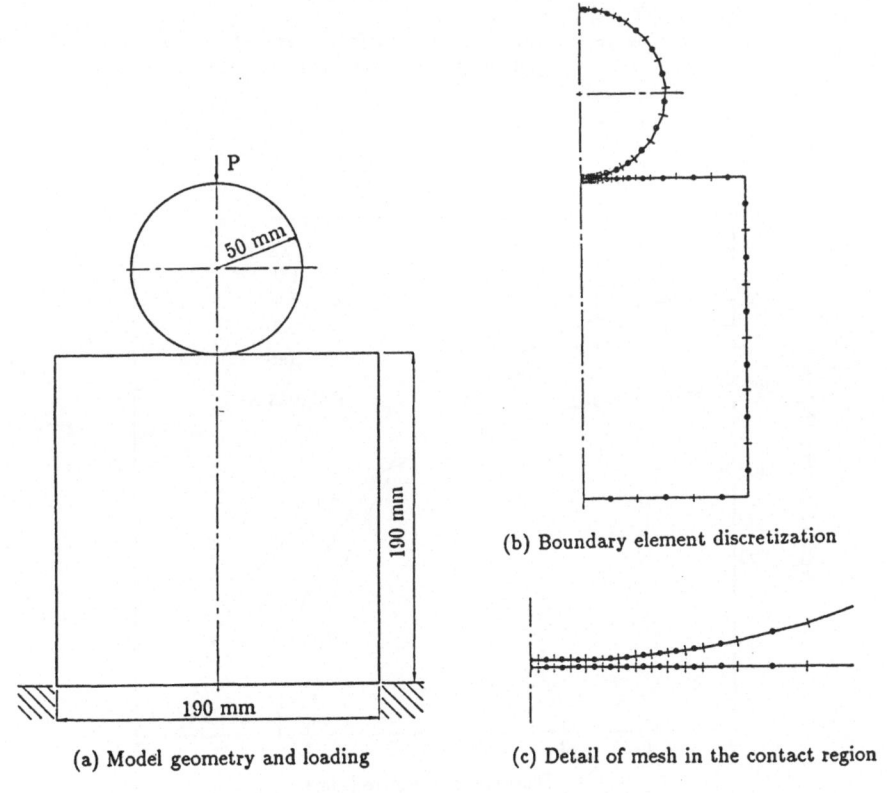

(a) Model geometry and loading

(b) Boundary element discretization

(c) Detail of mesh in the contact region

Figure 1. Model of cylinder on flat foundation

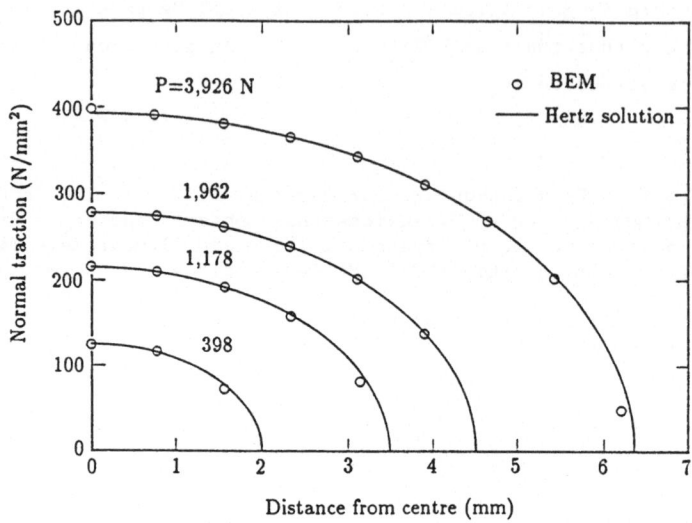

Figure 2. Normal traction distribution at contact
region of cylinder on elastic foundation

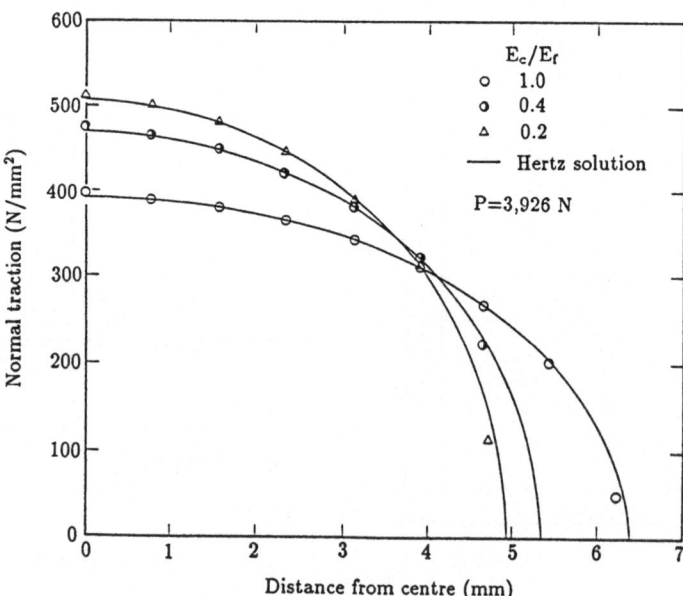

Figure 3. Normal traction distrubution at contact
region of cylinder on flat foundation

The Solution of Transient Heat Transfer Problems Using Boundary Elements

C.A. Brebbia

Computational Mechanics Institute, Ashurst Lodge, Ashurst, Southampton, UK.

Summary

This paper describes the solution of transient heat transfer problems using the Dual Reciprocity boundary element formulation. The approach eliminates the need of dividing the boundary into the type of internal cells which need to be applied in many transient boundary element problems. The technique can be extended to the case of material non-linearities by using a transformation. The present paper concentrates on the practical applications of the new technique and the way it has been implemented in the BEASY code.

Introduction

Boundary Element Methods have rapidly developed as a major tool of engineering analysis and design [1][2][3]. Their range of applications includes a large number of non-linear and time dependent problems. Their main drawback when solving these problems however is that the domain under consideration usually needs to be subdivided into a series of cells or internal elements. This destroys in part the attraction of the technique.

In 1982 a new formulation was proposed by Nardini and Brebbia [4][5] which referred the solution to the boundary and a series of internal points for the case of dynamic problems. The formulation was extended to transient conditions by Wrobel, Brebbia and Nardini [6] in 1986 and applied to solving a range of engineering problems by Wrobel and Brebbia [7]. A full explanation of this new technique for the case of elastodynamics (hyperbolic) can be seen in reference [8], and the same authors presented a description of the way in which the method could be used in parabolic problems [9]. More recently Niku and Brebbia [10] have applied the same idea to represent arbitrarily distributed sources.

The implementation of the new method in general purpose BEM codes was attempted in 1986 by Wrobel, Brebbia and Nardini [11] and the extension to non-linear transient thermal problems was presented in 1987 by Brebbia and Wrobel [12]. The same formulation was applied for the solution of axisymmetric problems in 1986 [13].

The present paper concentrates on the practical applications of the new technique and in particular its implementation into the BEASY code [14] for solving automotive type components. Different results were presented at the Conference but are not included here due to space limitation.

2. Governing Integral Equation

Transient heat transfer problems are governed by the following equation

$$\kappa \nabla^2 u = \frac{\partial u}{\partial t} \qquad \text{in } \Omega \qquad (1)$$

in which κ is the diffusivity of the medium ($\kappa = k/c\rho$, where k is the thermal conductivity, c is the specific heat and ρ the density of the material). The right hand side involves the derivative of temperature u with respect to time, t. u is the temperature which satisfies the following conditions

$$\text{Essential or temperature conditions:} \quad u = \bar{u} \qquad \text{on } \Gamma_1$$

$$\text{Natural or flux conditions:} \quad q = k \frac{\partial u}{\partial n} = \bar{q} \qquad \text{on } \Gamma_2 \tag{2}$$

where Γ is the total boundary and n denotes the normal to it. The bar indicates that the value of the corresponding variable is known. Conditions (2) can be easily extended later on to deal with convection type of problems and others.

The problem defined by equation (1) and (2) can be reduced to a boundary integral form using the fundamental solution u*, i.e.

$$\nabla^2 u^* + \Delta^i = 0 \tag{3}$$

where Δ^i is the Dirac delta function at a point x_i. One can also define the fundamental solution flux q*, such that $q^* = k \, \partial u^*/\partial n$.

The solution u* for two and three dimensional problems is well known, i.e. for 3D it is

$$u^* = \frac{1}{4\pi r} \tag{4}$$

and for two dimensional cases,

$$u^* = \frac{1}{2\pi} \ln\left(\frac{1}{r}\right) \tag{5}$$

Applying weighted residual or reciprocity principles [1][3] between the u and u* fields an integral equation valid for the fundamental solution at any point 'i' is obtained, i.e.

$$c^i u^i + \int_\Gamma q^* u \, d\Gamma + \int_\Omega c\rho \frac{\partial u}{\partial t} u^* \, d\Omega = \int_\Gamma q \, u^* \, d\Gamma \qquad (6)$$

This equation is the starting point for the boundary element method in transient heat transfer. Notice that all integrals are on the boundary with the exception of the one corresponding to the time derivative term. The constant c^i depends on the type of boundary point i under consideration. (i.e. $c = \frac{1}{2}$ for smooth bodies or otherwise proportional to the solid angle at i. $c^i = 1$ for internal points and $c^i = 0$ for the case of the point i being external to the domain).

3. The Dual Reciprocity Method

This method which was initially proposed by Nardini and Brebbia [4][5] for elastodynamics is essentially a generalized search for particular solutions using localized rather than global functions. Because of this the technique can also be used to represent body force distributions over the whole boundary or along a part of it as demonstrated by Niku and Brebbia [10]. The method has also been applied to solve transient diffusion problems [6] of the type presented in this paper. Papers related to transient thermal analysis were produced by Brebbia and collaborators and include the extension of the method to non-linear diffusion problems [7,11,12,13]. This paper instead concentrates on the practical applications of the dual reciprocity technique and in particular its implementation into the BEASY code [14] for solving automotive type components.

In order to describe the fundamentals of dual reciprocity let us consider the domain term in equation (6) and propose the following representation of the time derivative, i.e.

$$\dot{u} = \frac{\partial u}{\partial t} = \sum_{j=1}^N f^j \dot{\alpha}^j \qquad (7)$$

where f^j represents a set of N known coordinate functions. $f^j = f_j(\xi, x)$ is chosen as a function between the points ξ_j and x, where the 'j' point can be one of the boundary nodes. α^j are the unknown coefficients associated with each f^j. Notice that the f^j functions are considered to originate at 'j' different points, most of which are on the boundary. These functions are of the same type for all the points.

One can first define the functions f^j and then find their corresponding particular solutions, i.e. such that

$$k \ \nabla^2 \hat{u}^j = f^j \tag{8}$$

where the $\hat{u}^j$ field and its associated variables such as $q^j = k \ \partial \hat{u}^j / \partial n$ can be found by integrating the above equation.

Equation (6) can now be written as follows,

$$c \ u + \int_\Gamma q^* \ u \ d\Gamma - \int_\Gamma q \ u^* \ d\Gamma = - c\rho \sum_{j=1}^{N} \{\dot{\alpha}^j \int_\Omega (k \ \nabla^2 \hat{u}^j) u^* \ d\Omega\} \tag{9}$$

Notice that the superscript 'i' in the first term has been eliminated for simplicity's sake but the reader should keep in mind that these terms and the whole requation still refer to a particular position 'i' of the source. Each term on the right hand side can be integrated by parts resulting in only boundary integrals, i.e.

$$c \ u + \int_\Gamma q^* \ u \ d\Gamma - \int_\Gamma q \ u^* \ d\Gamma =$$

$$= c\rho \sum_{j=1}^{N} \{\dot{\alpha}^j [c^j \ \hat{u}^j + \int_\Gamma q^* \ \hat{u}^j \ d\Gamma - \int_\Gamma u^* \ \hat{q}^j \ d\Gamma] \ \} \tag{10}$$

Notice that the right hand side of this equation only involves boundary integrals and is formed by adding a series of terms, each of them localized at a particular point 'j'.

Equation (10) produces the following matrix system after the usual boundary element discretization (see for instance [5][11]).

$$\underset{\sim}{H} \ \underset{\sim}{U} - \underset{\sim}{G} \ \underset{\sim}{Q} = \underset{\sim}{S} \ \underset{\sim}{\dot{\alpha}} \tag{11}$$

where

$$\underset{\sim}{S} = c\rho \ [\underset{\sim}{H} \ \underset{\sim}{\hat{U}} - \underset{\sim}{G} \ \underset{\sim}{\hat{Q}}] \tag{12}$$

$\hat{U}$ and $\hat{Q}$ are now square matrices, $N \times N$ each (N is the number of points where the function f^j has been applied). The columns of these matrices represent the values of $\hat{u}^j$ and $\hat{q}^j$ at the different nodes for the case of the f^j function acting at a particular 'j' point.

Notice that the $\dot{\alpha}$ coefficients are different from the values of $\dot{u}$ at the points under consideration. They are however related throughout equation

The introduction of one more equation for the constant will require setting up another equation in terms of α, which can be done by defining an internal degree of freedom or 'pole'. In general the introduction of more degrees of freedom than simply those on the boundaries is recommended to obtain better results when the ü functions are difficult to represent in function of the boundary values only.

Other types of functions proposed by Nardini and Brebbia [9] included localized harmonic functions and polynomials in terms of x_k coordinates, but the best results were always reported using 'conical' type functions.

It has already been pointed out that the success of the DRM is strongly dependent on the choice of coordinate functions which ought to satisfy completeness. Further investigation of this was carried out by Tang and Brebbia [15] who proposed using generalized functions to represent the domain terms, including higher powers of the distance, i.e. terms of the type r^k, where k = 1,2,3 etc.

For axisymmetric cases the solution depends not only on r but also on the distance from the source and field points to the axis of revolution. Because of this Nardini, Telles and Brebbia [13] proposed the following function for these cases, i.e.

$$f^j = r \left[1 - \frac{R^j}{4R} \right] \qquad (19)$$

where R^j is the distance from the different poles to the axis of revolution and R is distance from any point on Γ(or Ω) to the same axis.

5. Computer Implementation

The above formulation has been implemented in the BEASY system [14] which carries out diffusion analysis using a two level time integration scheme.

The dual reciprocity formulation for diffusion problems preserves the main advantage of the BEASY code, namely the ease of data preparation, since only the boundary of the region under study needs to be discretized.

The BEASY codes offer a wide range of continuous and discontinuous boundary elements with constant, linear or quadratic variation for the functions u and q within the elements. Multizones can be studied using piecewise homogeneous regions and allows for many types of boundary and interface condi-

tions. Heat sources can be input as line, point or volumetric effects.
The BEASY codes permit using planes or lines of symmetry, on which no elements are required.

The method presents an efficient formulation for the solution of transient
thermal problems which is easy to implement in general purpose BEM codes.

References

1. Brebbia, C.A. The Boundary Element Method for Engineers, Pentech Press,
    London, Computational Mechanics Publications, Boston, 1978.

2. Brebbia, C.A., Telles, J.C.F. and Wrobel, L.C. Boundary Element Tech-
    niques - Theory and Applications in Engineering, Springer-Verlag,
    Berlin & NY, 1984.

3. Brebbia, C.A. and Dominguez, J. Boundary Elements - An Introductory
    Course, Computational Mechanics Publications, Southampton and Boston
     (This book includes a PC diskette with BE codes) 1988.

4. Nardini, D. and Brebbia, C.A. A New Approach to Free Vibration Analysis
    using Boundary Elements, in Boundary Element Methods in Engineering,
    (Ed. C.A. Brebbia) Springer-Verlag, Berlin & NY, 1982.

5. Brebbia, C.A. and Nardini, D. Dynamic Analysis of Solid Mechanics by
    an Alternative Boundary Element Approach, Int. J. Soil Dynamics and
    Earthquake Engg., Vol.2, No.4, pp.228-233, Computational Mechanics
    Publications, Southampton, 1983.

6. Wrobel, L.C., Brebbia, C.A. and Nardini, D. The Dual Reciprocity Bound-
    ary Element Formulation for Transient Heat Conduction, Proc. 5th Int.
    Conf. on FEM in Water Resources, Computational Mechanics Publications,
    Southampton and Boston, 1986.

7. Wrobel, L.C. and Brebbia, C.A. The Dual Reciprocity Boundary Element
    Formulation for Non-linear Diffusion Problems, Computer Methods in
    Applied Mechanics and Engg., Vol.65, pp.147-164, 1987.

8. Nardini, D. and Brebbia, C.A. Boundary Integral Formulation of Mass
    Matrices for Dynamic Analysis, in Topics in Boundary Element Research,
    Vol.2, (Ed. C.A. Brebbia), Springer-Verlag, Berlin & NY, 1985.

9. Nardini, D. and Brebbia, C.A. The Solution of Parabolic and Hyperbolic
    Problems using an Alternative Boundary Element Formulation, in Proc. of
    VIIth Int. Conf. on BEM, Computational Mechanics Publications, Southamp-
    ton & Boston, 1985.

10. Niku, S.M. and Brebbia, C.A. Dual Reciprocity Boundary Element Formula-
    tion for Potential Problems with Arbitrarily Distributed Sources, Tech.
    Note, Eng. Analysis Journal, Vol.5 No.1, 1988.

11. Wrobel, L.C., Brebbia, C.A. and Nardini, D. Analysis of Transient Ther-
    mal Problems in the BEASY System, in BETECH/86, (Eds J.J. Connor & C.A.
    Brebbia), Computational Mechanics Publications, Southampton & Boston, 1986.

12. Brebbia, C.A. and Wrobel, L.C., Non-linear Transient Thermal Analysis using the Dual Reciprocity Method, in Boundary Element Techniques: Applications in Stress Analysis and Heat Transfer (Eds C.A. Brebbia & W. Venturini), Computational Mechanics Publications, Southampton & Boston, 1987.

13. Wrobel, L.C., Telles, J.C.F. and Brebbia, C.A. A Dual Reciprocity Boundary Element Formulation for Axisymmetric Diffusion Problems, in Proc. VIIIth Int. Conf. on BEM, Tokyo, 1986, Computational Mechanics Publications, Southampton & Boston, 1986.

14. BEASY - Boundary Element Analysis System as a CIM Tool, in Computer Aided Engineering Systems Handbook, Vol.1 (Eds. J. Puig-Pey & C.A. Brebbia) CM Publications, Southampton, 1987.

15. Tang, W. and Brebbia, C.A., Critical Comparison of Two Transformation Methods for taking BEM Domain Integration to the Boundary, Submitted for publication to Engineering Analysis Journal.

A Simplified Technique for the Analysis
of Electrostatic and Magnetostatic Problems

J.TREVELYAN, S.NAGESWARAN, C.A. BREBBIA

Computational Mechanics, Inc.
Billerica, Massachusetts, U.S.A.

Summary

The analysis of engineering problems in electrostatics and magnetostatics is
a subject of growing importance as technological advances have introduced a
wide range of new devices. However, computational methods such as FEM have
not been as successful as expected in this field, mainly because traditional
numerical analysis techniques are not well suited to the type of geometry
typically encountered.

The boundary element method has developed over the last five years to offer
now a real alternative to the finite element method in various engineering
disciplines, notably in heat transfer and stress analysis. This paper describes how a large class of electrostatic and magnetostatic problems can be
treated simply and quickly using a commercial boundary element program.

Applications examples are given showing how the BEASY program handles general
electrostatic and magnetostatic field problems, calculates capacitance relationships between discrete conductors and simulates cathodic protection systems.

INTRODUCTION

Over the last twenty years, numerical analysis of engineering problems has
relied heavily on the finite element method, which has a wide range of applications in the engineering and scientific worlds. However, this is frequently
unsatisfactory in electrical field problems as the method does not lend itself
to typical geometries found. Commonly, such problems involve areas of intricate
detail, at which the field variables need to be found accurately, yet also
extend to "infinity" since the medium to be analysed may be the sea-water, air-
space or any other conducting medium.

There are two major problems associated with finite element modeling in these
cases. Firstly, the sharp element grading which is necessary to avoid large
numbers of elements, and hence large run-times, can lead to numerical ill-
conditioning and thus inaccuracy. Secondly, it takes a great deal of time and
effort to generate the the model data, and this can make the solution unecon-
omic for even comparatively simple cases.

A technique which overcomes these problems is the boundary element method [1]. By contrast to the finite element method, in which elements of volume (3D) or area (2D) are defined, boundary elements are only required on the surface of the model, and are therefore surface elements for 3D problems and line elements for 2D geometries (fig.1). This is the major reason for the benefit in data preparation time which is found when boundary elements are used.

The technique is currently in wide use in the analysis of heat transfer and stress analysis [2]. We shall consider in this paper the heat transfer type of analysis which is, in essence, a Laplace equation solver. The paper shows how electrostatic problems and some magnetostatic problems can be reduced to the Laplace form and can thus be solved using today's commercial boundary element software with great ease and efficiency.

ELECTROSTATICS AND CAPACITANCE PROBLEMS

For the electrostatic case, the governing equation is Laplace's, i.e.

$$\nabla^2 v = 0 \qquad \qquad \text{....(1)}$$

or, in the presence of sources, the Poisson form

$$\nabla^2 v = -\frac{1}{\varepsilon} \cdot \rho \qquad \qquad \text{....(2)}$$

where v is the (scalar) potential, or voltage, ε is the permittivity, and the term ρ expresses the charge density field. Clearly, there is no work involved in transforming this behavior to the Laplace form, so a commercial boundary element heat transfer code may be used to solve the equations directly [3].

A useful area of application is the determination of capacitance relationships between discrete conductors. This is important to assess with accuracy in much of today's electronic design. Cottrell and Buturla [4] described a finite element based algorithm, but the mesh generation time could easily be prohibitive, especially for 3D cases. However, the same approach has been used successfully using boundary elements, as demonstrated by Hiwasara and Nakamura [5], and completely overcomes the mesh generation problem. For a system of n conductors, the method involves considering a unit potential on one conductor with zero on the others, calculating the current densities (and thus the charges), and so finding one column of the capacitance matrix C by substitution into

$$\vec{Q} = C \vec{V} \qquad \qquad \text{....(3)}$$

where $\vec{Q}$ is a vector containing the n charges, and $\vec{V}$ is a vector containing the n voltages, which are known. Applying the method to the solution for the capacitance of two coaxial conductors (fig.2), the boundary element results compare well with the classical solution, as shown below.

$$\begin{bmatrix} 2.726 & -2.726 \\ -2.726 & 2.726 \end{bmatrix} \qquad\qquad \begin{bmatrix} 2.728 & -2.728 \\ -2.728 & 2.728 \end{bmatrix}$$

Classical solution   BEASY solution

CATHODIC (CORROSION) PROTECTION SYSTEM SIMULATION

Another application of the boundary element method in electrostatics is the simulation of cathodic protection systems for the prevention of corrosion [6]. This is ideally suited to boundary element analysis as the domain (the sea-water) may be assumed to be infinite. There is therefore only an internal boundary to the sea-water, and this is the outer surface of the body to be protected. So all that is required for an analysis of, for example, a pipe-joint is a surface representation of the joint (fig.3). Note that the model shown in the figure is a BEASY model of the water surrounding the pipe joint, and not of the joint itself.

Being in essence an electrostatic field problem, cathodic protection studies are governed by the Laplace equation. But the boundary conditions are non-linear, and this requires the use of an iterative scheme to proceed to a solution. The special cathodic protection option BEASY-CP incorporates such an iterative scheme, allowing complete definition of these non-linear "polarization" boundry conditions, allowing an accurate solution to be found quickly for both galvanic and impressed current systems (fig.4).

MAGNETOSTATIC PROBLEMS

The reduction of Maxwell's equations to the Laplace form is not as straight-forward as the electrostatic case. Assumptiopns need to be made, which restrict the range of applications currently solved by commercial boundary element software. It should be noted that the boundary element method can be used for a more complete solution to Maxwell's equations, but usually in conjunction with a finite element solver.

A subset of Maxwell's equations, neglecting time dependencies and currents (i.e. magnetostatics), may be expressed

$$\nabla \times \vec{H} = 0 \; ; \qquad \nabla \cdot \vec{B} = 0 \qquad\qquad ....(4)$$

where $\vec{B}$ is the magnetic induction (flux density) vector and $\vec{H}$ is the negative field intensity vector. Magnetic flux density and magnetic field intensity may be related by

$$\vec{B} = \mu \vec{H} \quad\quad\quad ....(5)$$

where μ is the magnetic permeability, assumed to be constant through the domain. One defines a magnetic vector potential $\vec{A}$ such that

$$\vec{B} = \nabla \times \vec{A} \quad\quad\quad ....(6)$$

For the 2D case, $\vec{B}$ and $\vec{H}$ have components only in the in-plane direction, and the vector potential $\vec{A}$ has components only in the out-of-plane direction, and may therefore be considered a scalar. Using equations (4) to (6) we can write

$$\vec{B} = \nabla \times \vec{A} = (\partial A/\partial x_1, \ \partial A/\partial x_2, \ 0) \quad\quad\quad ....(7)$$

and

$$\nabla \times \vec{H} = \nabla \times \vec{B} = \nabla^2 A = 0 \quad\quad\quad ....(8)$$

This is the familiar Laplace operation, and the equation (8) may therefore be solved using the BEASY boundary element package. The restrictions applied by the above assumptions mean that only magnetostatic cases can be considered at present, but the advantages offered by the boundary element method in terms of speed and accuracy often outweigh these restrictions in a design environment.

Figures 5 to 8 show various steps in the analysis of a typical rotor configuration found on electrical machinery. Both 2D and 3D cases are considered.

CONCLUSIONS

For electrostatic problems, the boundary element method provides an accurate and efficient means for solving a wide range of problems in diverse fields. There are many problems for which boundary elements offer great advantages over finite element programs, as the analysis domain frequently extends to infinity. Capacitance relationships in geometrically complex cases can be simply computed and non-linear boundary conditions, such as commonly found in corrosion problems, can be handled fully automatically.

For electromagnetic problems, the areas of application are currently limited to magnetostatics, but development of more sophisticated algorithms is under way. The boundary element method is overcoming some of the more serious drawbacks of finite element modeling, and makes the analysis of many geometries a practical option.

REFERENCES

1. Brebbia,C.A.;Dominguez,J. Boundary Elements - An Introductory Course, Computational Mechanics Publications, 1988.

184

2. Brebbia,C.A.,Adey,R.A. Integrated Design and Analysis using the Boundary Element Method, 9th Int Conf on BEM, Stuttgart, 1987, CM Publications.

3. Symm,G.T. Applications of Boundary Elements in Electrostatics, 1st BETECH Conf Adelaide, Nov. 1985, Computational Mechanics Publications.

4. Cottrell,P.E.;Buturla,E.M. VLSI Wiring Capacitance, IBM Journal of R&D, Vol.29,Number 3,May 1985.

5. Hiwasara,M.;Nakamura,M Analysis of Capacitive Transducer for Displacement Measurement by Boundary Element Method, 8th Int. Conf on BEM,Tokyo, 1966.

6. Brebbia,C.A.;Niku,S.M.;Bray,J.A. State of the art: Computer application of the boundary element method for cathodic protection of offshore structures, 7th Int. Conf. on Offshore Mechanics and Arctic Engineering, Houston, 1988.

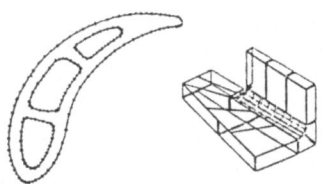

Fig.1. 2D and 3D BEASY models

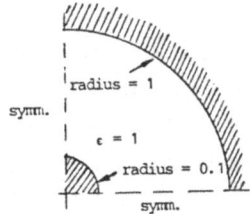

Fig.2. Coaxial conductors

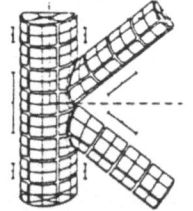

Fig.3. BEASY model of K-joint

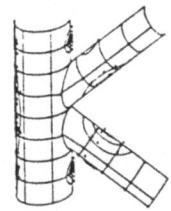

Fig.4. Potential results on joint

Fig.5. Rotor configuration

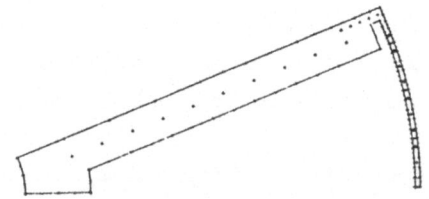

Fig.6. BEASY model for rotor air-gap

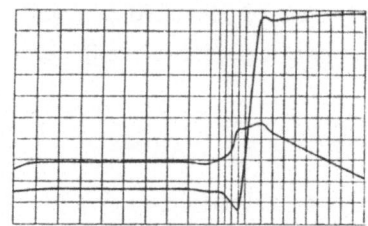

Fig.7. Flux density components (2D)

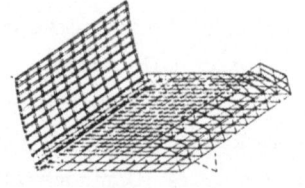

Fig.8. 3D BEASY model for rotor

Two Domain Heat Transfer Studies Using Coupled Finite-Difference and Boundary Element Programs

S. Dikmenli and P. J. Florio

Mechanical Engineering Department
New Jersey Institute of Technology,
Newark, New Jersey

1   Abstract

In order to investigate conjugate heat transfer problems on an arbitrary two dimensional or axi-symmetric geometry, the procedure chosen was the coupling of two existing programs known to accurately model the individual domains. The method used in coupling allows these programs to maintain their independence, and it essentially involves an iterative substitution of compatible boundary data (local heat transfer coefficient and surface temperature) between these programs until a relatively small change occurs (2%) at every location of conjugate heat transfer boundary. The result of the iterative coupling scheme is found to be a useful engineering tool that can be used to study conjugate systems consisting of boundary layer and conductive domains of arbitrary shapes with the advantages of total job-time and a data storage reduction and a substantial collection of system characteristics.

2   Leading Programs

The convective and conductive domains have been represented by STAN5 [2][1] and BEASY [3], respectively.

STAN5 is a general finite-difference program developed to solve parabolic two-dimensional boundary layer equations along the arbitrary surfaces. It is capable of solving upto five diffusion type boundary equations which are applicable to laminar and/or turbulent flows. It is also able to handle rough and smooth surfaces [4]. The individual test computations performed by STAN5 [1] showed that the accuracy of the result are well suitable compare to experiment and theory. Additionally, considering moderate computation time requirements, the generality of the program for the solution of two dimensional boundary layers,

---

[1]Numbers between [] refer to the items in bibliography.

suitable coupling with an boundary element program, where the stepwise solution pursued along the boundary and the additional modifications showed that STAN5 program shall be one of the most advantageous programs available.

BEASY: Boundary Element Analysis System [3], [5] which is developed by Computational Mechanics , Southampton, is used for the solution of a wide range of problems in many fields of engineering. The solution method of BEASY is an advanced technique called Boundary Element Method which, in spite of some early formulations derived at the beginning of this century, has in recent years come out of research into a reliable form. The method basically allows continuum modeling of 2D; 3D and axi- symmetric problems, involving solution of potential field[2] operated on discrëtized boundary field and field governed by the Laplace or Poission type equations. Consequently, considering a much smaller system of equation, less amount of computation, preparation, and checking time and introduced easy coupling with other numerical techniques, BEM shall be the best choice to represent conductive domain in coupling.

3   Coupling

For the solution of steady conjugate heat transfer problems, the thermal and fluid properties governed with boundary layer approximation are discretized in fluid domain by use of the method of STAN5 and the thermal properties of solid domain governed by boundary element method are approximated by discretizing the boundary by use of BEASY.

To accurately model the two leading programs: solid and fluid domains, two sets of data needed to be prepared. In that preparation, distribution of velocity is required for the solution of momentum equation in STAN5. Several methods were used to obtain the velocity for a series of data points on the arbitrary surface: perturbed potential velocity or stream function solution implemented by BEM, functions for some specific geometries or from experimental data. If well organized, it is found that the preparation time for the conjugate model was to the order of 60 minutes. This includes initial input files for BEASY and STAN5 and two data files which establish the compatibility of the coordinate systems.

The coupling of the two leading programs was achieved by three programs. Their essential role is to initiate the iteration, transfer the compatible data and recognize the convergence and accordingly, determine whether or not continue the iteration. One program reads the STAN5's output file and modifies the local heat transfer coefficient in

---

[2]static linear elastic and thermo-elastic analysis are also available

BEASY's input file and compares new and old values of HTC and gives and output to indicate the variation. The other data transfer program reads BEASY's output file and accordingly, modifies the surface temperature and initial temperature profile in STAN5's input file. The last program controls the simultaneous iteration (Fig.1) and allows both interactive and batch executions on VAX-DCL level.

4 Applications

In order to clarify some capabilities of the procedure, various models have been prepared and tested [1]. One of the models investigated was Model.3.1. In Fig.2, the surface of the model starts with 45° inclination from the forward stagnation point and maintains that to 10% of the cord, then declines to 0° which is maintained on the remaining section of the model. Agreement between inclination and thermal properties can be clearly seen in Fig.3. On the 45° inclined isothermal boundary (solid line), dimensionless HTC ($Nu_x/\sqrt{Re_x}$) should be 0.384 and on the flat surface (0° inclination) it should be 0.292. One can also see the smooth convergence of dimensionless HTC from 0.384 to 0.292 in the same figure and this also well simulates its behavior in nature. Prediction of flow parameters is well demonstrated by boundary layer development (illustrated in Fig.4). Fig.5 clearly shows the importance of the conjugate coupling and the high convergence rate in terms of heat flux variation. Fig.6 and Fig.7 demonstrate the compatible data variations transferred between two leading programs. In the figures 6 and 7, the agreement between HTC and temperature for each iteration is well established (increase in HTC decreases surface Temperature) as well. Furthermore, the thickness effect of solid domain (point A to B in Fig.2) might be seen on every thermal value obtained at the interface.

Other models studied include airfoil shapes. Typical of these is Model.2.1 (Fig.8). For this model, free stream velocity was obtained by perturbed solution of velocity potential which is implemented by BEM. Fig.9 shows temperature variation along the interface for various iterations and Fig.10 illustrates temperature distribution in solid domain for the last iteration.

For the remaining applications, which simulate effects of turbulent flow, different material zones in solid domain, variant fluid properties, convective inner regions, etc. have been exampled in [1].

5 Conclusion

Recent reports on the progress of conjugate heat transfer methods have been investigated [6],[7], but none were seem to be applicable to the various geometries as the present procedure does.

From the applications, it is found that the present set-up is very efficient in computation time as well as data preparation. Furthermore, the procedure allows a relatively simple method to extend the capabilities of domain or boundary type numerical methods.

For future consideration, the optimization techniques might be introduced into the procedure or a similar procedure might be used to account transient and radiation effects on two or three dimensional geometries.

References

1. Dikmenli, S., "Two Domain Heat Transfer Studies Using a Coupled Finite-Difference and Boundary Element Methods", M.S. Thesis, New Jersey Inst. of Tech., 1988.

2. Crawford, M.E. and Kays, W.M., "STAN5 - A Program for Numerical Computation of Two-Dimensonal Internal and External Boundary Layer Flows", NASA CR-2742, 1976.

3. "BEASY User's Manual", Computational Mechanics Inc., Mass., 1986.

4. Parthasarthy, R. I., "Modifications on Heat Transfer Program Including Icing Phenomena", M.S. Thesis, New Jersey Ins. of Tech., 1984.

5. "Boundary Element Research", Editor: C.A. Brebia, A Computational Mechanics Publ., 1985.

6. Aidun, C.K. and Lin, S.P., "Conjugate Heat Transfer from a Hollow Cylinder", Int. Heat Trans. Conf., San Francisco, 1987, pp. 373-378.

7. Karvinen, R., "Some New Results for Conjugated Heat Transfer in a Flat Plate", Int. J. Heat Trans., Vol. 21, 1978, pp. 1261-1264.

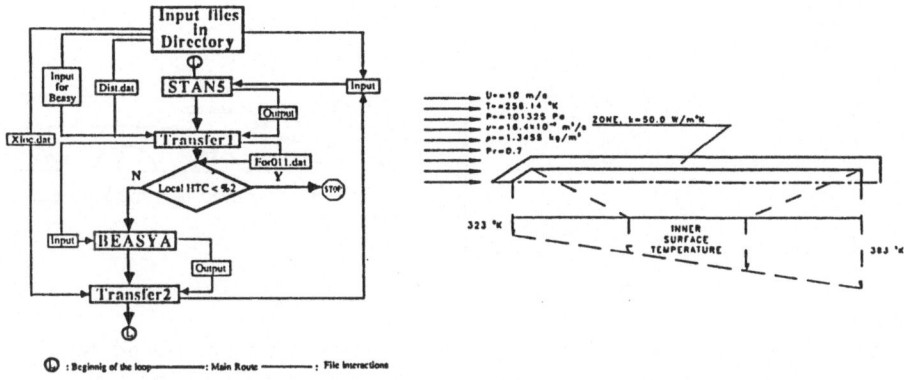

Fig.1: General Architecture of the
Iteration-Loop

Fig.2: Presentation of Boundary Conditions on Model.3.1 Geometry

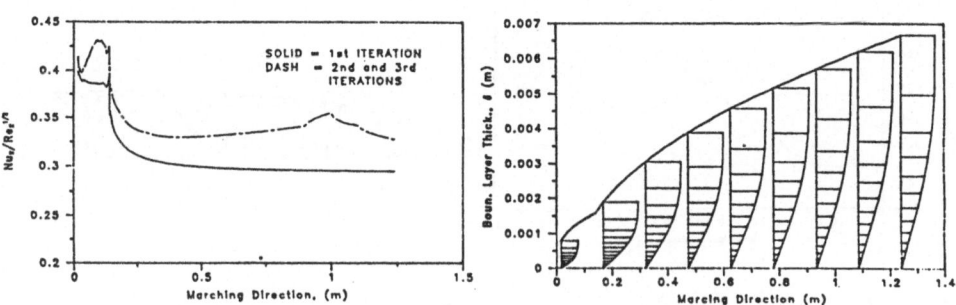

Fig.3: Dimensionless HTC Variation among the Iterative Solutions of Model.3.1

Fig.4: Boundary Layer Development along the Surface of Model.3.1

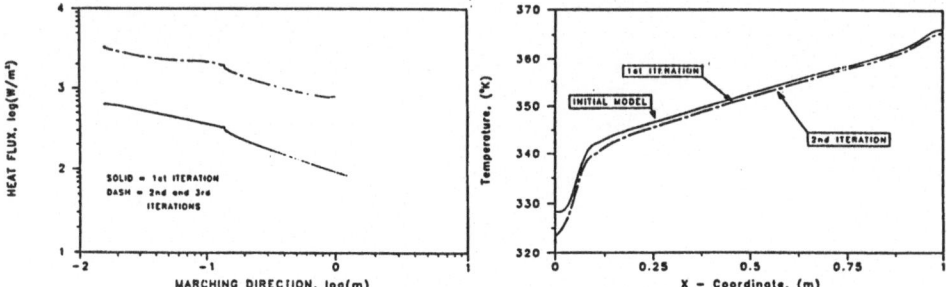

Fig.5: Heat Flux Variations Along the Model.3.1

Fig.6: Comparison of Temperature among the Iterative Solutions of Model.3.1

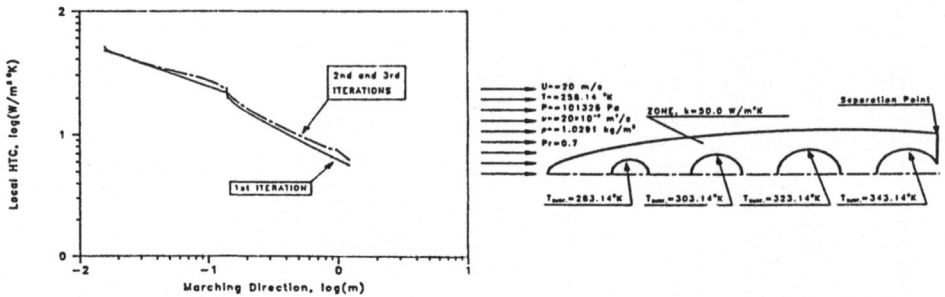

Fig.7: Comparison of Local HTC Variations among the Iterative Solutions of
Model.3.1

Fig.8: Presentation of Boundary Conditions on Model.2.1 Geometry

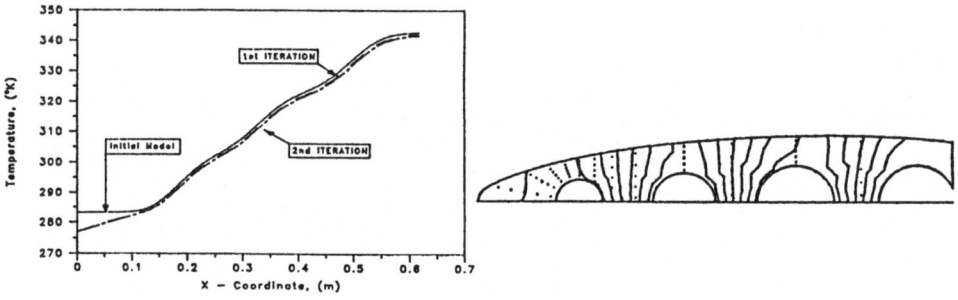

Fig.9: Comparison of Temperature among
the Iterative Solutions of Model.2.1

Fig.10: Temperature Distribution in Solid
Domain, Last Iteration

The Appropriate Use of Boundary Elements in the Aircraft Gas Turbine Business

Gareth H. Richards

General Electric Aircraft Engines
Cincinnati, Ohio

Summary

Boundary Element techniques have been introduced at General Electric
Aircraft Engines (GEAE) to improve productivity in the analysis of a large
class of structural problems. This paper examines the business reasons that
motivated this introduction, stressing the principle that the most
appropriate analysis tool should be selected for a given task. The
practical obstacles to achieving the maximum potential benefit from these
techniques are discussed.

How Analysis Drives the Business, and Vice-Versa

The classic challenge of structural engineering in the aerospace industry is
to produce maximum safe performance from the least weight. The useful
capability of a material is defined not only by its properties, but also by
the confidence the engineer places on his ability to predict its limiting
behavior in the real operating environment. When confidence in the analysis
is reduced, greater margins of safety must be assigned, and the amount of
material necessary increases. This basic motivation for methods improvement
has kept the aerospace industry at the forefront in thermal, structural and
fatigue analysis.

It is natural then that the success of a part or product becomes dependent
on high quality analysis. This happens in two ways. First, the design
goals may be achievable only with a certain caliber of analysis, and
secondly operational limits may be set or maximized using such high quality
work. Each of these has its own implications. The level of analysis in a
very real way determines the capability of the company and its competitive
position, and it usually becomes necessary to maintain or exceed that level
in the future development of that product. Further, when a product's safe
operational limits are determined using high quality analysis it becomes
essential to again employ similar techniques when studying variations to the
part or its environment. Locking in analysis like this can make parametric
studies, repairs and redesigns expensive and slow. For example, the expense
of analyzing a single non-conforming part, to determine whether it meets
design intent, can be greater than the cost of manufacturing a new
conforming part.

The above considerations provide a strong business need for analysis tools that combine high quality and low usage costs. Some comments on the components of usage cost are appropriate. In recent years the efficiency of both software and hardware has increased to the point where the cost of solution time is often not the limiting item in structural analysis. Note that this is certainly not true in other fields, such as computational fluid dynamics. Although some large non-linear 3D models are certainly expensive to run, the single biggest cost of performing a typical aerospace structural analysis task today is the engineering time required for pre- and post-processing. The requirement for reduced cost analysis quickly translates into a need for ease and speed of use, without compromising quality.

The Potential of BE Analysis

Boundary Element (BE) analysis answers the above requirement very well. It has an obvious advantage over Finite Element (FE) analysis in that much less model definition effort is required. The task of defining the surface geometry is common to both FE and BE, but the additional effort of specifying the mesh is dramatically reduced in the case of BE. A large class of problems require only that the maximum surface stress be determined; the full interior solution provided by FE models is normally needed only when analyzing complex problems having redundant load paths. In many cases only a few interior results will actually be used, such as in crack propagation problems. BE actually handles this better than FE, as internal points may be placed without the mesh transition struggles familiar to most analysts.

As an example, Figure 1 shows the BE model of a disk dovetail slot that is to be reworked by surface material removal, and then returned to operation. When the new surface is exposed the material will already have been subjected to operational stresses, unlike a new part. Therefore, to find the total life capability of the part, it is necessary to determine how much fatigue damage has been accumulated before the rework is performed. The sub-surface stresses are thus required. For this purpose, internal points have been placed where the new surface will be after rework. Note the large number of points and their proximity to the surface. To acquire the same information using Finite Elements would involve a lengthy and often tedious meshing exercise that, while entirely possible, would be very slow and expensive. Even with the best FE pre-processors, specifying fixed internal points on a generally curved line like this is a major task. By contrast, note how easily the old and new peak stress locations in the BE model have been linked with a series of points to study the stress decay.

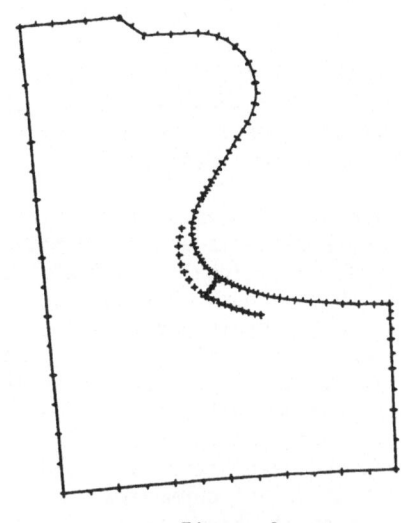

Figure 1

A 2D Boundary Element Model
Showing Ease of Placing Internal Points

When this type of rework or repair is designed, the optimization process
makes repetitive stress analysis an unpleasant fact of life for the
designer. In these problems the effort and cost of changing the model may
have a fundamental impact on the number of iterations carried out, and
therefore on the success of the final design. Consider the multiplication
of effort in the previous example when the whole process is repeated four or
five times. If each analysis takes three hours instead of three days, then
an extremely favorable impact on the business is achieved.

Another benefit of the method is the net quality of the results. This is a
consequence not only of the mathematics of the solution itself, but the way
the results are used and presented. The volume of results produced is so
much smaller, and mesh refinement so simple, that there is a definite
tendency for all users to impose and achieve higher standards of quality in
their work. One big driver of this trend is the presentation of unaveraged
stress results at the surface. Even seasoned analysts sometimes overlook
that stress discontinuities exist at every node in an FE model, and people
have an understandable habit of being seduced by contour plots that can
conceal major inadequacies. But presented with a graph that actually shows
a measurable jump in stress at a point, most analysts will recognize and
correct the deficiency immediately.

The Realities of Change

With all the above selling points, software manufacturers may have a hard time wondering why anyone would not install a system such as this. In fact at large sites having a staff of several thousand engineers, the inertia and expense of change is not to be underestimated.

The reality of whether a given new program will deliver its full potential is very site-dependent. At GEAE there is a complex network of mainframes from many of the major vendors. Additionally, CAD stations, PC's and workstations are attached to this network, giving rise to a wide variety of work habits among engineers. Some obtain geometry from the central drafting system, some generate it on PC's, and still others use mainframe packages such as PATRAN or in-house equivalents. In such an environment, the introduction of 'yet another program' will only succeed where it is made to dovetail with as many existing programs as possible, or it simply will not be used. The message to software manufacturers should be 'Interface or Die'. Stand alone packages that do not communicate are essentially worthless from an economic viewpoint, since they will be used only by a few enthusiasts, no matter what their capabilities.

Relative Strengths of BE and FE Analysis

There remains a large class of problems for which the flexibility of FE analysis is still required. GEAE performs a large amount of non-linear 3D analysis, and whole rotor axisymmetric analysis, using complex transient thermal distributions. The thermal analysis itself is a major undertaking and in most cases it is not practical to combine the analyses. The temperatures from the thermal analysis must be mapped to the stress analysis models. In these cases, internal nodes and elements are required in order to be able to apply the temperatures that result from the thermal analyses. BE is not the right tool for these analyses, since the advantage of not meshing the interior is lost. Contact problems and generalized constraint equation problems have been carefully validated over many years using FE analysis, and it is not anticipated that BE will offer any significant advantage here.

Conversely, individual non-conforming parts are easily evaluated using BE since the exact measured surface profile, which may be quite complex, is quickly transformed into a BE model. This type of analysis is also readily extendable into studies of complex profile tolerances, hole oversize/out-of-position, etc. that would be tedious or impractical in FE. Such studies are excellent ways of laying the groundwork for improved design practices and tolerance procedures.

The BE method should therefore not be considered a replacement for FE analysis, but rather a complement to it. Experience has shown that it is important to emphasize this fact when introducing the method to new users. Each technique has its own strengths that should be recognized and exploited as appropriate. It is time for the champions of both camps to accept this.

Future Direction

Methods are now in development to further automate the link between CAE geometry and analysis. With careful planning, it will be possible to radically improve the use of these CAE geometry definitions in BE analysis. The realities of file transfer and translation in very large multi-vendor computer installations need careful attention, or they can become real obstacles. Yet, if these aspects can be made completely transparent, there is potential for dramatic increases in capability and productivity. Perhaps someday not too far away a phone inquiry may be resolved, using high quality analysis, without hanging up.

Conclusion

Engineers spend a great deal of time generating numerical data and results. The aim of producing these numbers is to gain accurate information with which to make engineering and business decisions. However, in most cases, the time spent considering the action required by a given set of results is probably one hundredth the time consumed in producing them. In today's competitive environment that is a poor use of engineering talent. The use of Boundary Elements goes some way to restoring the balance, as it frees engineers to be more creative and more productive.

Methodology for Steam Valve Chest Assessment Using Boundary Elements

F.J.CUNHA[1]; J.TREVELYAN[2]

1 Public Service Electric and Gas Co, Newark,New Jersey
2 Computational Mechanics Inc., Billerica, Massachusetts

Summary

The operational reliability of steam valve chests subjected to cyclic loading conditions is of concern from the point of life prediction as well as life extension in power generating stations.

In many cases, the amount of time and effort required to describe the condition of a station component prevents a rapid and timely answer by most conventional numerical analysis techniques. The boundary element method has emerged as a promising new tool that can simplify and speed up the analysis. This paper deals with a methodology required to estimate the useful life of a typical steam chest configuration for continued and reliable operation using boundary elements.

Introduction

High temperature stainless steel valve chests with service life in excess of 20 years have been subjected to inner surface cracks. The severity of this cracking situation needs to be evaluated in order to determine how to maintain and operate the unit at rated conditions. Traditional design procedures focus on preventing creep rupture failure in structures that are nominally free of defects. However, some defects can be introduced during fabrication, repair and/or operation. Areas particularly prone to cracking are those having drastic changes in geometry such as transition sections to valve chest blind back ends and valve sitting areas. These can be subjected to not only primary stresses due to weight and pressure loading conditions, but also to very high secondary stresses due to operating differential thermal expansion. This paper illustrates the application of a currently available step-by-step procedure for fatigue and creep crack growth in stainless steel valve chests due to primary and secondary stresses. Even though primary streses are often sustained throughout the service life of the chest, it is generally recognized that secondary stresses are significant to crack growth.

A case history is presented for a 347 stainless steel steam chest operating at the Public Service Electric & Gas Sewaren Generating Station

Stress Analysis

The deformation controlled stresses such as those due to differential thermal expansion were not generally calculated before the advent of widespread numerical techniques. However, these stresses may be significant to crack growth. In many cases, the amount of time and effort required to describe the component geometry to a finite element program and interpret the results prevents a rapid and timely answer. The boundary element method has emerged as a promising alternative that can simplify and speed up the analysis since only the surface of the valve chest needs to be defined. A commercially available boundary element program BEASY was used to perform the required stress analysis of the steam chest. The analysis was performed in two separate parts. The first part considers an initial warming up period through transient states, and the second part considers a period when inflow heats up the chest walls to the final steady-state condition with full design pressure and temperature.

Warming up period

During the steam chest warming up period, the surrounding heating blankets will induce heating necessary to raise the metal temperature before admitting steam. This may induce a certain amount of stress, which depends on the initial thermal ramp. The stress analysis is obtained for the thermal gradient induced by different heating ramps of 50°F/hr, 75°F/hr and 100°F/hr. The thermal profile is calculated first, and then the corresponding stress response. The thermal profile was calculated from an assumed behavior through a semi-infinite medium [1], and the internal metal temperatures are shown in Table 1. These values are then specified at given distances x from the outside surface where the heating blanket elements are placed. As a result, the analytical model used for the BEASY code had to be broken down into subregions so that the internal metal temperatures could be specified at the interfaces shown on the 2D BEASY model in figure 1.

Steam inflow period

As the steam flows into the valve chamber, the heat is diffused into the metal and the temperature is raised from an initial heating blanket temperature to another temperature until an equilibrium state is reached. The stresses induced by the advancing thermal profile are obtained by examining the following conditions

Condition	Int temp.	Outside tem.	Pressure	Comments
A	900°F	400°F	1500 psia	Steam Admission
B	1050	650	1500	Heat Diffusion
C	1050	750	1500	Heat Diffusion
D	1050	850	1500	Heat Diffusion
E	1050	950	1500	Heat Diffusion
F	1050	1050	1500	Steady State

For the steam inflow period, a three-dimensional BEASY model was used for the valve back end section with the first valve inlet section (figure 2). Internal points were used to study the internal stress results for the above loading conditions. For simplicity, only the direct stress results for the points close to the internal surface were considered. It is believed that the triaxial state of stress would be representative of the working condition on the existing cracks and their stress values are tabulated in Table 2.

Life Prediction Methodology

In assessing the operational life of a steam valve chest it is important to consider the conditions at crack tips in a cracking situation. These conditions are characterized by the magnitude of the stress-intensity factor, which relates the applied nominal stress σ, the square root of the crack length a and geometry parameter factor F. Irwin [2] gave an expression for the stress intensity factor for a semi-elliptic surface crack similar to those studied for the Sewaren No.1 steam chest. The stress intensity KI is given by:

$$KI = F \sigma \sqrt{\pi a} \qquad \dots (1)$$

The stress results having been evaluated in the previous section, it is possible to calculate a critical crack length a_c at which time the flaw becomes unstable. This critical crack length can usually be approximated in the linear elastic regime by substituting KI=KIc which is the material fracture toughness.

A prediction of the operational life of the chest may be made by considering the crack growth due to both fatigue and creep. In fatigue, the well known Paris' law has been used extensively, relating crack length to cycles. The effect of non-linearities occurring locally around the crack tip can be considered using the J-integral parameter, which is an average measure of the energy release rate per unit crack extension in the elastic-plastic field of the material. Creep crack growth has been investigated by Saxena [3], and this can be added to the fatigue crack growth term to assess the likely number of remaining cycles for each stress condition. Care must be taken to include a creep/fatigue interaction factor (taken to be 3 in this study as the presence of each phenomenon accelerates the other.

Conclusions

As a result of the analysis performed, the following conclusions were
extracted:

1. The boundary element method proved a tool to simplify and speed up the
   numerical stress analysis of the component in both 2D and 3D cases.
2. With the exception of the steady state operation, the chest inner wall is
   subjected to compressive stresses when steam inflow heats up the chest
   which do not support an opening mode of cracking. However, during the
   warming up period or shutdown, tensile stresses will appear and the opening
   mode of cracking is assisted by the induced stresses.
3. At stress ranges of 25ksi, the predicted number of remaining cycles is
   reduced significantly.
4. At 1050°F and 25ksi stress levels, the rate of creep crack growth is higher
   than that of fatigue.

Based on the analysis performed, the cyclic stress conditions of 25ksi should
be avoided for older units similar to the subject chest. This may be achieved
by monitoring wall temperatures and restricting chest body thermal gradients.

References

1. Cunha,F.J. Methodology for Steam Valve Chest Assessment, Fossil Plant
              Cycling Conference, Princeton, NJ, Oct. 20-22, 1987.
2. Irwin,G.R. Crack-Extension Force for a Part-Through Crack in a Plate,
              Journal of Applied Mechanics, Vol 29, Trans ASME, Vol 84,
              Series E, No.3, Dec 1962, pp 651-654.
3. Saxena, A. Recent Advances in Predicting Creep Crack Growth in Steam
              Turbine Rotor Materials, EPRI CS-4160, 1985.

TABLE 1

METAL TEMPERATURES

X (in)	$t_1 = \frac{1}{2}$ hr	$t_2 = 1\frac{1}{2}$ hr	$t_3 = 2\frac{1}{2}$ hr
0.00	150°F	250°F	350°F
1.00	146.8	240.4	334.0
2.00	143.6	230.9	318.2
2.375	142.5	227.3	312.2
2.75	138.2	214.2	290.7

TABLE 2
DIRECT STRESS RESULTS

SIG (XX)

Int.pt	A	B	C	D	E	F
1	−17,140	−13,389	−9,768	−6,148	−2,527	1,093
2	− 7,532	− 5,912	−4,332	−2,753	−1,173	407
3	− 5,240	− 4,124	−3,035	−1,946	− 858	231

SIG (YY)

Int.pt	A	B	C	D	E	F
1	−14,087	−11,477	−8,528	−5,580	−2,632	316
2	− 7,914	− 6,210	−4,553	−2,896	−1,240	417
3	−11,738	− 8,961	−6,523	−4,084	−1,646	792

SIG (ZZ)

Int.pt	A	B	C	D	E	F
1	−22,912	−18,340	−13,522	−8,704	−3,886	932
2	−20,492	−16,393	−12,054	−7,715	−3,377	962
3	−17,279	−13,715	−10,055	−6,395	−2,734	926

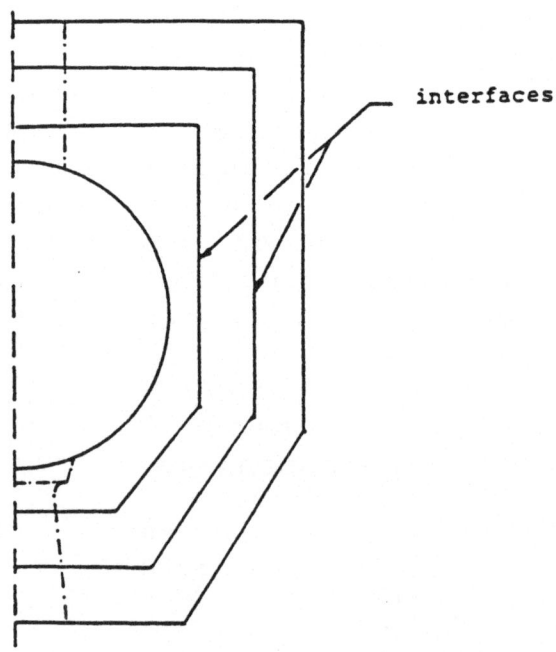

Figure 1.   2D BEASY model for warming up period

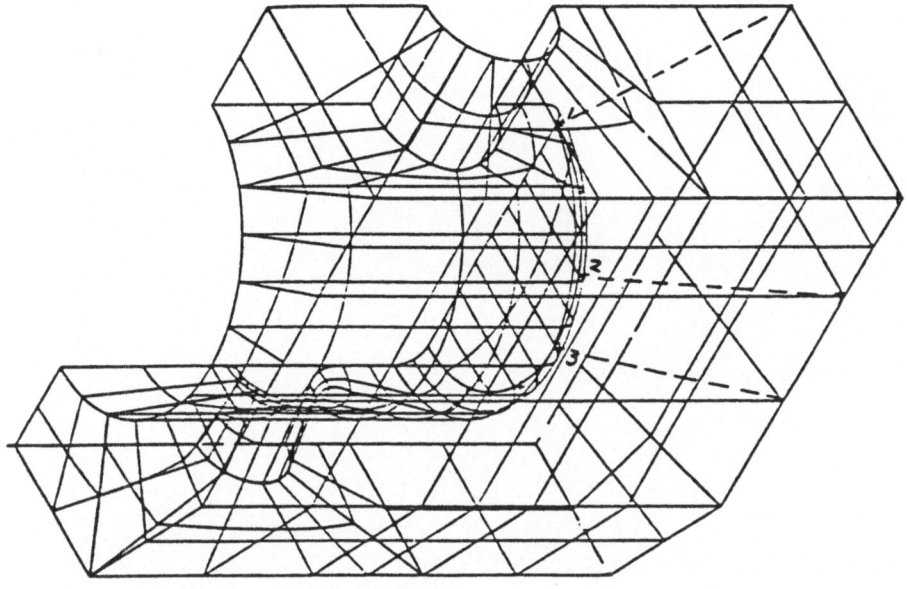

Figure 2. 3D BEASY model for steam inflow period

Other Numerical Methods

Numerical Analysis of Air-Borne Particle Movement in Industrial Clean Room

Yoshihide Suwa

Manufacturing Engineering Laboratory
Toshiba Corporation, Yokohama, Japan

Summary

A method to predict air-borne particle movement in an industrial clean room has been desired for high yield manufacturing of semiconductor devices. To analyse this particle movement, an equation for statistical density transportation was derived by the author. The analysis method to solve this equation and some of applications are reviewed in this paper. Using these systems, particle movement is able to be investigated in reference to each size, for each material and at each moment.

Introduction

Several kinds of industrial clean rooms are used to manufacture semiconductor devices. They must be kept at a high level of cleanness. For example, classes 10 to 100[1] are said to be required for most manufacturing processes[2]. To maintain such a cleanness level, particles must be prevented from being generated, and contamination by air-borne particles must be reduced. However, it is not so easy to measure distribution for air-borne particles, and it will be more difficult to measure it in the future. Now, the pattern width for 4Mbit DRAM (Dynamic Random Access Memory) is about 0.8μm. Under 0.5μm is going to be necessary for 16Mbit or 64Mbit DRAMs, in the near future[3]. It is said that a particle, which size is 1/10 of pattern width, is demanded for a device. On the other hand, in case larger sized particles may fall off quickly, due to the gravitation effect. Their distribution is transformed every moment, and any measurement system, such as a Laser Counter, becomes ineffective.

Using the computer simulation systems based on numerical analysis method, particle movement can be investigated in detail. So, the most suitable clean room structure or layout will be able to be designed.

Basic study for Particle Distribution

Factors acting on particle movement are considered to be as follows.

    i)   Diffusion by number density distribution
            (based on isotropy hypothesis of Brownian movement)
    ii)   Transportation by flow transfer
    iii) Falling phenomenon due to the gravitation effect

When, concerned with such small scale as that around air-borne particle, the Stokes approximation is able to be applied. On the assumption that particle shapes are roughly spherical, all resistance F_D values, which act on the sphere are described, using Cunningham correction factor C_c [4],

$$F_D = \frac{3 \pi \eta U d}{C_c} , \quad C_c = 1 + \frac{\lambda}{d} \{ C_1 + C_2 \exp \ (\frac{C_3 \ d}{\lambda}) \} \tag{1}$$

Where η, U, d, and λ are air viscosity, particle falling velocity, particle diameter, mean free path, respectively, and c_1, c_2, c_3 are experimental constants. Here, the diffusion force is settled by the distribution of particle number density. Considering the resistance is equivalent to the diffusion force, diffusion coefficient is obtained as follows, according to Van't Hoff's law.

$$D = \frac{\kappa T C_c}{3 \pi \eta d} \tag{2}$$

where κ is Boltzmann constant.

Equation for Statistical Density

Terminal settling velocity means the final velocity of particles in free fall. In case of air-borne particles, the falling phenomenon reaches a uniform velocity motion, for a moment. It is caused by the balance between buoyancy and gravitation. Relaxation time, until the force balances, is seriously short, around 3.1×10^{-2} second for 100μm particles and only 3.6×10^{-6} second for 1μm particles[5].

Resistance coefficient (resistance divided by acceleration), concerned with free fall, is divided into three regions, Stokes region, Newtonian region and the transient region. Considering the resistance is equivalent to gravitation, terminal settling velocity in the Stokes region is described as follows.

$$v_t \Big|_{\text{In Stokes Region}} = \frac{(\rho_p - \rho_g) \, d^2 g}{18 \eta} \tag{3}$$

where g is gravitational acceleration, ρ_p is particle density and ρ_g is air density. Here, resistance coefficient C_D is described using the Reynolds number R_e,

$$C_D = \frac{24}{Re} (1 + \frac{Re^{2/3}}{6}) \tag{4}$$

when the particle motion is in the Newtonian region, using Newton's resistance law,

$$F_D = C_D \frac{\pi}{8} \rho_g \, d^2 u^2 \tag{5}$$

Terminal settling velocity in the Newtonian region is obtained as follows.

$$v_t \Big|_{\text{In Newtonian Region}} = (\frac{4 (\rho_p - \rho_g) \, d g}{3 C_D \rho_g})^{1/2} \tag{6}$$

In the transient region, Davies has investigated experimental values for $C_D R_e$ for each Reynolds number[6]. The author tried incremental solutions for Eqs.(4),(6), using Eq.(3) as initial value. Using this simple method, terminal settling velocity in any region can be obtained. Figure 1 shows calculated terminal settling velocity for each material[7]. These materials are predicted to be generated in semiconductor manufacturing processes. When comparing terminal settling velocity for oil mist with that for the same sized Fe or Si particles, 10 to 20 times differences appear. Using Eqs.(1),(2), and calculated terminal settling velocity, the equation for statistical particle density n, concerned with time t and three dimensional coordinate x, is described as follows.

$$\frac{\partial n}{\partial t} = -u_1 \frac{\partial n}{\partial x_1} - v_t \frac{\partial n}{\partial x_3} + D \frac{\partial^2 n}{\partial x_1^2} \tag{7}$$

Here, velocity u_i is obtained from the Reynolds equation[8][9].

$$\frac{\partial u_1}{\partial t} + \frac{\partial u_1 u_1}{\partial x_j} = -\frac{1}{\rho}\frac{\partial P}{\partial x_1} + \frac{\partial}{\partial x_j} \{\nu (\frac{\partial u_1}{\partial x_j} + \frac{\partial u_1}{\partial x_1}) - \overline{u'_1 u'_j}\} \qquad (8)$$

Where p is pressure, ν is kinetic viscosity. To solve Eqs.(7)and (8), a two-step method related to the basic Crank-Nicholson scheme[10][11] was adopted for the Galerkin procedure definition, and the Broyden's update[12][13] was adopted for incremental solution.

Applications

Figure 2 shows one of the applications for transformation of particle distribution in a down flow type clean room. Figure 2 a) (contour line n=0.5 in each time step) indicates a gap between the HEPA (High Efficiency Particulate Air) filters on the ceiling partly hinders laminal down flow. The reason is that the flow under the gap is stirred by well known Karman vortices. Figure 2 b) shows the transformation of statistical density in three individual points under the gap. It indicates that the computed result corresponds with experimental measurement.

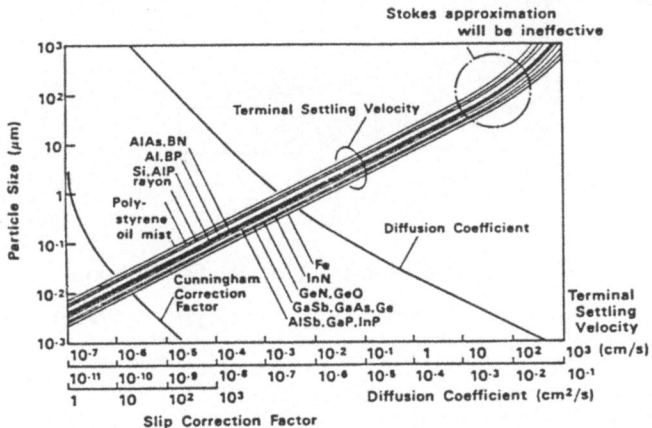

Fig.1 Calculated Terminal Settling Velocity

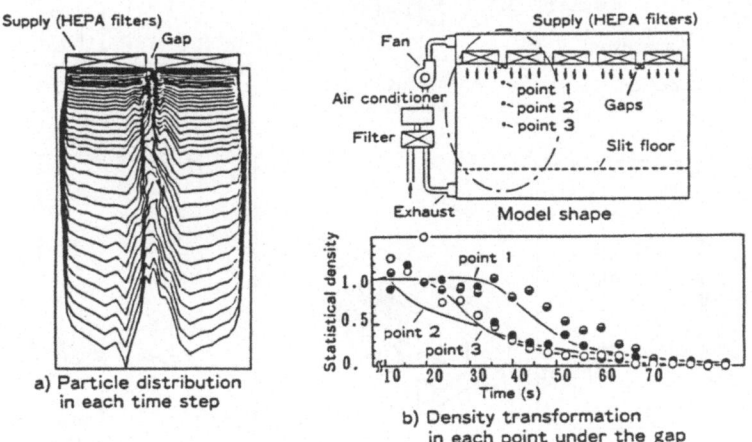

Fig.2 Particle distribution transient analysis

208

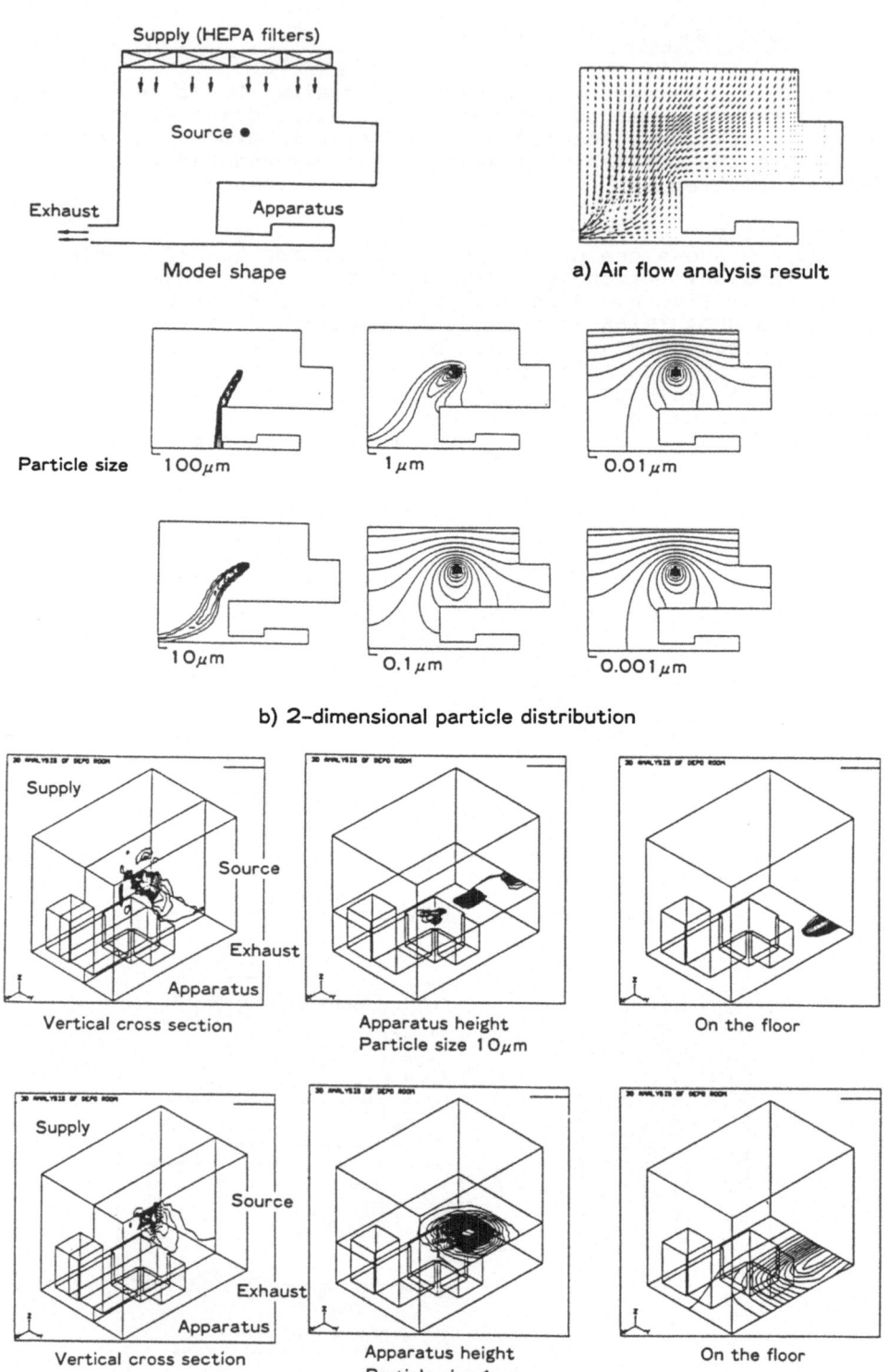

Model shape

a) Air flow analysis result

Particle size　100μm　　1μm　　0.01μm

10μm　　0.1μm　　0.001μm

b) 2-dimensional particle distribution

Supply

Source

Exhaust

Apparatus

Vertical cross section

Apparatus height
Particle size 10μm

On the floor

Supply

Source

Exhaust

Apparatus

Vertical cross section

Apparatus height
Particle size 1μm

On the floor

c) 3-dimensional particle distribution

Fig.3 Air-borne particle distribution analysis

Other applications are used for each sized particle movement.
Figure 3 shows results for two and three dimensional analysis
concerned with each sized amorphous silicon particle[7]. For 100
µm particle size, the gravitation effect is high. Comparing with
the flow state, a particle falls at a more acute angle than for
flow, and it is not diffused so much. According to the decrease
in the particle size, the falling angle becomes obtuse, and the
degree to be diffused becomes larger. When the particle size is
under 1µm, distribution does not depend on its size. This means
the assumption of a passive-contaminant is realized. Particles
sized under 1µm are stirred by Brownian movement of air mole-
cules, rather than by falling. So, it is understood that smaller
sized particles, under 1µm, can be reduced by flow transfer, and
larger sized particles are deposited on the floor, near the
source.

Conclusion
    For high yield manufacturing of semiconductor devices, it is
quite important to cope with several problems in regard to appa-
ratuses, processes, handling methods, materials, and environments. Up to this time, experimental methods, using smoke tracer
and Laser Counter, have been used to measure internal flow or
particle distribution. However, it has quite difficult to estimate macroscopic efficiency for an industrial clean room,
because of limited partial observation, prediction impossibili-
ty, and apprehension about causing tracer contamination. On the
contrary, using the computer simulation based on numerical
analysis methods, a macroscopic investigation for clean room
efficiency has become possible. These methods will be especially
effective, not only for problems in a clean room, but also for
apparatus problems.

References
[1] Fed.Std.209B:Clean Room and Work Station Requirements,
      Controlled Environment,(1973)
[2] Hayakawa,K.:Clean Room,(1985),Inoue Syoin,Tokyo(in Japanese)
[3] Maeda,K.:VLSI and Process Technology,Materials for
      Electronics, Vol.26,3,22-28(1987)(in Japanese)
[4] Cunninghum,E.:Definitive Equation for the Fluid Resistance
      of Spheres, Proc.Phys.Soc.,57,322(1945)
[5] Hinds,W.C.:Aerosol Technology,(1982),John Wiley & Sons,Inc.
[6] Davies,C.N.:Definitive Equation for the Fluid Resistance of
      Spheres, Proc.Phys.Soc.,57,322(1945)
[7] Suwa,Y.:Numerical Analysis of Air-Borne Particle Movement in
      Clean Room,The 7th Symp.on Clean Tech.and Contamination
      Controll,JACA,(1988)(in Japanese)
[8] Edited by Tani,I.:Turbulent Flow,(1980),Maruzen,Tokyo
      (in Japanese)
[9] Launder,C.W.,Spalding,D.B.:Lectures in Mathematical Models
      of Turbulence,(1972),Academic Press
[10] Edited by Washizu,K.et al.:FEM Handbook I,(1981),Baifukan,
      Tokyo(in Japanese)
[11] Roache,P.J.:Computation Fluid Dynamics,(1976),H.Hermosa
      Publishers,Inc.
[12] Engelman,M.S.,Strang,G.,Bathe,K.J.:The Applications of
      Quash-Newton Methods in Fluid Mechanics,Int.J.for
      Numerical Meth.in Eng.,Vol.17,707-718(1981)
[13] Bercovier,M.,Engelman,M.S.:A Finite Element for the
      Numerical Solution of Viscous Incompressible Flows,J.Comp.
      Phys.,30,181-201(1979)

Evaluation of Finite Difference Schemes for Solution to the Inverse Velocity and Acceleration Problem for Robot Manipulators

R. K. DEAN

Department of Mechanical and Aerospace Engineering
West Virginia University, Morgantown, WV

S. M. NESBIT

Department of Engineering Technology
Murray State University, Murray, KY

Introduction

The problem of getting a revolute jointed manipulator to plan and execute Cartesian straight line trajectories is analytically and computationally difficult [1]. A complete analytical solution to this problem would use the trajectory generator to calculate the Cartesian kinematic quantities, the inverse position kinematics to calculate joint positions, the inverse Jacobian matrix for joint velocities, and the inverse Jacobian matrix plus its derivative for joint accelerations [2]. A simpler way often used in practice is as follows: At the path update rate, use the inverse position kinematic solution to calculate the vector of joint angles. Numerical differentiation is then used to calculate the joint velocities and accelerations [3]. This paper will describe and evaluate several numerical differentiation schemes used to calculate the above joint velocity and acceleration vectors.

The RHINO XR-2 Robot is used to test the finite difference schemes. It was modified to allow it to follow Cartesian straight line trajectories. The path shape connecting the taught task points is linear with a parabolic blend region added at each task point. The velocity is constant except at the blend portion of the trajectory where constant acceleration is used to change velocity smoothly.

Finite Difference Schemes

The following expression is used to calculate the vector of joint velocities from the Cartesian specified end effector velocity

$$\{\dot{\theta}\} = [J(\theta)]^{-1}[V] \tag{1}$$

where $\{\dot{\theta}\}$ is the joint velocity vector, $J(\theta)$ is the Jacobian matrix, and [V] is the Cartesian specified end effector velocity. The transformation from Cartesian specified end effector accelerations to joint accelerations is

$$\{\ddot{\theta}\} = [J(\theta)]^{-1}[A - F(\theta,\dot{\theta})] \tag{2}$$

where $\{\ddot{\theta}\}$ is the joint acceleration vector, A is the Cartesian specified end effector acceleration, and $F(\theta,\dot{\theta})$ is the Coriolis and centrifugal acceleration components of A. Calculation of eq. 1 and eq. 2 for the RHINO manipulator requires that approximately 350 and 14,000 multiply times be performed per path update respectively using the computational equivalency convention specified by Craig [3]. The determination of the joint velocity and acceleration vectors using these equations is computationally lengthy and may overburden the controller when performing real-time calculations.

Since the joint angles must be known before the inverse velocity equation (eq. 1) can be calculated, and both the joint angles and joint velocities must be known before the inverse acceleration equation (eq. 2) can be calculated, it is possible to use finite difference formulations in place of eq. 1 and eq. 2 to solve for the joint velocities and accelerations. All of the numerical differentiation schemes presented here are a combination of forward, backward, and central difference equations. The forward and backward difference equations are necessary at the path end points. The central difference equation was used for all other points because it yielded better results than the forward and backward difference equations.

Two finite difference schemes are presented as alternate solutions to the inverse velocity problem. The first scheme uses forward, backward, and central difference formulas of error δt, δt, and δt^2 respectively, where δt represents the interval of time between points. The second scheme utilizes finite difference equations with a higher order error. This scheme employs forward, backward, and central difference equations of error δt^2, δt^2, and δt^4 respectively. The forward, backward, and central difference equations for the first scheme are

$$\{\dot{\theta}\}_i = \frac{\{\theta\}_{i+1} - \{\theta\}_i}{\delta t} \qquad \text{FOR } i=0 \tag{3}$$

$$\{\dot{\theta}\}_i = \frac{\{\theta\}_i - \{\theta\}_{i-1}}{\delta t} \qquad \text{FOR } i=N \tag{4}$$

$$\dot{\{\theta\}}_i = \frac{\{\theta\}_{i+1} - \{\theta\}_{i-1}}{2\delta t} \qquad \text{FOR } i=1 \text{ to } N-1 \qquad (5)$$

where N is the number of path update points. The forward, backward, and central difference equations for the second scheme are

$$\dot{\{\theta\}}_i = \frac{-\{\theta\}_{i+2} + 4\{\theta\}_{i+1} - 3\{\theta\}_i}{2\delta t} \qquad \text{FOR } i=0,1 \qquad (6)$$

$$\dot{\{\theta\}}_i = \frac{3\{\theta\}_i - 4\{\theta\}_{i-1} + \{\theta\}_{i-2}}{2\delta t} \qquad \text{FOR } i=N-1,N \qquad (7)$$

$$\dot{\{\theta\}}_i = \frac{-\{\theta\}_{i+2} + 8\{\theta\}_{i+1} - 8\{\theta\}_{i-1} + \{\theta\}_{i-2}}{12\delta t} \qquad \text{FOR } i=2 \text{ to } N-2 \qquad (8)$$

The above two finite difference schemes require 2.5 and 5.2 multiply times (for the central difference equation) per path update respectively.

There are four finite difference schemes presented as an alternate solution to the inverse acceleration problem. The first two solve for the joint acceleration vectors from joint velocity information using the same two finite difference schemes presented above for the calculation of joint velocities. This requires that the joint velocity vectors replace the joint position vectors in eq. 3 through eq. 8. The second two schemes solve for the joint accelerations directly from joint position information only. This has the advantage of using exact joint position data over approximate joint velocity data (obtained from finite difference equations) in the calculations. The first of these schemes employs forward, backward, and central difference formulas of error δt, δt, and δt^2 respectively. The second scheme uses forward, backward, and central difference equations of a higher order error of δt^2, δt^2, and δt^4 respectively. The forward, backward, and central difference equations for the first scheme are

$$\ddot{\{\theta\}}_i = \frac{\{\theta\}_{i+2} - 2\{\theta\}_{i+1} + \{\theta\}_i}{\delta t^2} \qquad \text{for } i=0 \qquad (9)$$

$$\ddot{\{\theta\}}_i = \frac{\{\theta\}_i - 2\{\theta\}_{i-1} + \{\theta\}_{i-2}}{\delta t^2} \qquad \text{for } i=N \qquad (10)$$

$$\{\ddot{\theta}\}_i = \frac{\{\theta\}_{i+1} - 2\{\theta\}_i + \{\theta\}_{i-1}}{\delta t^2} \quad \text{for } i=1 \text{ to } N-1 \tag{11}$$

The forward, backward, and central difference equations for the second scheme are

$$\{\ddot{\theta}\}_i = \frac{-\{\theta\}_{i+3} + 4\{\theta\}_{i+2} - 5\{\theta\}_{i+1} + 2\{\theta\}_i}{\delta t^2} \quad \text{for } i=0,1 \tag{12}$$

$$\{\ddot{\theta}\}_i = \frac{2\{\theta\}_i - 5\{\theta\}_{i-1} + 4\{\theta\}_{i-2} - \{\theta\}_{i-3}}{\delta t^2} \quad \text{for } i=N-1,N \tag{13}$$

$$\{\ddot{\theta}\}_i = \frac{-\{\theta\}_{i+2} + 16\{\theta\}_{i+1} - 30\{\theta\}_i + 16\{\theta\}_{i-1} - \{\theta\}_{i-2}}{12\delta t^2} \quad \text{for } i=2 \text{ to } N-2 \tag{14}$$

These two finite difference schemes require 3.9 and 7.5 multiply times (for the central difference equation) per path update respectively.

Results and Conclusions

The results presented here are for the RHINO XR-2 Robot following a Cartesian straight line trajectory operating at maximum speed from the initial to the final task point (2.25 sec). The path update rate is 100Hz and the blend time is 0.33 seconds for this trajectory. Fig. 1 shows the exact position, velocity and acceleration for joint 1 (waist rotation) for this trajectory.

Two finite difference schemes were enlisted as alternate methods for calculating joint velocity vectors from joint position information. Both finite difference schemes are shown compared to the exact velocity by Fig. 2. As can be seen from the graph, both difference schemes yield results which are graphically indistinguishable from the exact solution for both the smooth and discontinous portions of the curve. There were four finite difference schemes used to solve for the joint acceleration vectors. Two of the schemes use joint velocity information and are shown by Fig. 3 (error δt^2) and Fig. 4 (error δt^4). The other two schemes calculated the joint accelerations directly from joint position data only. These are shown by Fig. 5 (error δt^2) and Fig 6 (error δt^4). Analysis of the graphs reveals that all of the schemes were able to follow the exact solution during the smooth portion of the curve. However, only the finite difference scheme of error δt^2 that uses position data was able to match the exact solution at the discontinous portions of the curve.

This work has demonstrated that finite difference schemes which employ a combination of forward, backward, and central difference formulations are an accurate alternative solution to the inverse velocity and acceleration problem for robot manipulators. These schemes were also shown to yield a substantial reduction in the number of compuations per path update per desired joint quantity.

References

1. Nesbit, S.M.; Real-Time Safety and Collision Avoidance System for a Five Axis Robot. Ph.D. Dissertation for the Department of Mechanical and Aerospace Engineering, West Virginia University, 1988.

2. Gorla, B.; Renaud, M.: Robot Manipulateurs, Cepadus-Editions, Toulouse, 1984.

3. Craig, J.J.; Introduction to Robotics: Mechanics and Control. Reading, Mass. Addison-Wesley Publishing Company 1986.

4. James, M.L.; Smith, G.M.; Wolford, J.C.: Applied Methods for Digital Computation. New York: Harper & Row, Publishers 1977.

215

POSITION, VELOCITY, AND ACCELERATION

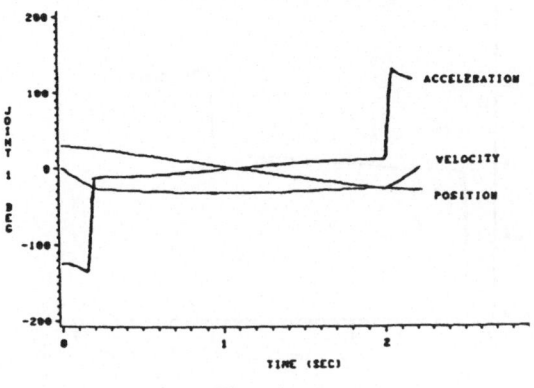

Fig. 1 Exact

JOINT VELOCITY USING FINITE DIFF EQU

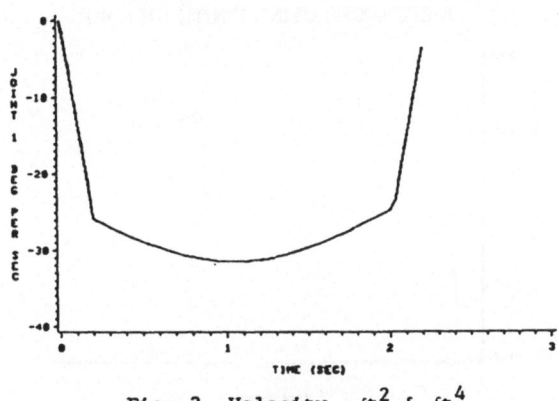

Fig. 2 Velocity $6t^2$ & $6t^4$

JOINT ACCEL USING FINITE DIFF EQU

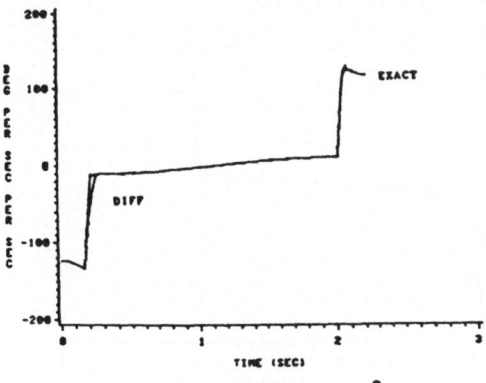

Fig. 3 Accel $6t^2$

JOINT ACCEL USING FINITE DIFF EQU

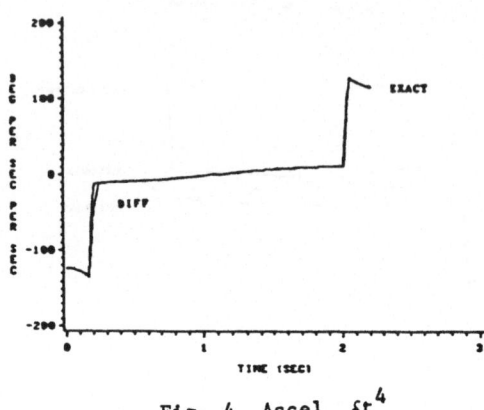

Fig. 4 Accel δt^4

JOINT ACCEL USING FINITE DIFF EQU

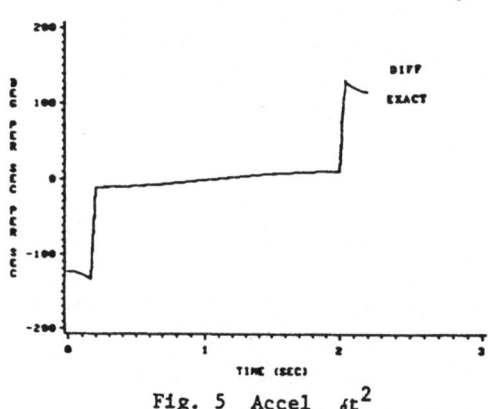

Fig. 5 Accel δt^2

JOINT ACCEL USING FINITE DIFF EQU

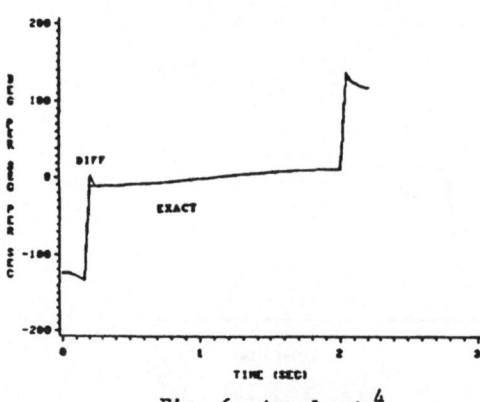

Fig. 6 Accel δt^4

Dynamics and Vibration

Computer Instrumentation of Robotic Test Arm

Dr. Joseph E. Barbay
Associate Professor

Southern Illinois University
Carbondale, Illinois 62901

ABSTRACT

A robotic arm vibration test station is being designed
and constructed by the students and faculty of the Department
of Technology, Southern Illinois University at Carbondale.
This unit will be used by both undergraduate students in the
Mechanical Engineering Technology program and by graduate students in the new MS program in Manufacturing Systems. This
paper will describe the PC based computer instrumentation
technology which will be an integral part of the vibration
test station.

INTRODUCTION

The continued decrease in the cost of personal computers,
coupled with the increased availability of data acquisition
hardware has provided new opportunities for engineering and
research uses of more sophisticated measurement techniques.
Low cost data acquisition hardware can be used with the PC to
configure a capable, but reasonably priced, computerized data
acquisition system. Such a system can be used for high speed
acquisition to collect transitory signals which in the past
could only have been captured by very costly equipment.

Not only can the computer be utilized as an instrument
for data acquisition, it can be effectively used to analyze
the data and then organize and display the results of an
experimenter's effort. In addition to playing a key role in
the collection and analysis of data, the computer can be used
in the theoretical analysis of complex systems. These techniques will be combined and applied to study laboratory motion
in mechanical systems. In particular, the mechanical systems
will be mock-ups of a variety of robotic arms. The mechanical
design aspects of the arm will be the responsibility of a
mechanical design team led by a faculty member in the Mechanical Engineering Technology Program.

The instrumentation design team will be responsible for
developing and selecting the components of the PC based instrumentation system. Transducers to measure physical variables representing force, stress, and acceleration will be

selected and integrated into the design of the mechanical arm. Signal conditioning equipment will be selected from existing product lines from the major manufacturers, and also designed and built by student experimenters.

SYSTEM DESCRIPTION (mechanical)

The mechanical system which will be instrumented will be presented to the students as a robotic arm which will be used as a vehicle to study vibrations in mechanical systems using PC based instrumentation technology. The robotic arm test station will consist of a rugged base unit. The base unit will be mounted to the table and will serve as the central platform for the vibration experiments. The base unit consists of a rotating barrel mounted on a metal shaft. The rotating barrel provides mounting hardware for the various robotic arms or beams to be mounted upon. The rotating base will have a starting position mounting bracket that holds the arm in initial position. The bracket will have provisions for a release mechanism which will be under the control of the experimenter. The release mechanism will be controlled by the computer and hence, by the operator through software commands. An adjustable mass will be used to supply the operational force to provide the initial motion of the mechanism. The force is applied by a pulley system. The arm is designed to impact upon a stop bracket a short distance from the starting bracket. The mechanical engineering technology students have designed and supervised the construction of the rotating base. A variety of robotic test arms or long beams will be constructed and built for use with the unit.

SYSTEM DESCRIPTION (electrical)

The electrical system which is used to instrument the rotating impulse beam tester will be designed and constructed by the electrical engineering technology students. The system will consist of the basic system transducers, signal processing circuitry, data acquisition hardware for the computer, the computer and associated peripheral equipment, and software required to operate the system.

It is anticipated that the measurement of actual system parameters from an operating mechanical system will enliven and challenge the electrical students. Experience has shown that although the use of electrical signal generation equipment to provide signals for testing for laboratory purposes is very useful, the collection of actual signals from a real operating system is much more stimulating for the students.

Of interest to the mechanical students will be the measurement of the vibrations set up in the beam when the beam is slammed against the stops of the rotating base unit. Accelerometers will be mounted on the beam to measure the acceleration initiated by the application of force on the beam when the mechanism is triggered by the computer. Strain gauges will be mounted along the beam to measure the vibratory action at various positions on the beam. It is planned that a

linear variable differential transformer (LVDT) will be added
to the system in order to provide a displacement signal. A
linear displacement transducer can be used to measure displacement as the actual travel length of the beam will be very
limited. In addition to measuring displacement by the use of
LVDT, the electrical students will be asked to design a rotating plate capacitor for measurements of position. The LVDT
output and measurement will be used as a standard to judge the
accuracy of the student designed and constructed rotating
plate transducers.

The mechanical system will be designed to provide relatively low frequency signals for the instrumentation system.
By selecting relatively low frequency signals, the cost of the
instrumentation system can be minimized. There are a number
of low cost single board data acquisition systems available
which will plug in to the IBM PC XT/AT bus. These systems can
provide sample rates of up to 100,000 samples per second.
Typically, many of these units provide the user with either
eight differential input channels, or sixteen single ended input channels. The multiple channels are obtained by multiplexing a single A/D converter. The sample rate is normally
specified for the A/D converter. Thus, the available sample
rate for an eight channel unit with a 100,000 samples per
second capability is 100,000/8 or 12,500 samples per second.
For a sixteen channel operation, the maximum available sample
rate would be 6,250 samples per second. The design parameters
of the beams will be adjusted so that the resonant frequency
of oscillation will be in the neighborhood of from one to
fifty hertz. For the maximum anticipated frequency of fifty
hertz, the sampling rate will allow a minimum of twenty
samples for the sixth harmonic of the fundamental frequency.
Since one of the goals of the project is to present visual
information to the students by means of the graphic capability
of the computer, (both the video screen and printer output)
high sample rates are required to insure quality presentations.

WORK STATION

The laboratory will consist of six individual work stations. The work stations will be developed in pairs with two
work stations being placed back to back and a work station extension shelf placed between the pairs of work stations. The
work stations will consist of 8/10 megahertz AT PC clones.
Each of the machines will be equipped with a DAS-16(F) high
speed analog/digital I/O board and a PIO-12 24 bit parallel
digital I/O board. The students will be introduced to the
problem of signal conditioning when they have to design the
circuitry to provide the proper level of signals for the A/D
I/O board mounted in the computer. The work stations will
have a complete set of electronic equipment consisting of oscilloscope, power supply, frequency counter, digital volt
meter and other equipment. This equipment constitutes a complete electronic lab which will be used during the design of
the signal conditioning circuitry.

It will be the responsibility of the students in the Electrical Engineering Technology program to develop suitable signal conditioning circuitry for interfacing the output of the transducers utilized on the mechanical portion of the system to the input of the multiplexer on the A/D board. This will greatly expand the student's exposure to real-world problems of noise, stability, drift, offset, and other problems associated with the design and construction of sensitive electronic measurement equipment. Students in both the mechanical and electrical programs will be paired to form development teams. This will give each person exposure to problems outside their area of study. In addition, it will develop the student's ability to work as part of a team applying new technology to real problems.

The signal range of the data acquisition boards will be set for plus or minus 10 volts. The data acquisition boards are 12 bits which will provide very adequate resolution of the signal, provided the signal has been properly conditioned and its amplitude is in the range of from -10 to +10 volts. Previous experience with students at the university and experience with problems a typical engineer has in the field with computerized data collection, indicates that the students need instruction designed to instill a basic understanding of the problems associated with signal conditioning. The work stations have been designed so that the students have access to standard measurement devices and will be able to view and measure the signal presented to the computer. This will ensure that they understand that the computer A/D board must be presented with a quality signal before the computer can digitize the signal to provide information about the process undergoing measurement. This is a simple concept, but the author has seen numerous instances where practicing people in the field have overlooked the need for proper conditioning of signals.

COMPUTER

A major component of the instrumentation work station is the AT style computer which will be used for the data acquisition and data analysis. The plummeting price of IBM PC's and their copycat cousins puts remarkable computer power in the price range of most instrumentation engineers. The cost of the basic systems is now such that the computer can be viewed as a low cost tool for just about anyone who wants one, even in economically strapped institutions such as universities.

Today's students can expect to find a growing use for personal computers wherever they may be working. In a typical program an engineering or engineering technology student may be required to take only a single semester course of a programming language such as Fortran. In today's environment, where the personal computer is fast becoming a standard tool, a single programming language course is no longer adequate. Both the mechanical technology and electrical technology students will be required to write their own data acquisition programs even though very excellent off-the-shelf programs are available for use with the PC in a data acquisition system. Experience has shown that beginning students turned loose with prepackaged data acquisition software do not properly comprehend the overall problems associated with gathering data. The beginners must be introduced to each step of the process. They must understand the significance of sampling rate, signal levels, timing considerations produced by multiplexing, noise problems, storage problems associated with the computer, etc.

The language for the work station will be Microsoft QuickBASIC. This language was selected after a survey of a large number of instrumentation engineers. In addition, hardware drivers for the boards are provided by the manufacturer. These drivers are written with the intention of interfacing to BASIC. The new versions of BASIC are a very powerful language which can be used to construct well written programs. They include many of the structure statements found in other languages such as Pascal and C. The integrated environment of QuickBASIC is ideal for student use. The language is more than adequate for the intended job and is relatively simple to use. QuickBASIC can be used for the development of the basic data handling and analysis programs required by the students.

The data acquisition and analysis programs developed by the students will be designed to interface with Lotus 123. In this way, both the electrical and mechanical engineering technology students will be encouraged to utilize the standard spreadsheet techniques for analysis and presentation of data collected by the system. An additional graphics capability will be provided by use of graphic programs developed by the faculty of the department.

CONCLUSION

The availability of low cost personal computers and numerous data acquisition hardware and expansion boxes, along with the large number of general purpose data acquisition and analysis software packages, makes it a foregone conclusion that the role of PC based data acquisition systems will greatly expand in the future. Many scientists, engineers, and technicians will be using computerized measurement systems. With the proper training and preparation, many people will become more productive and more competitive due to the use of such systems. The use of the robot arm vibration tester will provide very beneficial educational experiences for students in both the mechanical and electrical engineering technology programs at Southern Illinois University, Carbondale. The students will be exposed to real-world problems of data acquisition including transducer design, signal conditioning, programming, and data analysis.

Mechanical Design of a Robotic Arm Test Station

Dr. Marek L. Szary, Assistant Professor

College of Engineering and Technology
Southern Illinois University-Carbondale, Illinois 62901

Summary
This paper describes the design state of the mechanical portions of a vibration test station which is to be utilized in testing the vibration modes
of several robotic arm configurations. The robotic arm vibration demon-
stration and measurement unit will be used for research and student laboratories at Southern Illinois University at Carbondale.

Introduction
Every mechanical component of a robot has resonance frequencies which are
determined by the inertia and stiffness of the mechanical components. The
natural mechanical frequencies typically would lie in a range between very
low frequency and about 100 Hz for significant translational vibrations
and in range of much higher frequency for longitudinal and torsional vibrations. A long light-weight arm combined with an elbow and/or wrist equipped with a vibrating tool excited by variable speed motions will create
relatively high levels of vibrations. The amplitudes of the natural vibrations of the various mechanical components of a robot will increase when
the operating speed of the robot increases. The vibrations will be rich
in a variety of harmonics. Mechanical vibrations will play a role in the
trajectory accuracy of a robot. A good damping system and/or proper material of the arm with a well chosen coefficient of damping will improve accuracy, resolution, positioning, and repeatability.

The mechanical design of the robotic arm model will consist of a rigid
frame with a horizontal mounting system. Three different exchangeable
models of robotic arms will be developed. These arms will be designed to
test for and demonstrate the mechanical vibrations which would be present
in a robotic arm. The design will include provisions for computerized instrumentation that will be utilized to measure the vibratory motion. In
addition to the provisions made for instrumentation, the arms will be designed so that mathematical studies of the mechanical properties of the
arm can be readily accomplished by the researchers and the students who
will use the test station.

Initially, three arm simulators will be designed and constructed. The
first will be a single long arm with constant stiffness designed inten-
tionally to produce a large low frequency mechanical vibration. A second
long arm will be designed with variable stiffness to demonstrate a slightly
more complex mechanical system. The third model will be an arm with an
elbow and wrist which will be used to demonstrate a more complex mechanical
system. Provisions will be made to attach an unbalanced rotating mass to
the wrist or free end of the single arm. This rotating unbalanced mass
will be used to simulate variable driving forces which would normally be
developed by the gripper and the tools in use by the robot. The rotating
mass will be an adjustable weight traveling at a known rotational velocity,
thus, producing a force that can be calibrated and controlled. Mechanical

damping will be provided by adjustable air dampers. The intent will be to create a known mechanical system with known stimulation that can be readily modeled utilizing existing computer analysis programs.

Accelerometers and strain gauges will be attached at strategic locations along the arm. Suitable signal conditioning equipment will be developed to couple the output of the accelerometers and strain gauges to single board data acquisition hardware. PC based data acquisition hardware/software will be utilized. The test station will provide the user an opportunity to investigate free and forced vibration of linear and non-linear vibration systems with varying degrees of damping. The computer will be utilized both as the means for theoretical analysis and for the measurement of experimental data.

<u>Mechanical Design</u>
A classical solution of the robot with prismatic joint and a long flexible arm has been chosen to design a model of robotic vibrating arm station.

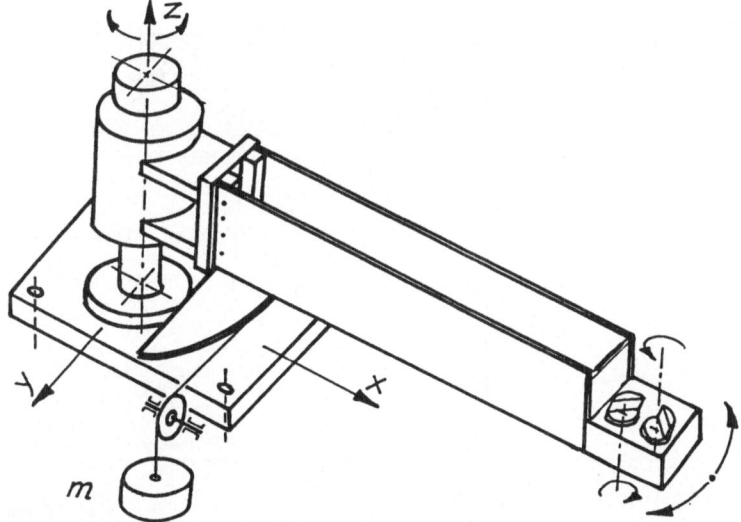

Figure 1. The model of a robotic arm with vibrating tool on the free end.

Figure 1 shows the three-dimensional model of a robotic's arm with prismatic joint on the one end and working-vibrating tool on the other. Figure 2 shows mechanical design of the mounting system of the arm simulators. Very high regidness of the base plate and associated parts may develop natural resonance frequencies from frequencies of the arm simulators. The thick steel base plate with a pressed vertical shaft used in this design should have this characteristic. Two ball bearings which have been chosen for minimum friction are carrying barrel with heavy and rigid arms mounting system. Five bolts on the end of the arm and extra clamps are performing rigid "cantilever" joint for clamped-free mode of vibration. On the bottom of the arm mounts is a pulley which can be used to perform initial rotation of the arm around shaft using known mass. The mass determines initial energy applied to the arm and also initiates vibrations. Figure 3 shows the basic configuration of the robotic arm simulators which will be used in research and teaching.

226

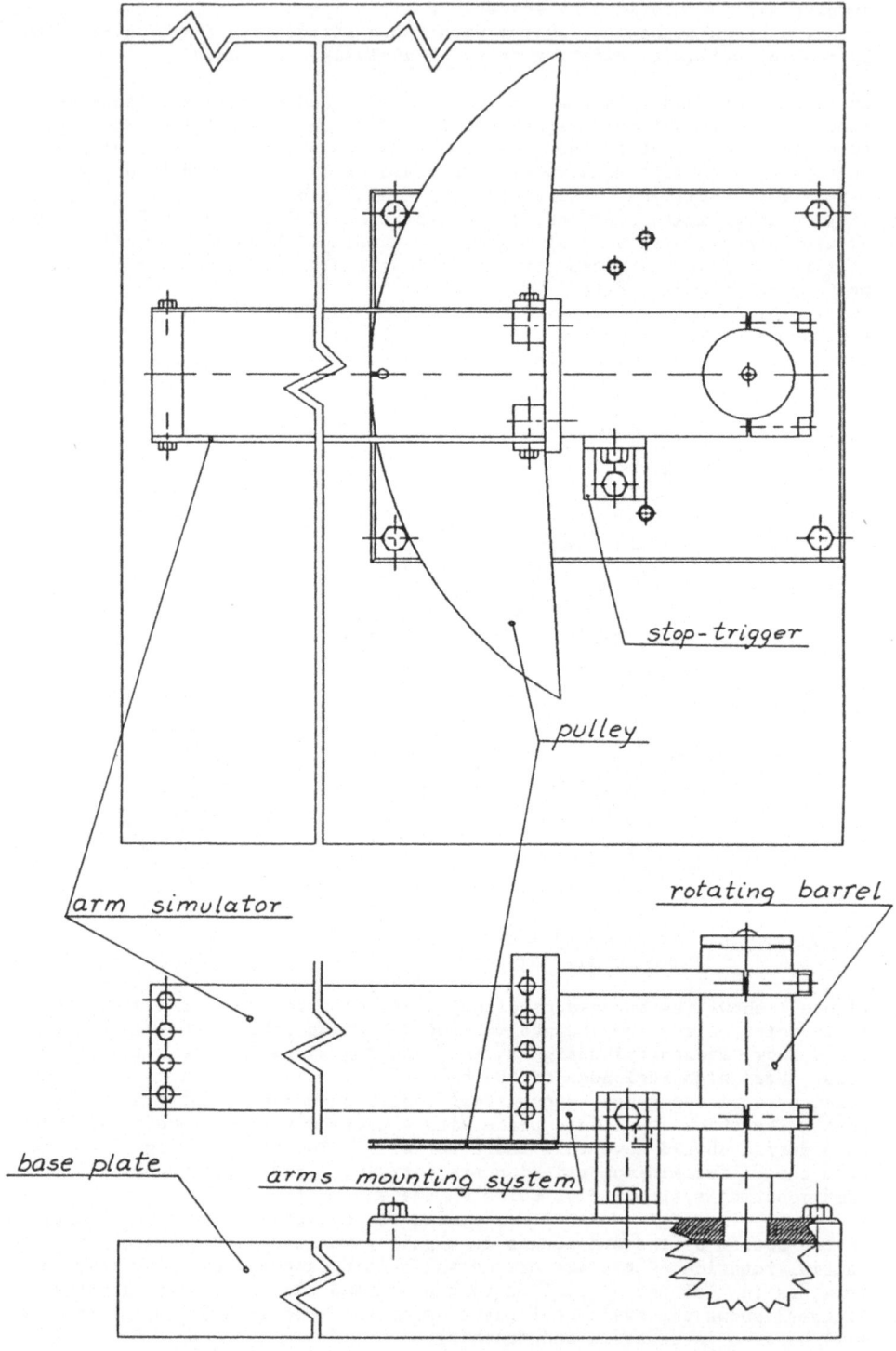

Figure 2. Mechanical design of the arm simulators mounting system.

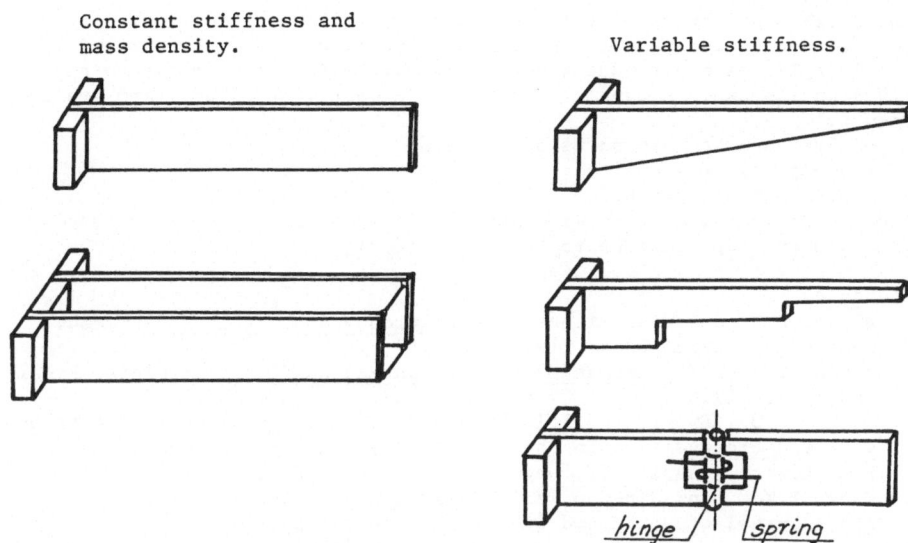

Constant stiffness and mass density.

Variable stiffness.

Figure 3, Basic configurations of the robotic arms simulators.

Future Work

The light-weight long robot arm for relatively high-speed operation requires the consideration of the vibrational effects on manipulation (Ref. 4). The characteristic vibrations of the beam can be explained using three models.

Model A. When the characteristic length of a test beam is comparable to stress wavelength (corresponding to a chosen frequency). In this case standing waves are generated in the test sample and the moment of inertia becomes an important parameter (Ref. 1).

Model B. When the characteristic length of a test specimen is much larger than stress wavelength. In this case the waves are propagated in the beam (Ref. 1).

Model C. When only transfer vibrations, together with torsional and longitudional, are significant for the arm. A properties described by models A and B will be subjects of a future research.

In laboratory practice using different models of the arm and models of the vibrating tools the different mathematical models and experiments can be expected:
1. A harmonic vibration (the simplest arm and tool or only free vibration of the arm).
2. A non-harmonic periodic motion. (The Fourier Series Mathematical Theorem is giving possibility to find a number of harmonically related frequencies. The amplitude vs. time and amplitude vs. frequency are describing time and frequency domains.)
3. A random stationary vibrations.
4. A random non-stationary vibrations.

The variable stiffness of the robotic's arm will bring non-linearity to the motion equations and frequency response functions. (Ref. 2 and 3) In principle, each particular non-linear vibration problem has to be solved on its own. A popular non-linear property of robotic arm are related to:
- clearance between parts of the arm;
- asymptotic elasticity;
- pretentioned springs.

Research and parallel work with mathematical apparatus can bring more information and applications to the future design.

References
1. Blake, P.M. and W.S. Mitchell, Vibration and Acoustic Measurement Handbook, Spartan Books, 1972.
2. Broch, Trampe J., Mechanical Vibration and Shock Measurements, Bruel and Kjaer, 1980.
3. Ferry, J.D., Experimental Techniques for Rheological Measurements on Viscoelastic Bodies. In Rheology Theory and Applications edited by F.R. Eirich, Chapter II, New York: Academic Press, 1958.
4. Wang, P.K.C. and Jin-Duo Wei, Vibrations in a Moving Flexible Robot Arm, Journal of Sound and Vibration, 116(1), pp. 149-160, 1987.

Flexural Vibrations of a Tip Loaded Stepped Robot Arm

K.K. PUJARA, VIKRAM R. JAMALABAD, SUKESH MOHAN

Mechanical Engineering Department
Indian Institute of Technology, New Delhi, India

Summary

This paper obtains natural frequencies and mode shapes in flexure of a
tip loaded robot arm when the arm is made of two steps, each being of
uniform cross-section but of different dimensions. Effects of shear
deformation and of rotary inertia are included. The technique used is
to equate the vibrational responses at the junction of the two steps.
The procedure is general and can be extended to arms having a number of
steps. The technique is recommended for the analysis of robotic arms
of telescopic nature. Computed results for natural frequencies are
presented. Required mode shapes can be computed by using equations included in the paper.

Introduction

Sanger [1] and Goel [2] have attempted dynamic analysis of variable section beams by cumbersome methods which have the further inadequacy of
not including the effects of shear deformation and of rotary inertia.
This assumption is not justified in robot applications where high manipulator speeds are required or when greater positional accuracy is needed.

Huang [4] using [5] and Bruch & Mitchell [6] have presented results with
analyses that include shear deformation and rotary inertia effects.
These references [4] & [6] are however limited to uniform section beams
which do not adequately simulate actual arms which may be telescopic.

In this paper the beam is considered to be composed of two segments for
each of which the Timoshenko equations are used. The technique used
is to equate the vibrational response by constraints enforced at the junction. This analysis is valid for a beam with several steps. However
analysis presented here is for a single step cantilevered beam.

Basic Equations

Fig. 1 shows the beam under analysis where x_1 & y_1 are co-ordinates w.r.t. the fixed frame of reference and x_2 & y_2 w.r.t. a frame at the junction.

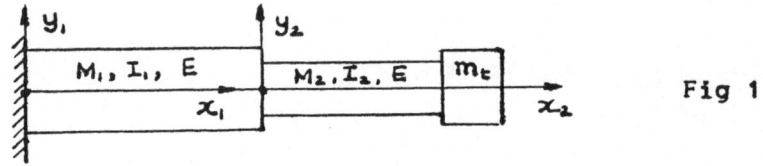

Fig 1

Timoshenko's coupled equations given in references [5] & [7] applied to either segment are, for i=1,2 :

$$EI_i\frac{\partial^2\psi_i}{\partial x_i^2} + KGA_i(\frac{\partial y_i}{\partial x_i} - \psi_i) - \rho I_i(\frac{\partial^2\psi_i}{\partial t^2}) = 0 \tag{1}$$

$$\rho A_i\frac{\partial^2 y_i}{\partial t^2} - KGA_i(\frac{\partial^2 y_i}{\partial x^2} - \frac{\partial\psi_i}{\partial x_i}) = 0 \tag{2}$$

with notations as in reference [6].

The end conditions for the beam under consideration are :

$$y_1 = 0 \quad \& \quad \psi_1 = 0 \qquad \text{at } x_1 = 0 \tag{3 & 4}$$

$$EI_2\frac{\partial\psi_2}{\partial x_2} + I_t\frac{\partial^2\psi_2}{\partial x^2} = 0 \qquad \text{at } x_2 = L_2 \tag{5}$$

$$KGA_2(\psi_2 - \frac{\partial y_2}{\partial x_2}) - m_t\frac{\partial^2 y_2}{\partial t^2} = 0 \qquad \text{at } x_2 = L_2 \tag{6}$$

where $I_t = \frac{1}{2}m_t k^2$ is the mass moment of inertia of the tip mass m_t about the axis of bending and k is the radius of gyration.

At the step the following conditions must be met :

$$\frac{\partial y_1}{\partial x_1} = \frac{\partial y_2}{\partial x_2} \quad \& \quad \frac{\partial\psi_1}{\partial x_1} = \frac{\partial\psi_2}{\partial x_2} \qquad \text{at } x_1=L_1 \& x_2=0 \tag{7 & 8}$$

$$y_1 = y_2 \quad \& \quad \psi_1 = \psi_2 \tag{9 & 10}$$

To reduce equations (1) & (2) and boundary conditions (3) through (10) to non-dimensional form, the following dimensionless factors are introduced. (Some of these are adapted from references [4] & [6]) : for i=1,2 :

$$\epsilon = d_2/d_1 \quad , \quad \eta = L_2/L_1 \quad , \quad \xi_i = x_i/L_i$$

$$b_i = (\rho A_i L_i^4 \omega^2)/EI_i \quad , \quad \frac{b_1}{b_2} = \epsilon/\eta^2$$

$$r_i^2 = I_i/A_i L_i^2 \quad , \quad s_i^2 = Er_i^2/KG \quad , \quad r_1/r_2 = s_1/s_2 = \eta/\epsilon$$

$$\bar{M}_2 = \frac{m_t}{\rho A_2 L_2} \quad , \quad \sigma_2 = \frac{k}{L_2} \quad , \quad \bar{M}_1 = \eta\epsilon^2 \bar{M}_2$$

Before the equations are written in the non-dimensional form, they will be simplified further by using the modal vibration in the form :

$$y_i = Y_i(\xi_i)e^{j\omega t} \tag{11}$$

$$\psi_i = \Psi_i(\xi_i)e^{j\omega t} \tag{12}$$

where $j = \sqrt{-1}$.

The non-dimensional ratios and the above simplification results in reducing the equations (1) through (10) to :

$$s_i^2 \frac{d^2 \Psi_i}{d\xi_i^2} - (1 - b_i^2 r_i^2 s_i^2)\Psi_i + \frac{1}{L_i}\frac{dY_i}{d\xi_i} = 0 \tag{13}$$

$$\frac{d^2 Y_1}{d\xi_i^2} + b_i^2 s_i^2 Y_1 - L_i \frac{d\Psi_i}{d\xi_i} \tag{14}$$

$$Y_1 = 0 \quad \& \quad \Psi_1 = 0 \quad \text{at} \quad \xi_1 = 0 \quad \text{(15 \& 16)}$$

$$\left.\begin{array}{l} Y_1 = Y_2 \quad \& \quad \Psi_1 = \Psi_2 \\[2mm] \dfrac{dY_1}{d\xi_1} = \dfrac{dY_2}{d\xi_2} \quad \& \quad \dfrac{d\Psi_1}{d\xi_1} = \dfrac{d\Psi_2}{d\xi_2} \end{array}\right\} \quad \text{at} \quad \xi_1 = 1 \ \& \ \xi_2 = 0 \quad \text{(17-20)}$$

$$\frac{d\Psi_2}{d\xi_2} - \tfrac{1}{2}.\bar{M}_2 b_2^2 \sigma_2^2 \Psi_2 = 0 \quad \text{at} \quad \xi_2 = 1 \tag{21}$$

$$\Psi_2 - \frac{1}{L_2}\frac{dY_2}{d\xi_2} + \frac{1}{L_2}\bar{M}_2 b_2^2 s_2^2 Y_2 = 0 \quad \text{at} \quad \xi_2 = 1 \tag{22}$$

Solutions

Decoupling equations (13) & (14) gives two fourth order equations in Y_i and ψ_i. The solutions [3] are taken in the following standard format

$$Y_1 = C_1\cosh(b_1\alpha_1\xi_1) + C_2\sinh(b_1\alpha_1\xi_1) + C_3\cos(b_1\beta_1\xi_1) + C_4\sin(b_1\beta_1\xi_1) \tag{23}$$

$$Y_2 = C_5\cosh(b_2\alpha_2\xi_2) + C_6\sinh(b_2\alpha_2\xi_2) + C_7\cos(b_2\beta_2\xi_2) + C_8\sin(b_2\beta_2\xi_2) \tag{24}$$

$$\Psi_1 = C_1' \sinh(b_1\alpha_1\xi_1) + C_2'\cosh(b_1\alpha_1\xi_1) + C_3'\sin(b_1\beta_1\xi_1) + C_4'\cos(b_1\beta_1\xi_1) \tag{25}$$

$$\Psi_2 = C_5'\sinh(b_2\alpha_2\xi_2) + C_6'\cosh(b_2\alpha_2\xi_2) + C_7'\sin(b_2\beta_2\xi_2) + C_8'\cos(b_2\beta_2\xi_2) \tag{26}$$

where for i=1,2 ;

$$\begin{Bmatrix} \beta_i \\ \alpha_i \end{Bmatrix} = [\tfrac{1}{2}(\pm(r_i^2+s_i^2)+((r_i^2-s_i^2)+4/b_i^2)^{\frac{1}{2}})]^{\frac{1}{2}}$$

Inserting equations (23) through (26) in (13) & (14) gives :

$$C_1' = B_1 b_1 C_1/L_1 \; , \quad C_2' = B_1 b_1 C_2/L_1 \; , \quad C_3' = -B_2 b_1 C_3/L_1 \; , \quad C_4' = B_2 b_1 C_4/L_1$$

$$C_5' = B_3 b_2 C_5/L_2 \; , \quad C_6' = B_3 b_2 C_6/L_2 \; , \quad C_7' = -B_4 b_2 C_7/L_2 \; , \quad C_8' = B_4 b_2 C_8/L_2$$

where

$$B_1 = \frac{\alpha_1^2 + s_1^2}{\alpha_1} \quad B_2 = \frac{\beta_1^2 - s_1^2}{\beta_1} \quad B_3 = \frac{\alpha_2^2 + s_2^2}{\alpha_2} \quad B_4 = \frac{\beta_2^2 - s_2^2}{\beta_2}$$

Computation & Results

The solution equations (23) to (26) are inserted into the four boundary conditions (15) through (18) and the four enforced conditions (19) through (22) resulting in a matrix of the following form :

$$\begin{bmatrix} 1 & 0 & 1 & 0 & 0 & 0 & 0 & 0 \\ 0 & B_1 & 0 & B_2 & 0 & 0 & 0 & 0 \\ P_1 & P_2 & P_3 & P_4 & -1 & 0 & -1 & 0 \\ R_1 & R_2 & R_3 & R_4 & 0 & R_6 & 0 & R_8 \\ S_1 & S_2 & S_3 & S_4 & 0 & S_6 & 0 & S_8 \\ T_1 & T_2 & T_3 & T_4 & T_5 & 0 & T_7 & 0 \\ 0 & 0 & 0 & 0 & V_5 & V_6 & V_7 & V_8 \\ 0 & 0 & 0 & 0 & W_5 & W_6 & W_7 & W_8 \end{bmatrix} \begin{Bmatrix} C_1 \\ C_2 \\ C_3 \\ C_4 \\ C_5 \\ C_6 \\ C_7 \\ C_8 \end{Bmatrix} = \begin{Bmatrix} 0 \\ 0 \\ 0 \\ 0 \\ 0 \\ 0 \\ 0 \\ 0 \end{Bmatrix} \tag{27}$$

The solutions to the above give frequencies (and mode shapes). The details of various elements of the matrix and vector have to be omitted owing to space constraints.

Computed results are for stock segment of circular cross-section based on Huang's [4] definition and are presented in figure 2. Thus $r_1=0.02$ and $s_1=0.04$. The variation for ϵ and η is attributed to the scion beam.

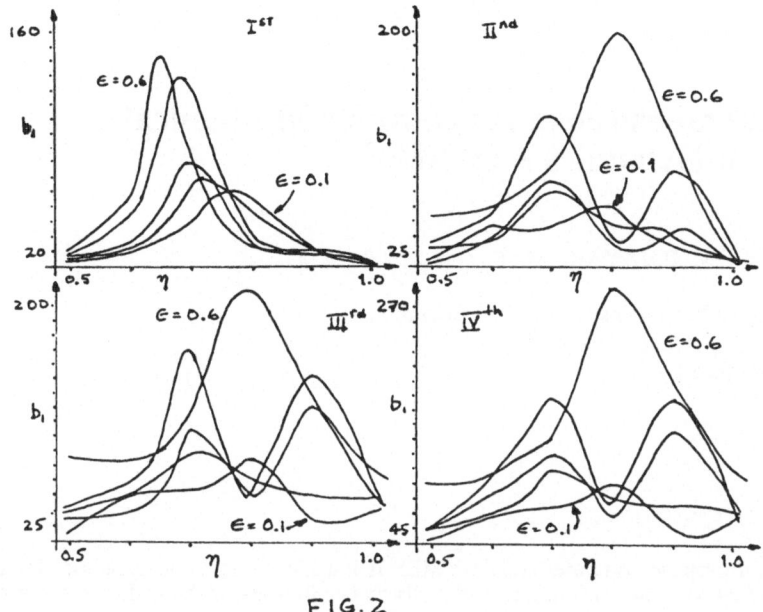

FIG. 2

References

1. Sanger D.J.:Transverse vibrations of a class of non-uniform beam. Jnl. of Mechanical Engineering Science 10 (1968) 111-120.

2. Goel R.P.:Transverse vibrations of tapered beams. Jnl. of Sound & Vibration 47 (1976) 1-7.

3. Gorman D.J.:Free vibration analysis of beams & shafts. New York: John Wiley & Sons 1975.

4. Huang T.C.:The effect of rotary inertia and of shear deformation on the frequencies and normal mode equations of uniform beams with simple end conditions. Jnl. of Applied Mechanics 28 (1961) 579-584.

5. Timoshenko S.P.:Vibration problems in engineering. New York: D. Van Nostrand & Co. 1955, pp.329-331.

6. Bruch Jr, J.C. & Mitchell, T.P.:Vibrations of a mass-loaded clamped-free Timoshenko beam. Jnl. of Sound & Vibration (1987) 341-347.

7. Meierovitch L.:Analytical methods in vibration. New York: The Mac-Millan Co. 1967, pp. 126-135.

A General Purpose Program for Mechanical System Dynamics – DYMES

O. K. KWON, P. JAYAKUMAR and R. KODALI

Engineering Mechanics Research Corporation
1707 W. Big Beaver Road
Troy, MI 48084

Summary

A general purpose computer code DYMES is introduced as an analysis tool for DYnamics of MEchanical Systems. Modeling and analysis capabilities of the code are described. A robot arm control example and simulation of passenger car handling demonstrate the code performance.

Introduction

The system being treated in DYMES is characterized as a constrained, multi-body, spatial mechanical system, in which body elements are connected through mechanical joints such as spherical, revolute, and translational joints and force elements such as nonlinear springs, dampers, and actuators. The fundamental difference of mechanical system dynamics from conventional structural system dynamics is the presence of a high degree of geometric nonlinearity associated with large displacement kinematics. Governing equations for conventional structural system dynamics are linear differential equations, while those for mechanical system dynamics are nonlinear differential equations coupled with nonlinear algebraic equations of kinematic constraints. Based on recently developed theory on nonlinear mechanical system dynamics and state of the art numerical solution techniques, a highly sophisticated program, DYMES, has been developed as a powerful tool for engineers.

Major areas of application of the code are analyses of machine mechanisms, robotics, vehicle dynamics, controlled systems, and spacecraft dynamics. Most of these mechanical dynamic systems can be conveniently modeled and analyzed using a library of standard elements written in the form of individual modules. The code also incorporates modules to represent special purpose subsystems and interdisciplinary effects of integrated mechanical systems. In addition, it allows the user to derive and incorporate equations that govern nonstandard effects and to enter FORTRAN description of such equations.

Kinematic and dynamic analyses of a spatial mechanical system is more complicated than the analysis of a planar system. The main complexity is due to large rotational motion of body elements in the system. The traditional way of describing 3-dimensional rotation is the use of a set of orientation coordinates and its time derivatives. Euler angles, Bryant angles, and Euler parameters are among the choices available for this set. However, the use of a set of orientation coordinates and its time derivatives for governing equations of motion may result in a singular or variable mass matrix [1]. DYMES resolves this difficulty by employing Euler parameters as a set of orientation coordinates and Cartesian angular velocity as a set of angular velocity of the system. This results in a constant mass matrix in time without any singularity, and improves both numerical efficiency and reliability. References 1 and 2 discuss the theoretical background of this approach in detail.

Modeling Capabilities

Standard mechanical joints and force elements can be used to model most multi-body mechanical systems. Examples of such elements are shown below. Figure 1 shows some basic joints which interconnect two bodies directly. If the inertia of a coupler is negligible, two bodies can be indirectly interconnected by using composite joints as shown in Fig. 2. Translational and rotational spring-damper-actuators are shown in Fig. 3. DYMES uses various curve elements to represent nonlinear stiffness, damping or actuator properties.

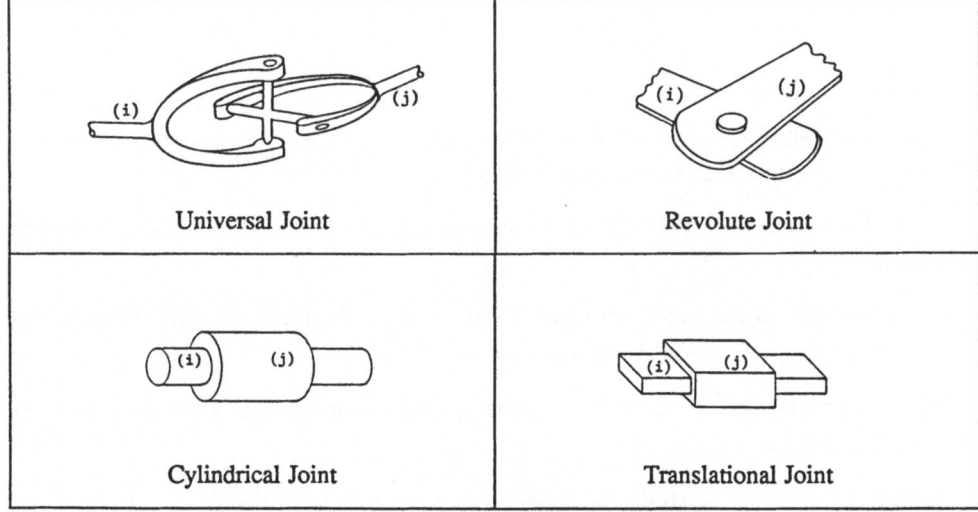

| Universal Joint | Revolute Joint |
| Cylindrical Joint | Translational Joint |

Fig. 1 Some Standard Mechanical Joints

In addition to the standard elements for general use, ground vehicle modeling capabilities such as tire, bushing, road profile, steering, and axles are implemented for special use of the code in vehicle system analysis. Realistic and effective modeling of various tires is achieved by a combination of analytical and empirical modeling techniques.

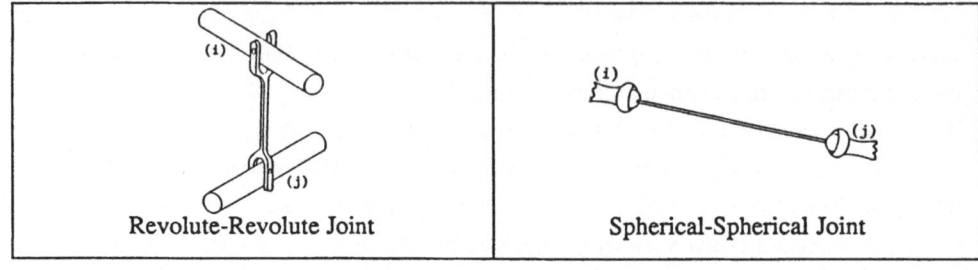

Fig. 2 Composite Joints

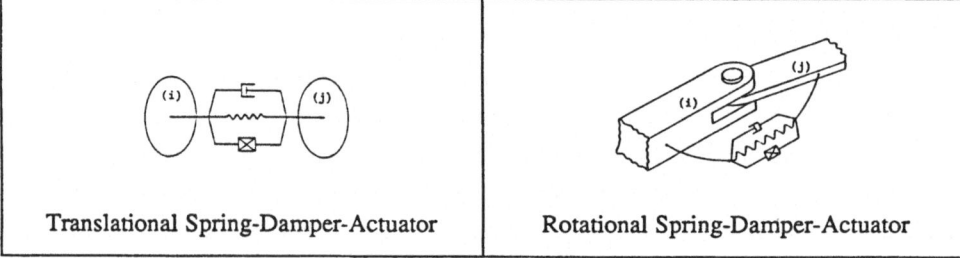

Fig. 3 Force Elements

Analysis Capabilities

- Assembly analysis - Assembly of a mechanical system, modeled from a drawing board or design draft, within a given assembly tolerance.

- Redundancy analysis - Remodeling an overconstrained system by eliminating redundant constraints.

- Static analysis - Static equilibrium analysis of a nonzero degree-of-freedom (DOF) system after the assembly and redundancy analyses.

- Quasi-static analysis - A sequence of static equilibrium analyses of a mechanical system for different applied forces.

- Dynamic analysis - Analysis for time-histories of position, velocity, acceleration, and forces of a nonzero DOF system.

- Kinematic analysis - Analysis for time-histories of position, velocity, and acceleration of a zero DOF system whose motion is partly prescribed.

- Inverse dynamic analysis - Analysis for time-histories of forces in addition to kinematic analysis of a zero DOF system.

Example Problems

Control of a Spherical Robot It is demonstrated that DYMES can be used for feedback control in mechanical system dynamics, using a three DOF spherical robot [3]. Figure 4 shows the three-DOF spherical robot consisting of four bodies with two revolute joints and a translational joint. The end-effector of the robot is located at point A. The intended function of the robot is to weld in such a way that the end-effector makes a circular path of 1 meter radius in X-Z plane at a constant Y position. The purpose of the simulation is to see how accurately the path is generated and to get the time-histories of the joint actuating forces.

DYMES performs dynamic analysis based on control forces supplied at the three joints. These joint actuating forces are calculated in a control loop of DYMES by detecting the error between the reference and actual paths. Simulation results show that the path is generated within maximum errors of 1.1 cm in the circle radius and 2.2 cm in Y position of the end-effector as shown in Fig. 5. Figure 6 shows the time-histories of the control forces required at the three joints. Note that TORQUE3 is initially large due to the effect of gravitational forces on BODY3 and BODY4.

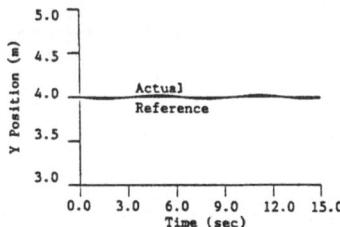

Fig. 5 Y Position of End-Effector

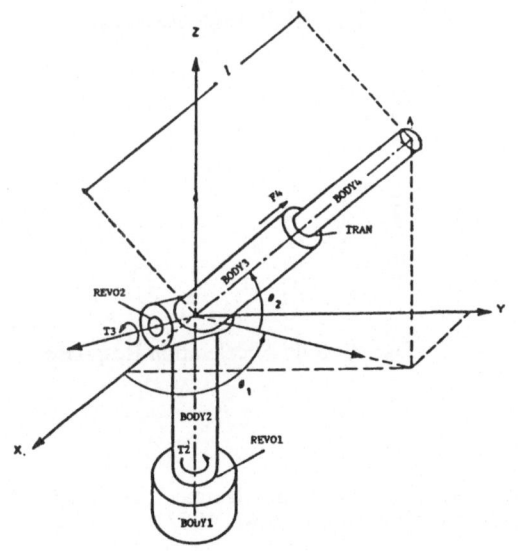

Fig. 4 Three-DOF Spherical Robot

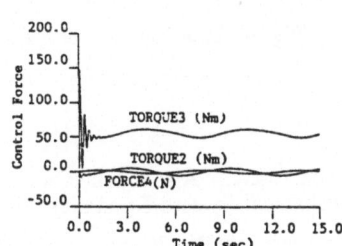

Fig. 6 Control Forces

Dynamic Response of a Passenger Car The dynamic response of a typical passenger car [4] subjected to a step steering input and moving on a flat ground is analyzed. The vehicle is modeled with chassis, steering rack, McPherson strut front suspension, semi-trailing arm rear

238

suspension, front and rear roll stabilizer bars, and four wheel/tire assemblies. Detailed modeling and inertial properties of the car can be found in Ref. 4. After initial assembly and static equilibrium analyses, dynamic analysis is performed with an initial forward speed of 60 mph.

The front tire steering angle is shown in Fig. 7. Figures 8, 9, and 10 show transient responses of chassis roll angle, yaw rate, and side acceleration, respectively. Roll gain of the vehicle is calculated as 4.4°/g. Response times, for the transient responses to reach 90% of their steady-state values, are calculated as 0.15 sec for the yaw rate response time and 0.43 sec for the side acceleration response time.

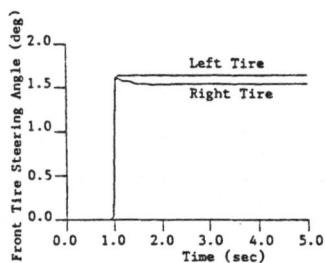

Fig. 7 Front Tire Steering Angle

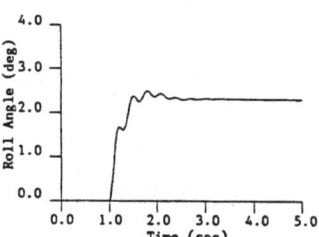

Fig. 8 Roll Angle Response

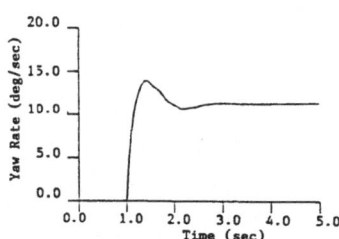

Fig. 9 Yaw Rate Response

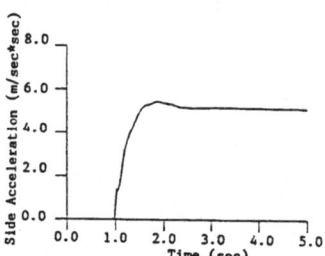

Fig. 10 Side Acceleration Response

References

1. Kwon, O.K.: An Index One Formulation of Differential-Algebraic Equations for Mechanical System Dynamics. PhD thesis, The University of Iowa, 1987.

2. Nikravesh, P.E.; Kwon, O.K.; Wehage, R.A.: Euler Parameters in Computational Kinematics and Dynamics, Part II. ASME Journal of Mechanisms, Transmissions, and Automation in Design, Vol. 107 (1985) 358-369.

3. Shoham, M.; Koren, Y.: Motion Control Algorithms for Sensor-Equipped Robots. ASME Journal of Dynamic Systems, Measurement and Control, Vol. 109 (1987) 335-344.

4. Haug, E.J.: Computer Aided Kinematics and Dynamics of Mechanical Systems, Volume 1 Basic Methods. Boston: Allyn and Bacon, 1988.

A Multiprocessor System for Multiblade Coordinate Transformation

S. Ganesan, Department of Computer Science and Engineering, Oakland University, Rochester, MI 48309-4401

T. S. Balasubramanian, National Aeronautical Laboratory, Bangalore, India 560017

M. O. Ahmad, Department of Electrical Engineering, Concordia University, Montreal, P.Q., Canada, H3G 1M8

ABSTRACT

The multiblade coordinate transformation is widely used in the helicopter industry for the studies of rotor dynamics. A multimicroprocessor based instrument has been developed which transforms measured parameters of the rotor blades into parameters in the transformed coordinate system, which are of greater utility to the experimenter. One microprocessor is used to collect the input data and output the computed results at periodic intervals, while the second microprocessor is used to compute the results. Details of the tightly coupled multimicroprocessor system and the computations involved in the multiblade coordinate transformations are described here.

INTRODUCTION

Helicopter dynamics in forward flight involve complex aerodynamic forces dependent on time and on the azimuth angles of the blades. Investigation of these phenomena leads to the solution of multidegree parametric instability problems for which the computer time required is exorbitant. One effective method used by helicopter dynamists to reduce the computer time required is the multiblade coordinate transformation. The transformation is based on multiblade summations, and was first introduced by Coleman [1], who used it to solve ground resonance problems for helicopter rotors. Later on Hohenemser and Yin [2] have used N generalized multiblade coordinates to devise a general method for helicopter dynamics problems, which they call the "Method of Multiblade Coordinates". This has reduced the computational time drastically, since it converts the parametric instability problem into a simple eigenvalue/forced vibration problem.

METHOD OF MULTIBLADE COORDINATES

This method is based on multiblade summations. If there are b equally spaced blades, and if

$$C = \sum_{K=0}^{b-1} \cos\{N(\psi + 2\pi K/b)\} \qquad (1)$$

and
$$S = \sum_{K=0}^{b-1} \sin\{N(\psi + 2\pi K/b)\} \qquad (2)$$

then

$$C = \frac{\sin(\pi N)}{\sin(\pi N/b)} \cos\{N(\psi + \frac{b-1}{b}\pi)\}, \quad \text{if } N \text{ is not an integer,}$$

$$= 0, \text{ if } N \text{ is an integer which is not a multiple of } b,$$

$$= b \cos N\psi, \text{ if } N \text{ is an integer which is a multiple of } b. \qquad (3)$$

and
$$S = \frac{\sin(\pi N)}{\sin(\pi N/b)} \sin\{N(\psi + \frac{b-1}{b}\pi)\}, \quad \text{if } N \text{ is not an integer,}$$

$$= 0, \text{ if } N \text{ is an integer which is not a multiple of } b,$$

$$= b \sin(N\psi), \text{ if } N \text{ is an integer which is a multiple of } b. \qquad (4)$$

Also
$$\sum_{K=0}^{b-1} \sin^2\psi_K = \sum_{K=0}^{b-1} \cos^2\psi_K = b/2 \qquad (5)$$

and
$$\sum_{K=0}^{b-1} \sin\psi_K \cos\psi_K = 0 \qquad (6)$$

where ψ_K takes the values ψ, $\psi + 2\pi/b, \ldots, \psi + 2\pi(b-1)/b$.

Coleman first used these summations in a ground resonance problem of the form

$$\ddot{X}_K + 2\Omega F \dot{X}_K + \Omega^2 X_K = p(t) \sin\psi_K + q(t) \cos\psi_K + \cdots \qquad (7)$$

where $p(t)$ and $q(t)$ are functions of time, and F is a damping function. This is a parametric instability problem involving the variable quantity X_K which is measured with respect to the rotating K^{th} blade.

To find the total effect of all the blades Coleman introduced new coordinates such that

$$u = - (\frac{2}{b}) \sum_{K=0}^{b-1} X_K \cos\psi_K \qquad (8)$$

$$v = - (\frac{2}{b}) \sum_{K=0}^{b-1} X_K \sin\psi_K \qquad (9)$$

Differentiating u and v with respect to time and noting that $d\psi/dt = \Omega$, we get:

$$\sum_{K=0}^{b-1} \dot{X}_K \sin\psi_K = b(\Omega u - \dot{v})/2 \qquad (10)$$

$$\sum_{K=0}^{b-1} \dot{X}_K \cos\phi_K = -b(\Omega v - \dot{u})/2 \tag{11}$$

$$\sum_{K=0}^{b-1} \ddot{X}_K \sin\phi_K = -b(\ddot{v} - 2\Omega\dot{u} - \Omega^2 v)/2 \tag{12}$$

$$\sum_{K=0}^{b-1} \ddot{X}_K \cos\phi_K = -b(\ddot{u} + 2\Omega\dot{v} - \Omega^2 u)/2 \tag{13}$$

Multiplying Equation 7 by $\cos\phi_K$ and $\sin\phi_K$, summing over the blades, and using the multiblade summations, Equation 7 transforms into

$$\ddot{u} + 2\Omega F\dot{u} + 2\Omega\dot{v} + 2\Omega^2 Fv = -q(t) \tag{14}$$

and
$$\ddot{v} + 2\Omega F\dot{v} - 2\Omega\dot{u} - 2\Omega^2 Fu = -p(t) \tag{15}$$

Thus the transformation has effectively resolved a rotating quantity into components along fixed axes in the helicopter body, removing the periodic terms from Equation 7, i.e., this transformation converts a parametric equation into a simple eigenvalue/forced excitation problem.

Hohenemser and his team improved upon this method [2], and have developed a general method of solving rotor dynamics problems, called the "method of multiblade coordinates". This method is widely used in the helicopter industry. The method has the following characteristics: If X_K is a parameter of the Kth rotating blade, it can be expressed in the following form:

$$X_K = X_0 + X_d(-1)^K + X_I\cos\phi_K + X_{II}\sin\phi_K + X_{III}\cos2\phi_K + X_{IV}\sin2\phi_K + \cdots \tag{16}$$

where X_0, X_d, X_I, X_{II}, X_{III}, X_{IV}, are parameters independent of ϕ_K, representing coning, differential coning, tilting, and warping.

The stationary coordinate parameters are given by (from multiblade summations):

$$X_0 = \frac{1}{N} \sum_{K=1}^{N} X_K \tag{17}$$

$$X_d = \frac{1}{N} \sum_{K=1}^{N} X_K(-1)^K \tag{18}$$

$$X_I = \frac{2}{N} \sum_{K=1}^{N} X_K \cos\phi_K \tag{19}$$

$$X_{II} = \frac{2}{N} \sum_{K=1}^{N} X_K \sin\psi_K \qquad (20)$$

$$X_{III} = \frac{2}{N} \sum_{K=1}^{N} X_K \cos2\psi_K \qquad (21)$$

$$X_{IV} = \frac{2}{N} \sum_{K=1}^{N} X_K \sin2\psi_K \qquad (22)$$

If the X_K's are known then X_o, X_d, X_I, etc., can be evaluated from these transformations. This principle is used in the Multiblade Coordinate Transformation Processor. The parameter X_K can be the blade flapping angle, pitching angle, lead-lag angle or bending deflection in flap or lag direction, etc..

DETAILS OF THE INSTRUMENT

Figure 1 shows the block diagram of the instrument, which has two tightly coupled microprocessors. The local bus of each microcomputer is connected through a tristate interface circuit to the common bus and common memory. A bus arbitration circuit grants access to common memory taking into account the priority of the requesting microprocessor and mutual exclusion [3]. The multi-blade coordinate transformation equations for 3 blades are given by

$$\beta_o = \frac{1}{3} \sum_{i=1}^{3} \beta_i \qquad (23)$$

$$\beta_c = \frac{2}{3} \sum_{i=1}^{3} \beta_i \cos\psi_1 \qquad (24)$$

$$\beta_s = \frac{2}{3} \sum_{i=1}^{3} \beta_i \sin\psi_1 \qquad (25)$$

It may be seen that the transformation involves multiplication by $\sin\phi_i$ and $\cos\phi_i$ terms and summation only.

The value of sine and cosine for various angles at 10° intervals are obtained from the look-up table stored permanently in a read only memory integrated circuit. The computed values are stored in memory or displays it on an oscilloscope.

CONCLUSIONS

A specialized processor has been described which has been proven to very useful in supporting experimental studies of helicopter rotor dynamics. This multimicroprocessor based system transforms measured parameters of the rotor blades into parameters in the transformed coordinate system. The complex individual blade signals become simpler in the transformed form.

REFERENCES

1. R. P. Coleman, "Theory of Self-Excited Mechanical Oscillation of Hinged Rotor Blades," NACA Advanced Restricted Report 3G29. Republished as NACA Report 1351, 1943.

2. K. H. Hohenemser and S-K. Yin, "Some applications of the method of multiblade coordinates," Journal of the American Helicopter Society, 17, 3:3-12, 1972.

3. D. Sundararajan, M. O. Ahmad and S. Ganesan, "MULTI-PROCESSORS," IEEE Trans. on Circuits and Systems, p 620-622, June 1985.

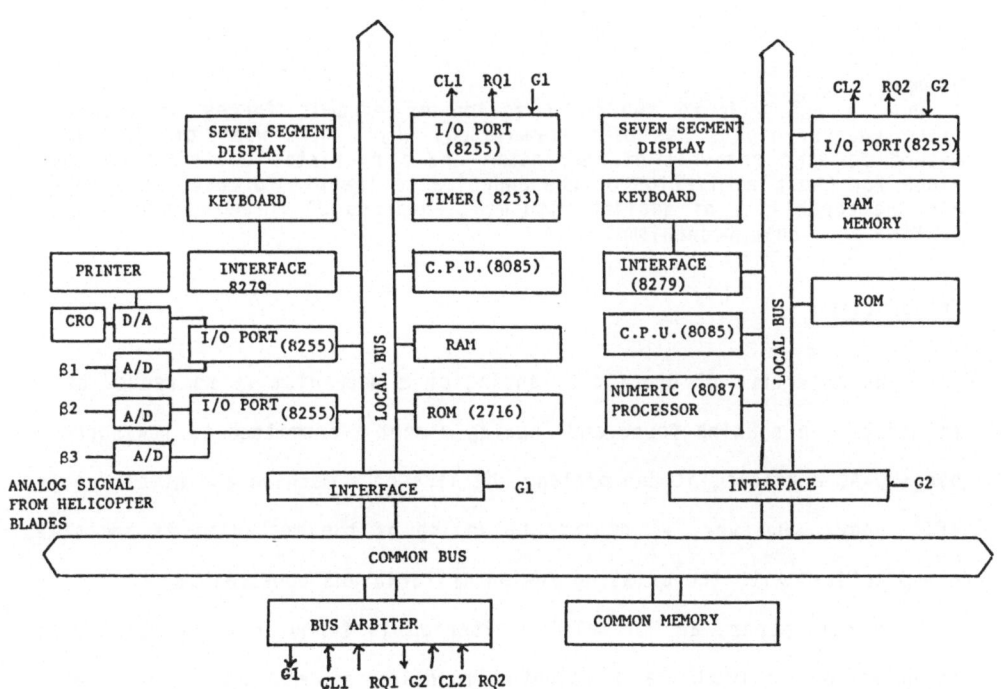

FIG.1 BLOCK DIAGRAM OF MULTIBLADE COORDINATE TRANSFORMATION PROCESSOR

A General Approach to the Dynamic Balancing of Mechanisms

Tin-Lup Wong

Department of Mechanical Engineering
Florida Atlantic University
Boca Raton, Florida

Summary
A general approach to the minimization of dynamic forces of mechanism using counterweights has been formulated, which make use of the matrix method. for the analysis of mechanisms and a heuristic optimization technique for the minimization of the undesirable inertia forces. This method has the capability of dealing with multi-degrees of freedom, multi-loops, spatial or planar mechanisms.

Introduction

The objective of dynamic balancing of a mechanism is to reduce or to eliminate the shaking force and shaking moment transmitted to its surroundings. As a result, it can prolong the life of a machine and attain better efficiency. However, a complete balancing of the mechanism is sometimes not practical with the existing design criteria and constraints.

In this paper, an optimality criterion is formulated for constrained optimization to minimize a weighted function of shaking force, shaking moment and input driving power. The computation of shaking force, shaking moment, and input power utilize the matrix method for the analysis of mechanisms [1] and the generalized Uicker and Sheth notation [2] of representing linkages in 3-D space, as a result of that the dynamic balancing approach should have the capability of dealing with multi-degrees of freedom, multi-loops, spatial or planar mechanisms, likely to be applicable to open chain mechanisms.

Calculation of Shaking Force and Moment

In formulating the problem of dynamic balancing, a four-bar linkage is used as an example for its simplicity. As shown in Fig. 1, firstly, in order to compute the shaking force and shaking moment produced by a mechanism, it is necessary to find the forces and moments transmitted to the surrounding. For each of the joints which are connected to the ground, the magnitudes of the transmitted forces or moments are equal to the bearing reactions in magnitude but opposite in direction. The ground bearing reactions in the local coordinates $(x_1y_1z_1, x_4y_4z_4)$ can be computed using the method of variation of constraints which is an extension of the principle of virtual work developed by Uicker [3].

Therefore, if one of the ground bearings, joint i is considered, the force F_{ti} or moment M_{ti} transmitted to the surrounding can be expressed as

$$F_{ti} = \{0 \;\; F_{\Delta xi} \;\; F_{\Delta yi} \;\; F_{\Delta zi}\}^t \qquad M_{ti} = \{0 \;\; F_{\Delta Li}, \;\; F_{\Delta Mi}, \;\; F_{\Delta Ni}\}^t \quad (1)$$

The components of F_{ti} or M_{ti} can be found from the following equations: [4]

$$F_{\Delta i} = -Tr \sum_{k=1}^{n} (W_{k\Delta} + D_\Delta U_{\Delta k}) \; [T_{0k}J_kT_{0k}] \; (\alpha_k + \omega_k\omega_k)^t$$

$$(2)$$

$$+ \sum_{k=1}^{n} m_k g^t (W_{k\Delta} + D_\Delta U_{\Delta k}) T_{0k}\bar{r}_k$$

where $\Delta = \Delta_x, \Delta_y, \Delta_z$ = linear variations along the x, y, z axis

$\Delta = \Delta_L, \Delta_M, \Delta_N$ = angular variations about the x, y, z axis

T_{0k} = Transformation matrix $W_{k\Delta}, D_\Delta U_{\Delta k}$ = kinematic derivatives

J_k = inertia matrix ω_k, α_k = velocity and acceleration matrix

m_k = mass of link k $\bar{r}_k$ = position of the C. G.

If these forces are expressed in the global or reference coordinates,

$$GF_{ti} = T_{0i}F_{ti} \quad \text{and} \quad GM_{ti} = T_{0i}M_{ti} \tag{3}$$

where T_{0i} is the transformation matrix.

Therefore the shaking force SHF can be expressed as

$$SHM = \sum_{i=1}^{a} GF_{ti} \qquad or \qquad SHF = \sum_{i=1}^{a} T_{0i}(0 \ F_{\Delta xi} \ F_{\Delta yi} \ F_{\Delta zi})^t \qquad (4)$$

and a = number of joints connected to the base.

The shaking moments transmitted through the joints with respect to a particular reference coordinates can be found by the summation of GM_{ti} and the moment effect of the shaking forces. Therefore the shaking moment SHM can be expressed as

$$SHM = \sum_{i=1}^{a} (GM_i + \bar{p}_i \ X \ GF_i) \qquad (5)$$

and $\bar{p}_i$ is the position vector between the joint connected to the ground and reference coordinates.

As can be recognized, the addition of counterweights will not affect the kinematic properties of the mechanism except the inertia matrices as long as the position versus time curves (or the work cycle) remains unchanged. Here, $\overset{*}{J}_k$ and $\overset{*}{m}_k$ denote the new inertia matrix, and the total mass of link k and the countermass respectively. $\bar{r}_k^*$ is the corresponding position vector of the new center of gravity. Since the original inertia matrix J_k and the inertia due to the additional mass ΔM_k at a position r_k^* are all measured in the local coordinates system, thus the new inertia matrix can be found by the following equations:

$$\overset{*}{J}_k = J_k + \Delta J_k \qquad and \qquad \overset{*}{J}_k = J_k + \Delta M_k \overset{*}{r}_k \overset{*}{r}_k^t \qquad (6)$$

The equation for the computation of the forces transmitted to the base by a counterweighted mechanism becomes

$$\overset{*}{F}_{\Delta i} = - Tr \sum_{k=1}^{n} \{(W_{k\Delta} + D_{\Delta}U_{\Delta k}) \ [T_{0k}(J_k + \Delta M_k \overset{*}{r}_k \overset{*}{r}_k^t)T_{0k}^t] \ (\alpha_k + \omega_k \omega_k)^t\}$$

$$+ \sum_{k=1}^{n} m_k^* g_k^t (W_{k\Delta} + D_{\Delta} U_{\Delta k}) T_{0k} \bar{r}_k^* \tag{7}$$

where

$$m_k^* = m_k + \Delta M_k \qquad \bar{r}_k^* = (m_k \bar{r}_k + \Delta M_k r_k^*) / m_k$$

Optimal Dynamic Balancing Formulation

The objective function F_{obj} to the optimal dynamic balancing problem can be expressed mathematically as

$$F_{obj} = \sqrt{\frac{1}{\tau} \int_0^\tau \left[W_1 \left| \frac{SHF}{F_o} \right| + W_2 \left| \frac{SHM}{M_o} \right| + W_3 \left| \frac{PWD}{P_o} \right| \right] d\tau} \tag{8}$$

F_o M_o and P_o are the values of the shaking force, shaking moment and input power respectively of a mechanism before the addition of counter-weights. τ is the time required to sweep through a period of the work cycle and PWD is the input power.

The constraints on the countermass ΔM_i and its position r_i^* can be set as

$$a_i \leq \Delta M_i \leq b_i \qquad c_i \leq r_i^* \leq d_i \tag{9}$$

The upper and lower constraints (a_i, b_i, c_i, d_i) of ΔM_i and r_i^* are numerical constants. The selection of numerical weighting factors W_1, W_2, and W_3 are generally dependent on the design purpose and may be chosen to be equal to each other.

The design spaces for mechanism parameter optimization are usually highly nonlinear and irregular. Heuristic search techniques are preferred in this application. Among the heuristic methods, a method called Complex [4] has been used in this application and shown reasonable results.

Result and Discussion

A four-bar linkage with the dimensions and parameters is shown in Fig. 1. Link 2 is chosen as the input crank and is set to rotate at a speed of 1000 rpm for 360 degrees. If all the links are allowed to be counterweighted simultaneously and all the weighting factors are set equal to one. A 40% reduction in peak shaking force and moment can be achieved after 1000 evaluations of the objective functions. In addition, there is a 20% decrease in peak input power after balancing. The optimum sizes and positions of the counterweights are listed in Table 1.

References

1) Uicker, J. J. Jr., J. Denavit, and R. S. Hartenberg, "An Iterative Method for the Displacement Analysis of Spatial Mechanisms," Journal of Applied Mechanics, Vol. 31, Trans. ASME, Vol. 86, Series E, 1964, pp 309-314

2) Sheth, P. M., and J. J. Uicker, Jr., "A Genealized Symbolic Notation for Mechanisms," Journal of Engineering for Industry, Trans. ASME, Series 13, Vol. 93, No. 1, Feb. 1971

3) Uicker, J. J. Jr., "Dynamic Force Analysis of Spatial Linkage," Journal of Applied Mechanics, Vol. 34, Trans. ASME, Series E, Vol. 89, No. 2, June, 1967

4) Wong, T. L. "Systems Design of Walking Robot," Ph.D. Dissertation, University of Wisconsin-Madison, 1986

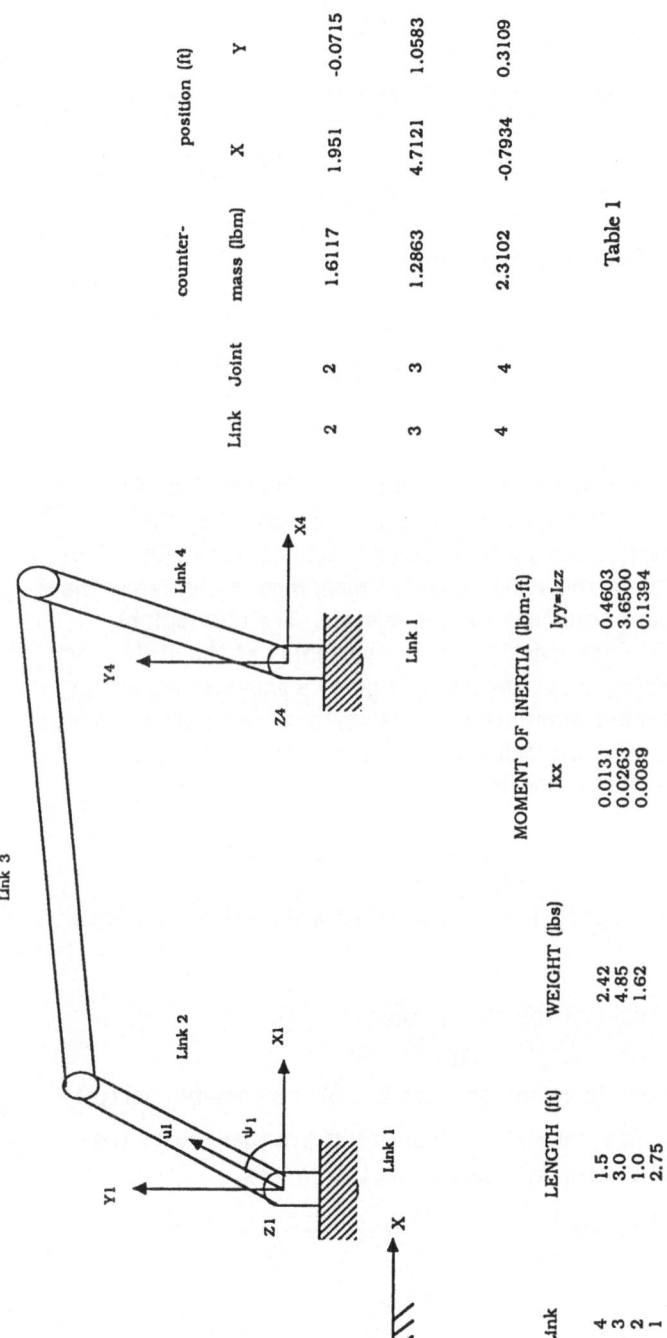

Figure 1

Link	Joint	counter-mass (lbm)	position (ft) X	position (ft) Y
2	2	1.6117	1.951	-0.0715
3	3	1.2863	4.7121	1.0583
4	4	2.3102	-0.7934	0.3109

Table 1

Link	LENGTH (ft)	WEIGHT (lbs)	MOMENT OF INERTIA (lbm-ft) Ixx	MOMENT OF INERTIA (lbm-ft) Iyy=Izz
4	1.5	2.42	0.0131	0.4603
3	3.0	4.85	0.0263	3.6500
2	1.0	1.62	0.0089	0.1394
1	2.75			

• Weight and inertia are measured at the center of mass which is assumed to be at the middle of a link.

On Spreadsheet Simulation of Nonlinear Dynamic Systems

T. G. Windeknecht
Computer Science And Engineering Department
Oakland University
Rochester, MI 48063

Abstract A discrete-time dynamic system is an interconnection of static elements and unit delays. If the system has no static loops then the variables in the static part of the system can be topologically sorted. This, together with the domains of the static element mappings, establishes the existence and uniqueness of solutions for the system. The topological sorting is also a computational ordering for the variables of the static part of the system. This ordering is not abstract; it gives a physical layout for the variables in a spreadsheet simulation of the system. Thus, spreadsheet programs may be used to simulate dynamic systems of practical interest including those with significant nonlinearities.

Introduction A *dynamic variable* is a real-valued sequence
$$v = \langle v(0), v(1), \ldots, v(t), \ldots \rangle$$
i.e., $v: \Omega \to \mathbb{R}$ where $\Omega = \{0, 1, \ldots, t, t+1, \ldots\}$ is the set of natural numbers and $\mathbb{R}$ is the set of real numbers.

A *static element* is a functional relation of the form
$$v(t) = f(v_1(t), v_2(t), \ldots, v_n(t)), \quad t \in \Omega$$
where $v, v_1, v_2, \ldots v_n$ are dynamic variables and $f: \mathbb{R}^k \to \mathbb{R}$ is a possibly partial real function. (Usually, f is total and continuous.) If $k < n$ then $n-k$ of the arguments $v_1, v_2, \ldots, v_n$ are *dummy* variables. In other words,
$$f(v_1(t), v_2(t), \ldots, v_n(t)) = f(v_{h(1)}(t), v_{h(2)}(t), \ldots, v_{h(k)}(t))$$
where $1 \leq h(i) < h(i+1) \leq n$ for $i = 1, 2, \ldots, k-1$.

A *unit delay* is the relation
$$v(t+1) = w(t), \quad t \in \Omega$$
connecting two dynamic variables v and w. A *dynamic system* is a relation over r dynamic variables v_i of the form:
$$v_i(t) = f_i(v_1(t), v_2(t), \ldots, v_r(t)) \quad (i = 1, 2, \ldots, q)$$
$$v_i(t+1) = v_{g(i)}(t) \quad (i = q+1, q+2, \ldots, q+n)$$

where $q+n \leq r$ and $1 \leq g(i) \leq r$. In other words, there are q static elements and n unit delays relating r dynamic variables. Here, the variables $v_1, v_2, ..., v_q$ are called *endogenous* variables, $v_{q+1}, v_{q+2}, ..., v_{q+n}$ are *state* variables, and $v_{n+q+1}, v_{n+q+2}, ..., v_r$ are *exogenous* (or *input*) variables.

A typical dynamic system is

$$u(t) = (x(t)+l(t))/2 + ((x(t)-l(t))/2)*SGN(s(t))$$
$$v(t) = x(t) + ((x(t)-l(t))/2)*SGN(s(t))$$
$$y(t) = f(x(t))$$
$$s(t) = f(x(t)+d)-f(x(t))$$
$$l(t+1) = u(t)$$
$$x(t+1) = v(t)$$

which models a well-known computational scheme for finding the maximum of a unimodal[1] function f. In this case, $n+q=r$ so the system is *autonomous*, i.e., it has no exogenous variables.

Existence And Uniqueness Of Solutions Suppose the static element functions f_i of the above general dynamic system are all total. Then, as is well-known [1-5], a sufficient condition for the existence and uniqueness of solutions for all variables is there exist no static loops of functional dependence in the static part of the system

$$v_i(t) = f_i(v_1(t), v_2(t), ..., v_r(t)) \quad (i=1,2,...,q)$$

In other words, there exists some ordering of the variables $v_1, v_2, ... v_q$ (say $\langle v_1, v_2, ... v_q \rangle$ itself) such that, by eliminating dummy variables, the static part of the system is triangular in functional dependence:

$$v_1(t) = f_1(v_{q+1}(t), v_{q+2}(t), ..., v_r(t))$$
$$v_2(t) = f_2(v_1(t), v_{q+1}(t), v_{q+2}(t), ..., v_r(t))$$
$$...\qquad ...$$
$$v_q(t) = f_q(v_1(t), v_2(t), ..., v_{q-1}(t), v_{q+1}(t), v_{q+2}(t), ..., v_r(t))$$

The ordering $\langle v_1, v_2, ... v_q \rangle$ is a *topological sorting* of the graph of functional dependence for the static part of the system [1-2]. In the case of the above example system, the graph of functional dependence is

---

[1] A possibly partial real function $f: \mathbb{R} \to \mathbb{R}$ is *unimodal* if it is defined and continuous over an interval [a,b], has a unique maximum at a point x* with a<x*<b, is increasing for $a \leq x < x*$, and decreasing for $x* < x \leq b$.

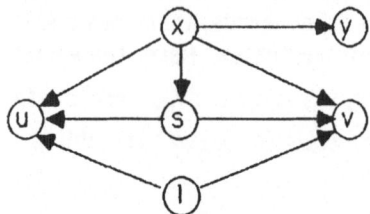

From the graph of functional dependence, assuming no cycles (circular paths in the direction of the arrows) exist, a topological sorting may be found by iteratively choosing a source (a variable with no indirected arrows) and then deleting it together with its outdirected arrows. In the example, this yields:

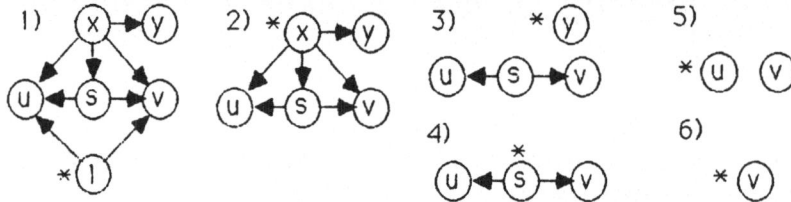

Overall, this gives the topological sorting $\langle 1,x,y,s,u,v \rangle$. Eliminating the state variables, we get the topological sorting $\langle y,s,u,v \rangle$ of the endogenous variables. The corresponding listing of the static part of the system is

$$y(t) = f(x(t))$$
$$s(t) = f(x(t)+d)-f(x(t))$$
$$u(t) = (x(t)+1(t))/2 + ((x(t)-1(t))/2)*SGN(s(t))$$
$$v(t) = x(t) + ((x(t)-1(t))/2)*SGN(s(t))$$

which is, of course, a triangularization. Clearly, both in the example and in general, the topological sorting $\langle v_1,v_2,...v_q \rangle$ also gives an *order in which to compute the variables of the static part of the system.*

Spreadsheet Simulation The spreadsheet layout consists of a sequence of horizontal blocks, corresponding to the successive time instants $0,1,...,t,t+1,...$ at which the system variables are defined. In each horizontal block are cells for storing the value at time t of each variable of the system. Each of those variables is either an exogenous variable, an endogenous variable, or a state variable.

In the tth horizontal block, the cells for storing the values $v_i(t)$ $(i=q+n+1,q+n+2,...,r)$ of the exogenous variables either contain numerical values or replicated formulas for successively generating them.

In the tth horizontal block, the cells for storing the values $v_i(t)$ $(i=1,2,...,q)$ of the endogenous variables contain replicated formulas describing the functions f_i of the static elements in terms of (a) the built-in functions of the spreadsheet and (b) the cell addresses (in the same horizontal block) of the values of the nondummy variables $v_j(t)$ on which f_i depends. The physical arrangement of these latter cells is determined by the order of calculation of the spreadsheet program:

1. If the order of calculation is row-by-row, then the cells for storing $v_i(t)$ $(i=1,2,...,q)$ are arranged in one or more rows in the order top-to-bottom, left-to-right of the topological sorting $\langle v_1, v_2, ..., v_q \rangle$. They are preceded by the cells for storing the values $v_i(t)$ $(i=q+1,q+2,...,r)$ of the exogenous and state variables. The relative arrangement of the latter cells is arbitrary.

2. If "natural" order of calculation is available (as in Lotus 1-2-3), then *any* arrangement of the cells for storing the exogenous, state, and endogenous variables may be used within the horizontal blocks. (The topological sorting is performed internally by the spreadsheet program.)

In the 0th horizontal block, the cells for storing the values $v_i(0)$ $(i=q+1,q+2,...,q+n)$ of the state variables contain given numbers (the initial system state). In the tth horizontal block $(0<t)$, the cells for storing $v_i(t)$ $(i=q+1,q+2,...,q+n)$ contain the replicated cell addresses of the values $v_{g(i)}(t-1)$, respectively, in the $(t-1)^{th}$ horizontal block.

Example For the above example system, the Lotus 1-2-3 spreadsheet layout for row-by-row calculation using the given topological sorting is:

```
        A       B      C       D           E                   F
 1  HILL-CLIMBING THE FUNCTION f(x)=1-x^2
 2      t      X(t)   x(t)    y(t)         s(t)                u(t)
 3      v(t)
 4
 5      0      -7     -2      1-C5^2       C5^2-(C5+.001)^2     @IF(E5>0,C5,B5)
 6      @IF(E5>0,C5+(C5-B5)/2,(C5+B5)/2)
 7
 8      A5+1   F5     A6      1-C8^2       C8^2-(C8+.001)^2     @IF(E8>0,C8,B8)
 9      @IF(E8>0,C8+(C8-B8)/2,(C8+B8)/2)
10
11      A8+1   F8     A9      1-C11^2      C11^2-(C11+.001)^2   @IF(E11>0,C11,B11)
12      @IF(E11>0,C11+(C11-B11)/2,(C11+B11)/2)
13
14      A11+1  F11    A12     1-C14^2      C14^2-(C14+.001)^2   @IF(E14>0,C14,B14)
15      @IF(E14>0,C14+(C14-B14)/2,(C14+B14)/2)
...
32      A29+1  F29    A30     1-C32^2      C32^2-(C32+.001)^2   @IF(E32>0,C32,B32)
33      @IF(E32>0,C32+(C32-B32)/2,(C32+B32)/2)
```

254

The given spreadsheet computes the special case of the system equations in which $f(x)=1-x^2$, d=0.001, I(0)=-7, x(0)=-2, and 0≤t≤9.

Using Wrap-Around One possible problem is the limited (though large) number of rows of a typical spreadsheet program. How do we proceed when the range of t is *very* large? Using a trick, it is possible to compute the system equations forward in time whenever recalculation (or iteration) is selected. For example, to employ wrap-around in the example spreadsheet, modify the spreadsheet as follows:

cell	old	new contents		
E3	empty	the label ˆINIT(1/0)=		
F3	empty	1 (then 0)		
A5	0	@IF(F3,0,A32+1)	*or*	(F3*0)+(1-F3)*(A32+1)
B5	-7	@IF(F3,-7,F32)	*or*	(F3*-7)+(1-F3)*F32
C5	-2	@IF(F3,-2,A33)	*or*	(F3*-2)+(1-F3)*A33

To start the (row-by-row) calculation, 1 is stored manually in cell F3 and then recalculation is selected once. This computes all variables for 0≤t≤9. Then, 0 is stored manually in cell F3 and recalculation is selected as many times as desired. After the first recalculation, the variables are displayed for 10≤t<19. After the second, we get the values for 20≤t<29, etc.

Bibliography

1. H. D'Angelo & T. G. Windeknecht, "On the Computation of State Variable Equations," Proc. Conf. Information Sciences and Systems, Johns Hopkins Univ., Baltimore, 1976.

2. H. D'Angelo, R. J. Faudree, & T. G. Windeknecht, "The Mathematics of Topological Sorting," Proc. Conf. Information Sciences and Systems, Johns Hopkins Univ., Baltimore, 1976.

3. T. G. Windeknecht & H. D'Angelo, "System Theoretic Implications of Numerical Methods Applied to the Solution of Ordinary Differential Equations," Trans. IEEE, Vol. SMC-7(11), 1977.

4. H. D'Angelo & T. G. Windeknecht, "A Simple Program for General System Simulation" *in* Modeling and Simulation, Vol. 6 (Proc. Sixth Annual Pittsburgh Conf.), W. G. Vogt and M. H. Mickle (eds.), Univ. Pittsburgh, 1975.

5. T. G. Windeknecht & H. D'Angelo, "A System Graph and Canonical State Equations," Proc. Sixth Annual Southeastern Symp. System Theory, Louisiana State Univ., Baton Rouge, 1974.

Normal Modes Analysis of Structures Using an Out-of-Core Component Mode Synthesis Technique

Walter L. Wolf
Cray Research, Inc.
Dearborn, MI 48126

Alan E. Duncan
Automotive Analytics, Inc.
Troy, MI 48084

Arthur R. Solomon
Analytical Engineering & Research, Inc.
Washington, MI 48094

ABSTRACT

A method is developed for the normal modes analysis of large structures using an out-of-core component mode synthesis technique. With the out-of-core technique, no more than one row or column of any matrix is required to be in memory at any one time.

In this approach, the structure is partitioned into smaller components and the vibration modes for each component are determined. The requirement for displacement compatibility at the connections generates a set of constraint equations among the component generalized coordinates. These constraint equations are put in a modified form using Gaussian elimination to identify an independent set of component generalized coordinates and to allow reduction of the system mass and stiffness matrices using the out-of-core technique. The result is a new set of dynamic equations for the connected structure which has fewer degrees of freedom than would exist if the structure had been analyzed in its entirety. These equations form an eigenvalue problem for the connected structure and can be solved for the system natural frequencies and mode shapes.

An example using the out-of-core technique is presented and comparison made with analysis results from the complete structure.

NOMENCLATURE

$[G]$ = constraint equation coefficient matrix

$[M],[K]$ = system mass and stiffness matrices

$[\hat{M}],[\hat{K}]$ = reduced system mass and stiffness matrices

$[m],[k]$ = component mass and stiffness matrices

$[\tilde{m}],[\tilde{k}]$ = component generalized mass and stiffness matrices

$[\phi]$ = component modal matrix

$\{q\},\{\ddot{q}\}$ = generalized coord. displ. and accel. vectors

$\{u\},\{\ddot{u}\}$ = component displ. and accel. vectors

Subscripts:

   f   = free dofs after constraint equation elimination

  i,j   = i-th and j-th components respectively

   m   = no. of constraint equations

   n   = total number of generalized coords. for all components

INTRODUCTION

As micro-computers increase in computational efficiency it has become economically desirable to apply micro-computers to larger and more complex C.A.E. computations. One such application is the determination of axisymmetric and non-axisymmetric modes of brake rotors. The technique described herein was developed to synthesize the modal properties of large structures from the modal properties of their components. This technique was designed to execute effectively in a micro-computer environment.

THEORY

The approach for the analysis of a structure assembled from a number of modal components follows the development of Hurty [1] with some modification. In Hurty's approach, the equations of motion for the free vibration of a structure assembled from a number of modal components is:

$$[M]\{\ddot{q}\} + [K]\{q\} = \{0\} \tag{1}$$

A similar equation can be written for each of the components. For the j-th component:

$$[m]_j\{\ddot{u}\}_j + [k]_j\{u\}_j = \{0\} \tag{2}$$

In terms of the component generalized coordinates, Eq. (2) becomes:

$$[\tilde{m}]_j\{\ddot{q}\}_j + [\tilde{k}]_j\{q\}_j = \{0\}$$

where:
$$\{u\}_j = [\phi]_j\{q\}_j$$

Displacement compatibility is required at the component connection or interface points. For two components, i and j,:

$$\{u\}_i = \{u\}_j$$

In terms of the component generalized coordinates:

$$[\phi]_i\{q\}_i = [\phi]_j\{q\}_j \tag{3}$$

Rearrangement of Eq. (3) gives:

$$[\phi_i | -\phi_j] \{ \frac{q_i}{q_j} \} = \{0\}$$

This last equation is a constraint equation among the components' generalized coordinates when assembled for all components in the structure and is of the form:

$$[G]\{q\} = \{0\} \tag{4}$$

The matrix [G] is of size (mxn) with (n>m). These constraint equations are modified by using Gaussian elimination with a maximum pivot strategy [2] and the [G] matrix has the form shown in Fig. 1.

The modified mass, [$\hat{M}$], and stiffness, [$\hat{K}$], matrices are formed from the system mass and stiffness matrices of Eq. (1) using the reduction method of Curiskis and Valliappan [3]. The reduction process eliminates one row and column (the last row and last column) for each constraint equation processed through a series of row and column manipulations when the constraint equations are taken one at a time. [$\hat{M}$] and [$\hat{K}$] are coupled matrices and the equations of motion form a new eigenvalue problem which can be solved for the system frequencies and mode shapes:

$$[\hat{M}]\{\ddot{q}\}_f + [\hat{K}]\{q\}_f = \{0\}$$

The system mode shapes are then recovered from the constraint equations given by Eq. (4) and:

$$\{u\} = [\phi]\{q\}$$

RESULTS OF NUMERICAL TEST CASE

A simple cantilever beam example shown in Fig. 2 was used to test the algorithms that were developed. A finite element normal modes analysis was performed on the full structure using a PC based finite element program for comparison. For the full model the bending modes computed in the XZ plane are lower in frequency than those in the YZ plane due to the mesh difference in the Y direction. The average first and second bending mode frequencies found from the full model are underestimated compared to beam theory by 2.3% and 18.% respectively.

Next the full beam was split along the XZ plane bounded by grids 2, 3, 23, and 22, resulting in two identical 5 inch beams (Parts 1 and 2). The normal modes were only required for Part 1 since both parts were identical. A total of 60 modes were computed for Part 1 since there were three degrees of freedom per grid. Finally the synthesis technique was applied to assemble Parts 1 and 2 and the results were compared to the full model.

When all modes from Parts 1 and 2 were used in the synthesis it was found that the frequencies were identical to the finite element solution. The results of several additional synthesis studies are shown in Table 1. Because the beam was split in the XZ plane the XZ modes found with the synthesis were unaffected by the number of modes used. However, the bending in the YZ plane achieves higher accuracy as more modes are used. It appears that 30 modes (50%) from Parts 1 and 2 are required for an accurate first YZ bending mode while 40 modes are required for the second YZ bending mode.

CONCLUSION

An out-of-core component mode synthesis technique has been developed for the normal modes analysis of large structures and an example was presented to show the application of the technique. The out-of-core technique permits the analysis of large structures on computers, such as personal computers, which may be otherwise unsuited for this type of analysis due to main memory limitations. Results from the example show that accurate results can be obtained with this technique even when a truncated set of component modes are used.

$$[G] = \begin{bmatrix} g_{1,1} & g_{1,2} & \cdots & g_{1,n-m-1} & -1 & 0 & \cdots & 0 & 0 \\ g_{2,1} & g_{2,2} & \cdots & & g_{2,n-m} & -1 & 0 & \cdots & 0 & 0 \\ \cdot & & & & & & & & \\ \cdot & & & & & & & 0 & 0 \\ \cdot & & & & & & & -1 & 0 \\ g_{m,1} & g_{m,2} & \cdots & & & & & g_{m,n-1} & -1 \end{bmatrix}$$

Fig. 1. Modified form of the constraint equation matrix, [G].

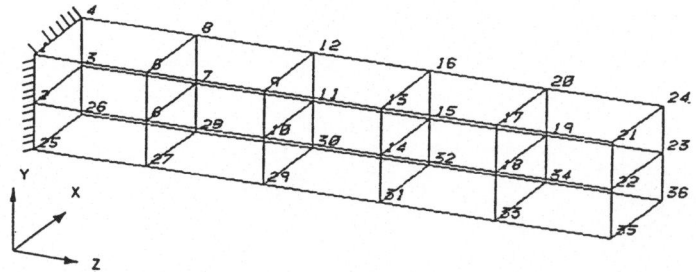

Fig. 2.   Cantilever beam model - 1.0 x 1.0 in. square steel bar
            5.0 in. long.

TABLE 1
MODESHAPE FREQUENCY COMPARISON FOR SEVERAL STUDIES
CANTILEVER BEAM MODEL

| | MODESHAPE AND FREQUENCY | | | | |
	1st XZ Bending	1st YZ Bending	Twist	2nd XZ Bending	2nd YZ Bending
Baseline synthesis using all modes	1260.	1290.	4170.	6480.	6890.
Study #1 40 modes each part	1260.	1300.	4170	6480.	7280.
% Deviation	0.0%	0.8%	0.0%	0.0%	5.7%
Study #2 30 modes each part	1260.	1310.	4730.	6480.	7730.
% Deviation	0.0%	1.6%	13.4%	0.0%	12.2%
Study #3 20 modes each part	1260.	—	—	6480.	—
% Deviation	0.0%			0.0%	

REFERENCES

1.   Hurty, W. C., "Dynamic Analysis of Structural Systems Using
     Component Modes," _AIAA Journal_, Vol. 3, No. 4, April, 1965,
     pp. 678-685.

2.   Carnahan, B., Luther, H. A., and Wilkes, J. O., _Applied
     Numerical Methods_, John Wiley & Sons, Inc., New York, New
     York, 1969.

3.   Curiskis, J. I. and Valliappan, S., "A Solution Algorithm
     for Linear Constraint Equations in Finite Element
     Analysis," _Computers & Structures_, Vol. 8, Pergamon Press,
     1978, pp. 117-124.

Chapter III:
Manufacturing and Production

Introduction

The papers in Chapter III have a substantial interest in the manufacturing and production processes. Many of these papers illustrate the integration of design and manufacture. In Section III.1 we have the papers dealing with manufacturing in a more general way followed by those on flexible manufacturing in Section III.2. Automation and robotics is given next in Section III.3 and the chapter concludes with Section III.4 and III.5 on production processes and process planning respectively.

The improvement of the manufacturing and production processes is essential for any nation to compete in world markets. A substantial part of that improvement will depend on the degree of success in the implementation of computer integrated manufacturing (CIM). The critical attention that current research is giving to the production and manufacturing processes, as exemplified by the papers in this chapter, illustrates the importance of this endeavor.

General Manufacturing

CIMI – Whole New Ballgame

Patrick J. Fallon, P.E. CMfg(E)

Applied Life Extension Resources and Technologies
17255 Ridge Road
Northville, Michigan 48167

Background: CIM or Computer Integrated Manufacturing is a hot topic in today's manufacturing and management circles. The driving force behind these discussions and debates is the declining market share and the degrading balance of payments that U.S. companies/U.S. industry are currently experiencing. CIM is being explored as a technological means of enhancing the U.S. industrial competitive position relative to its worldwide competition. This is currently being viewed as a means of increasing market share and reducing the balance of payments. Vendors of advanced manufacturing and automation equipment, computer vendors and consultants are attacking this need with a vengence not seen since the early days of nuclear power. This paper will attempt to outline CIM implementation from an end user perspective, that of the consumer. The method for this will be an analogy of the world market to the great American game of baseball.

The first thing to develop in this analogy is the scene. The below sets our scene and relationships.

The Scene

Ballpark is the U.S. and World Market

Fans are the consumers

Teams are Japan Inc. and U.S. Inc.

Hitting is manufacturing related items.

i)	Cost of finished goods and semi–finished goods
ii)	Quality of product
iii)	Flexibility in manufacturing
iv)	Inventory and control of inventory
v)	Engineering design

Pitching and fielding are marketing related ideas.

i)	Market survey ability
ii)	Public image
iii)	Advertising and Marketing
iv)	Service
v)	Sales
vi)	Product differentiation

Score is based on balance of Trade and Market Share.

The most important aspect of any game is understanding the rules. For our game the rules of free market trade apply. In the analogy restrictions could be viewed as pre-free agent days where restrictions on mobility of players caused the club with the most money to obtain the preponderance of available talent.

The current status of the game is also important so that the perspective of history is included. The principle areas of "hitting" and "pitching" are compared below:

Status:
Japan is clearly ahead in terms of balance of trade and gaining in market share.

Strengths and Weaknesses:
Category: Hitting (Manufacturing Related)
Japan Inc. is ahead in cost of finished goods, cost of semi-finished goods, flexibility in manufacturing, inventory and control of inventory. U.S. Inc. is ahead in engineering design.

Category: Pitching (Market Related)
U.S. Inc. is ahead in market survey ability, advertising/marketing, and sales. Japan Inc. is ahead in public image, service and product differentiation.

The second thing to develop in this analogy is the respective team strategies. This will help to show what is important in the perspective of each team and help to explain the current status.

Japan Inc. – The fundamental policy (strategy) of this team in the design area is to design products at all levels of complexity with simple logical designs. The risks of employing new or developing new technology do not support the philosophy of cost and design effectiveness that appears to be crucial to the Japan Inc. strategy. In baseball this is similar to drawing walks or making contact singles. The results of this strategy are incremental vice quantum.

In manufacturing the process function is the central focus. This functionality is focused toward the goal of optimal timing/cost. The baseball analogy to this manufacturing strategy is the hit and run offense. In summary, the strategies of

Japan Inc. In the "hitting" area center around logical design and manufacturing flow to take advantage of limited resources.

The "pitching" or market related area appears to be an area where Japan Inc. is not so strong. The success of Japanese products is due mainly to good timing. Japan Inc.'s image of quality and value have been built up over years. The major driving force in Japan Inc.'s marketing plan appears to be market penetration using accepted advertising techniques. Their products' superior image for quality and value coupled with the range of products they offer give Japan Inc. a market or "pitching" capability that keeps them in the "game".

U.S. Inc. - The policy (strategy) of this team in the design area is to achieve new products from new technologies. The corporate world does not supply the direction needed to focus available research funding into areas where technology could be easily developed into viable products. A second strategy focuses on solving problems with the product or expanding the product's capability through design measures. In baseball terms, U.S. Inc. is going for the long ball (home run/double) in the design area.

The policy (strategy) in the manufacturing area is twofold. First, the driving corporate desires of short term profitability and high return on investment constrain the manufacturing process. Cost, efficiency and schedule are the most controllable items in the manufacturing process and their control has a direct impact on profits. The second policy revolves around implementation of advanced manufacturing techniques. These methods will allow better use of the manufacturing process to meet management goals. These two strategies can be equated to a "squeeze play" or "double steal" for the first policy and to the use of advanced composite bats trying to hit home runs for the second policy.

U.S. Inc. team's "pitching" strategy for advertising utilizes psycological techniques that have been developed through years of study. Market survey techniques are quite well developed and prediction of consumer response is nearly a science.

The status of the game and the teams' respective strategies show the past, present and expected future trade picture between teams U.S. Inc. and Japan Inc. The teams' respective strategies and relative strengths/weaknesses explain the true nature of the competition. The nature of the competition also reveals the ways in which U.S. Inc. can improve its performance in the game.

The New Ballgame:

The normal way that baseball teams improve is by acquisition of better players or by changes in the team's fundamental strategies (typically inspired by a new manager). The new ballgame referred to in this paper will be a result of one of the aforementioned improvements in team capability or strategy.

Capability Improvements –

"Pitching"   The first area of improvement is in the area of "pitching".   Market assessment and evaluation needs radical improvement if U.S. Inc. is to turn around the flood of imports. A chief advantage of the U.S. Inc. is its market survey ability. We as producers understand the mood and buying habits of our consumers. We can use our consumer attitude to predict markets. Our market survey techniques need to be directed in order to obtain the optimum results with respect to market share. Our market analysis capability needs to view the market as a fluid environment. We can employ test and analysis of market response in order to properly direct our marketing/advertising efforts. This analysis in baseball terms is reading the batter. Once again proper direction of our capability to analyze the market will allow full utilization of our abilities.

A principle strength that is also available in the Team U.S. "bullpen" is the ability to "create" a market.   We need to more adequately evaluate our timing and the feasibility of "creating" markets.   This added capability coupled with proper tools/flexibility can help to provide again our ability to produce for the market.

Proper direction in the above three areas, market survey, market analysis and market creation, is absolutely necessary if we are to ever see large gains in market share. This direction can be likened to a pitching coach providing rundowns on opposing batters to help the pitcher plot his game strategy.

"Hitting"   The second aim of improvement is in the area of "hitting" or advanced manufacturing.   The functional areas of manufacturing are product design, process design, logistics, quality and inspection, rework, assembly, formation, monitoring and control of product flow.   The following matrix shows these areas and the area of advanced manufacturing that can be used to improve performance.

Area	Advanced Process
Product Design	MAP, LAN, MCAE, Flexis, CAD/CAM, CIM, Simultaneous Engineering, Robust Engineering, Design Process Analysis
Process Design	Robust Engineering, CIM, Programmable Controllers, Flexis, Bar Code Scanning
Logistics	Just-in-Time, CIM, MRP, CAM, Bar Code Scanning, Supplier Certification, SPC on Suppliers, Robotics
Quality Control and Inspection	Vision Systems, SPC, Total Quality Control, Robotics, CIM, Flexis, Ultrasonics
Rework	Total Quality Control, Robotics, Vision systems, CIM
Assembly	Hard Automation, Robotics, Vision Systems, Bar Code Scanning, SPC, Flexis, Total Quality Control, CIM, CAD/CAM, Just-in-Time, Robot Welding, Laser Welding, Water Jet Cutting
Formation	Robotics, Vision Systems, Total Quality Control, CIM, CAD/CAM, Thermography, Flexis, CNC/DNC
Monitoring	Vision Systems, Bar Code Scanning, CIM, Thermography
Control of Product Flow	Just-in-Time, CIM, CAM, Bar Code Scanning, Total Quality Control

The above matrix shows potential applications of advanced techniques, or in baseball terms more effective hitters. A good hitter that is used improperly or that is not fully integrated into the team is effectively useless. The use and integration aspect will be covered later under the "coaching" section.

Management –

"Coaching/Managing" This is a critical area for a baseball team. Management can be the difference between a pennant winner and an also ran. This fact has been proven time and time again. Management in the manufacturing area is just as critical. A producer needs to have proper teamwork and direction in all aspects of the product cycle in order to be effective. Management has many areas of responsibilities, the most critical of which is the long term viability of the company. The long term viability of producers in today's environment relies on management's commitment to properly use available resources. Short term items that enhance management's image with the stockholders are improper use of available resources for long term viability as a competitive world class producer. Overseas competition is well sponsored and in business for the long term. The improvement available in the management area consists of two principle areas: market oriented production and marketing, and maximum flexibility to enable timely response to the market.

How can management accomplish these areas? The following guidelines could help lay the groundwork for the programs.

1. **Involvement** – Involve employees/managers at all levels in helping to find/gauge the "score" of the company with respect to consumers/market.

2. **Understanding** – Understanding of the market and the product cycle is a must to determine status and direction. Management and workers must continually strive to understand the image and market requirements of their product. The manufacturing cycle must also be understood to capitalize on the knowledge gleaned from product image and market requirements for the product.

3. **Rewards** – The organization must reward progress toward the goal of market orientation and manufacturing flexibility.

4. **Organization** – Management must make the organization support this new philosophy. Middle level management rewards systems must use compliance to the plan as a basis. All impediments to free flow of ideas must be removed.

Conclusions: The "whole new ballgame" concept hinges on management. Embracing the ideals of market orientation and flexible response put forth in this paper could provide the required means to obtain large gains in market share. Of particular importance is product/producer image. Management needs to provide assurance to the consumer that his/her interests are foremost in the minds of the producer. This entails response to and proper identification of consumer requirements on new products and service in the aftermarket. When any new manager comes to a baseball team the first thing looked at is the "fundamentals". Manufacturing must also return to fundamentals – customer satisfaction.

Multicriteria Evaluation of Manufacturing Systems

TOM M. WEST and SABAH U. RANDHAWA

Department of Industrial and Manufacturing Engineering
Oregon State University
Corvallis, Oregon 97331

Summary

A major problem in the justification of advanced manufacturing systems is the prerequisite evaluation process. Replacing current technology with new technology often cannot be justified by traditional economic-based procedures alone. Implementation and acquisition decisions should take into account both tangible and intangible factors. A framework for assessing the acquisition decision through performance evaluation of competing designs is presented below.

Introduction

Today's manufacturing systems are being challenged by requirements which may change many aspects of the manufacturing scene. The trend is to incorporate many individual concepts and technologies into a single production system. This integrated system may include automatic storage and retrieval, automatic material handling, robots, numeric control machine tools, group technology, and hierarchical computer control.

Certain characteristics of these advanced manufacturing technologies make their justification process more complex than production equipment has required in the past. Newer technologies require full integration of all manufacturing functions, and component interfaces through extensive information networks. What is important in the justification process is a combination of factors rather than single elements. Furthermore, the exact relationship between these factors will vary over time, and as market conditions change. The integration process may demands major changes in management strategies and organizational structure.

Thus the problem of evaluating manufacturing systems is a
multi-attribute decision problem. The attributes or factors
that may effect the evaluation process include tangible or
quantitative factors such as manufacturing costs, work-in-
process inventories, equipment utilization, set-up times and
maintenance requirements, and intangible or qualitative factors
such as compatibility, flexibility, vendor integrity and the
ability of the management to understand the technology, or at
least its broad implications.

Purely economic-based procedures (such as payback, return on
investment and net present value) are inadequate for these
higher level systems. This is not to say that justification
studies should not be based on sound principles of discounted
cash flow analysis, but that additional implications, both
positive and negative, must be presented in a comprehensive and
consistent manner.

Evaluation Process

The first step in the evaluation process is to identify a set
of attributes applicable to the manufacturing system being
evaluated. These attributes may be classified into three
categories.

1. Production attributes measuring the operational performance
of the system. Example of production attributes are equipment
accuracy, tooling compatibility, dimensions control, and control
software.

2. Economic attributes reflecting the economic performance of
the system. The two primary cash flows in cost analysis are
acquisition costs and net operating profits. Net operating
profits include the dollar values resulting from material
savings, set-up time reductions, programming efficiency, inventory reduction and additional capacity hours available due to
improved efficiency of the newer technology less the direct
operating costs incurred.

3. Intangible attributes representing factors that are difficult to quantify. Two of the more important intangible attributes are (a) process flexibility, or adaptability of the manu-

facturing technology to changes in product mix and demand, and
(b) vendor's integrity reflected through his past performance,
technical ability, service reputation and service response
time.

Evaluation of alternatives follows the identification of attributes. One model that may be used for the evaluation process
is the linear additive model. This use of this model is summarized below.

Production model. Evaluate the alternatives (technologies competing to replace the current system) on production attributes.

1. Develop attribute weights reflecting the relative importance of the attributes in the decision environment. Pairwise
comparison (illustrated in Table 1) can be used for developing
attribute weights. The number of attributes should be adequate
to fully describe all system characteristics, but carefully
selected to minimize overlap. A typical descriptive array
could contain twenty or more attributes. The comparison values
in Table 1 show the importance of row attributes relative to
column attributes. Thus accuracy is twice as important as
speed. Also, the values below the principal diagonal are
complement of the values above the diagonal (0 is the complement of 2 and 1 is the complement of itself). This simplifies
the assessment procedure as only half the matrix values need to
be filled.

Table 1. Attribute weights through pairwise comparison
          (Importance scale 0 to 2; 0=less important, 1=same
          importance, 2=more important)

	Accuracy	Speed	Compatibility	Sum	Weight
Accuracy	0	2	1	3	0.500
Speed	0	0	1	1	0.167
Compatibility	1	1	0	2	0.333

2. Score the alternatives on each attribute. These scores
represent the performance of the alternative systems on how
well they satisfy the forecasted production requirements and
product specifications. A 0-10 quality-point scale can be used
to assess these scores (Table 2).

3. Combine the attribute scores and weights for each alternative into an aggregate "utility" value using the linear additive model. The additive model is given by

$$U_j = \sum_{i=1}^{n} w_i s_{ij}$$

where U_j is the aggregate value for the j-th alternative, w_i is the weight for the i-th attribute, s_{ij} is the score of the j-th alternative on the i-th attribute, and n represents the number of attributes (Table 2).

Table 2. Computing aggregate value scores
(Scores of alternatives on attributes based on 0 to 10 scale; 0=unsatisfactory performance, 10=superb performance)

Attribute	Attribute Weight	Alternative		
		A	B	C
Accuracy	0.500	6	5	9
Speed	0.167	8	6	9
Compatibility	0.333	9	7	3
Aggregate score on production attributes		7.33	5.83	7.00

Preliminary screening of alternatives. Alternatives that score poorly on production attributes can be removed from further consideration. In addition, alternatives that fail to satisfy certain critical criteria such as vendor integrity can also be eliminated from further analysis.

Cost model. At this stage cost estimates need to be developed for alternatives that have passed the initial screening. Life cycle or asset life costing may be used, but in most incidences corporate policy determines the study period. The cost analysis may use an evaluation criterion such as after-tax rate of return or after-tax net present value.

Final evaluation. The final evaluation requires integrating the production model, the cost model and intangible factors. The relative importance of production parameters, cost implications and intangible factors can be established using pairwise comparisons. The overall score for alternatives can then be obtained using the linear additive model. The approach results in a single point in the attribute space that describes the overall merit of the alternative.

Sensitivity Analysis

Sensitivity analysis consists of varying some of the parameter values that went into the initial analysis over some range of interest, and observing the effect on the final ranking of alternatives. The most important sensitivity to consider is sensitivity to changes in attribute weights. This is because attributes are the essence of value judgments, and because weights, being purely subjective numbers about which decision makers disagree, are more likely to be in dispute than other parameters. If the rankings remain unaffected as the weights are changed, this can be established as evidence that small errors in the estimation of attribute weights are not important. If one or more of the weights prove to be very sensitive, the analysis will indicate the accuracy of the estimating weights and the stability in the ranking of alternatives.

Conclusions

The model presented in this paper deals with performance evaluation and selection of manufacturing systems. The methodology is useful for a number of reasons. First, the approach is based on combining different facets of the manufacturing environment into a decision making framework. Second, the procedure allows the incorporation of various levels of expertise. For example, engineering experts and financial analysts can contribute in their respective areas of expertise, thus ensuring that all important concerns are properly addressed in the evaluation process. Third, questions of particular interest to management such as sensitivity issues and relative per-formance of the systems can be answered effectively, and with almost no additional effort. Finally, the proposed methodology is easy to understand, and provides a high degree of control to the decision maker for evaluation of manufacturing systems.

References

1. Abrel, A. and Seidmann, A.: Performance evaluation of flexible manufacturing systems. IEEE Transactions on Systems, Man and Cybernetics SMC-14 (1984) 606-617.

2. Meredith, J.R. and Suresh, N.C.: Justification techniques for advanced manufacturing systems. Int. J. Prod. Res. 24 (1986) 1043-1057.

A Justification Method for Advanced Manufacturing Systems

H. R. PARSAEI
W. KARWOWSKI
M. R. WILHELM

Speed Scientific School
Department of Industrial Engineering
University of Louisville
Louisville, KY 40292

Abstract

This paper presents a systematic approach (model) for evaluation and
selection of Advanced Manufacturing Systems. The suggested model is a top
down approach which measures the non-monetary benefits resulting from
implementation of Advanced Manufacturing Systems.

Introduction

The current trends in national and international markets are forcing
the manufacturers to carefully consider the benefits resulting from
implementing advanced manufacturing systems [1, 2]. The technological and
economic justification of the "high tech" hardware necessary to implement
modern manufacturing philosophies has been the subject of voluminous bodies
of literature. Some of the common barriers to widespread adoption of
automation may be attributed to the high cost of capital, the risks
associated with advanced technological decisions, technical uncertainties,
inadequacies in cost accounting methods, and difficulties in quantifying
the qualitative outcome which results from the new manufacturing technology
systems [4, 7].

Today, factory automation is considered by many executives as the
ultimate solution for the current dilemma of competition experienced by
U.S. manufacturers in the world market. However, it is believed that
automation should be implemented one piece at a time, making sure that
detailed planning and justification is complete before funding. Then,
after one piece of an automated system has been successfully implemented,
the next piece is selected and the process repeated.

Organizations often structure two types of plans. These are
commonlyreferred to as long-term (or strategic) and short-term (or
tactical) plans [6, 8]. Strategic plans configure the ultimate form of the
firm in future markets based on the external elements surrounding them,
such as consumer taste, and enviromental changes. Short term plans (or
tactical decisions) are referred to those which assist the firm in

accomplishing its short term goals. The short term goals may be interpreted as the systematic steps taken by the firm to achieve the long term strategic objectives [2, 4, 6, 8].

Automation decisions have some features which make the economic justification process using traditional techniques hard, if not impossible. These features may include, higher productivity, better quality, and higher flexibility, etc. The justification methods currently used by the corporate decision makers, such as net present value analysis, internal rate of return, payback period, benefit/cost ratio are not suitable for justification of flexible manufacturing systems since intangible benefits can not be easily formulated and incorporated into the cash flow definition [2, 5, 6, 7, 8].

According to a 1984 survey done by the National Electrical Manufacturers Association, 91 percent of the replying businesses use traditional financial justification methods to analyze the desirability of automated manufacturing systems [2]. These traditional justification methods when used along with a high Minimum Attractive Rate of Return (MARR) to reflect today's uncertain investment environment will normally result in the rejection of strategically vital automation proposals [6].

A Proposed Methodology

A methodology is proposed which enables the decision maker to measure and to evaluate the desirability of the firm's long-term and short-term investment plans. This methodology is implemented in two phases. Phase I examines the effectiveness of the firm's various strategic (long-term) factory automation proposals. In Phase II, the methodology evaluates each tactical alternative available for implementing the strategic proposal selected in Phase I. Figure 1 is an illustration of the two phases of the proposed method.

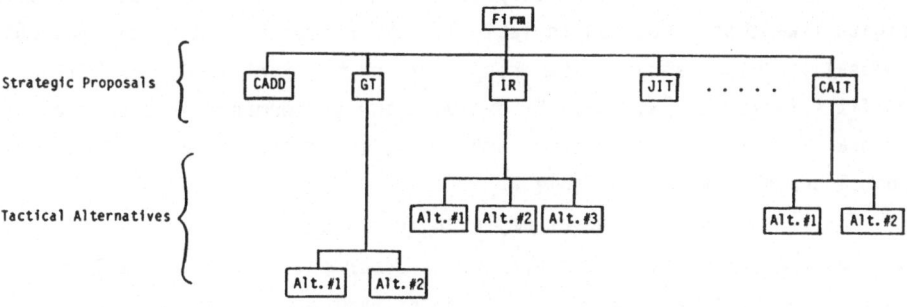

Figure 1. A graphical representation of the proposed methodology.

To implement the methodology, two sets of ordinal scale weights must be developed by the decision maker. These are then employed in a simple scoring technique known as a linear additive model [2, 4, 6, 10]. Examples of the ordinal scales which may be used in implementing this methodology are represented in Tables 1 and 2.

Table 1. Example Ordinal Scale weights for ranking the attributes of the available strategic proposals or tactical alternatives.

Very Important	1.00
Important	.75
Necessary	.50
Unimportant	.25

Table 2. The Ordinal Scale used to evaluate the performance of each strategic proposal or tactical alternative with respect to each attribute.

Superior	1.00
Good	.80
Average	.60
Below Average	.40
Poor	.20

Phase I

A hierarchy of the firm's strategic investment proposals is structured in Phase I. The benefits (attributes) resulting from these strategic proposals are ranked based on corporate long term objectives, using ordinal weights like those provided in Table 1. The performance of each strategic proposal is then evaluated with respect to each prioritized, resulting benefit (attribute) [8, 10]. To evaluate the performance of each strategic proposal with respect to a given attribute, the methodology uses the ordinal scale provided in Table 2.

The linear additive model employed in Phase I of this methodology compiles the information from distinct independent criteria in a linear fashion and facilitates overall scoring of each proposed strategic investment decision.

The linear additive model employed in this study can be expressed mathematically by the relationship:

$$\text{Max}_j \ N_j = \sum_{i=1}^{m} P_i \ X_{ij}, \quad \text{for } j = 1, 2, \ldots n$$

where:

N_j = The total score earned by the j-th strategic proposal in satisfying the corporate goals for factory automation.

P_i = A rank (ordinal weight from Table 1) assigned to the i-th strategic decision attribute, which expresses the relative importance of the corporate factory automation goal achieved by implementing each proposal, assuming $1 \leq i \leq m$.

X_{ij} = The rating (ordinal value from Table 2) assigned to the i-th strategic decision attribute, which reflects the anticipated performance of the j-th strategic proposal with respect to strategic decision attribute i.

m = The number of decision attributes common to all strategic proposals.

n = The number of strategic factory automation proposals under consideration by the corporate decision maker.

Phase II

In Phase II of the methodology the desirability of the available alternatives for the recommended proposal in the Phase I is evaluated. For example if the recommended proposal in the Phase I is to install a computer aided inspection system, in Phase II, the available systems/manufacturers are evaluated by the decision maker. The components of the linear additive model used in Phase II may be described by:

$$\text{Max}_j \ V_j = \sum_{i=1}^{m} W_i \ X_{ij}, \quad \text{for} \quad j = 1, 2, \ldots q$$

where:

V_j = The total score representing the ability of the j-th alternative to achieve the goals of the strategic proposal.

W_i = An ordinal scale weight assigned to the i-th characteristic (attribute) to reflect its importance in the strategic proposal, assuming $1 < i < p$.

X_{ij} = The ordinal scale weight assigned to the i-th characteristic, which reflects the anticipated performance of tactical alterative j with respect to the i-th characteristic (attribute).

p = The number of characteristics (attributes) common to all alternatives under consideration.

q = The number of alternatives under consideration.

An illustrative example of the two phases of this proposed methodology is demonstrated by Tables 1 and 2.

Table 3. Phase I Selection of the best strategic automation Proposal.

i	P_i	Attribute	Proposal 1 X_{i1}	P_iX_{i1}	Proposal 2 X_{i2}	P_iX_{i2}	Proposal 3 X_{i3}	P_iX_{i3}
1	.75	Reduced leadtimes	.60	.15	.80	.60	.80	.60
2	1.00	Improved Quality	.80	.80	.80	.80	.80	.80
3	1.00	Cost Reduction	1.00	1.00	.80	.80	.80	.80
4	.75	Product Flexibility	.80	.60	1.00	.75	.80	.60
5	.75	Schedule Flexibility	.60	.45	.80	.60	1.00	.75
6	.75	Safety improvement	.60	.45	.60	.45	1.00	.75
		N_J		3.75		4.00		4.30
		Normalized N_j		.87		.93		1.00

Table 4. Phase II Selection of the best tactical Alternative.

i	W_i	Attribute	Alternative 1 X_i	W_iX_{i1}	Alternative 2 X_{i2}	W_iX_{i2}	Alternative 3 X_{i3}	W_iX_{i3}
1	.75	Initial Cost	.80	.60	.60	.45	.40	.30
2	.75	Software Flexibility	.80	.60	.80	.60	.60	.45
3	.50	Required Floor Space	.60	.30	1.00.	.50	1.00	.50
4	1.00	Compatibility	.80	.80	.60	.70	.80	.80
5	1.00	Future Expansion	.80	.80	1.00	1.00	.80	.80
		V_j		3.10		3.15		2.85
		Normalized V_j		.98		1.00		.90

Conclusion

This paper has proposed a methodology which can be used for evaluation and selection of advanced manufacturing systems. The methodology takes into consideration the intangible factors which can not easily be formulated and placed into the traditional cashflow definition. The two phases of this methodology systematically identify the best strategic proposal and also evaluate and recommend the most desirable alterative for this selected strategic proposal. It should also be remembered that the attributes (characteristics) developed for the two phases of this methodology must be independent: that one attribute should not increase or decrease the relative importance of the other attributes.

Bibliography

[1] Son, Y. K. and C. S. Park, "Economic Measure of Productivity, Quality and Flexibility in Advanced Manufacturing Systems," Journal of Manufacturing Systems, Volume 6, No. 3, 1987.

[2] Sullivan, W. G., "Models IEs Can Use to Include Strategic, Non-monetary Factors in Automation Decision," Industrial Engineering, Volume 18, No. 3, March 1986.

[3] Parsaei, H. R., and R.E. Allen, "Decision-Support Expert System: Selection and Economic Justification of Flexible Automation Alternatives," Proceedings of the 9th International Conference on Production Research, Cincinnati, Ohio, August 17-20, 1987.

[4] Meredith, J. R. and N. C. Suresh, "Justification Techniques For Advanced Manufacturing Technologies," International Journal of Production Research, September/October, 1986.

[5] Varney, M. S., W. G. Sullivan, and J. Cochran, "Justification of Flexible Manufacturing Systems with the Analytic Hierarchy Process," Proceedings of the 1985 Annual Industrial Engineering Conference, Los Angeles, California, May 1985.

[6] Canada, J. R., "Non-traditional Methods for Evaluating CIM Opportunities Assign Weights to Intangibles," Industrial Engineering, Volume 18, No. 3, March 1986.

[7] Suresh, N. C. and J. R. Meredith, "Justifying Multimachine Systems: An Integrated Strategic Approach," Journal of Manufacturing Systems, Volume 4, No. 2, 1985.

[8] Parsaei, H. R., and M. R. Wilhelm, and W. Karwowski, "A Decision Making Methodology for Selecting Among Flexible Automation

Justification of CIM: Strategic Considerations for Economic Analysis

R. SONI; D. H. LILES and G. T. STEVENS, JR.
Department of Industrial Engineering
The University of Texas at Arlington

INTRODUCTION

The decline in the competitive strength of United States manufacturing is much discussed and well documented [1], [2]. In an attempt to successfully meet global competition, American manufacturing companies are modernizing their manufacturing operations. Many advanced manufacturing technology projects are, however, difficult to justify using traditional economic evaluation methods.

Traditional investment evaluation methods have three characteristics that tend to inhibit investments in advanced manufacturing. *First*, most economic evaluations of advanced manufacturing are based upon a comparison of the initial investment to the estimated cost savings. This approach ignores possible revenue enhancement which is one of the primary benefits of advanced manufacturing. *Secondly*, evaluations of the "existing technology alternative" usually assume a continuation of the current market share. Kaplan [3], however, states that the "correct existing technology alternative" should assume a declining market share, resulting in a decrease in revenues. *Finally*, high hurdle rates and short capital recovery periods have traditionally been used for investment evaluations as a means of protection against the risk associated with an uncertain future. This paper presents a procedure that attempts to address these three problems.

DEVELOPMENT OF THE MODEL

The decision model, presented in this paper, is based on a year by year comparison of a series of minimum annual *revenue requirements* $(R_j, j=1,2,\ldots,N)$ to a corresponding series of *expected revenues* or gross incomes (G_j, j=1,2...,N). The revenue requirement is the minimum gross income required to cover all project costs including capital recovery, return to capital, cost of goods sold, and taxes. It is determined using the following equation [4].

$$R_j = Db_j + Fe_j + I_j + C_j + t_j \; ; \; j=1,...,N. \qquad (1)$$

In the above equation, Db_j is the total capital (debt and equity) recovered in year j; Fe_j and I_j are returns to equity and debt capital, respectively, in year j; C_j is annual costs in year j; t_j is the taxes paid in year j; and N is the capital recovery period. A comparison of the R_j values to the expected gross incomes requires that the analyst explicitly consider all of the benefits of the projects. The benefits include both cost savings and revenue enhancements that result being more competitive. This paper defines expected gross incomes as a function of market share and market share, in turn, as a function of technological leadership and of market growth pattern.

Technological leadership can be divided into three categories. A technological leader is one of the first companies in a particular market to adopt advanced manufacturing concepts. Its market share (Figure 1) will increase with time due to improved quality, increased flexibility, reduced cost and price, shorter lead times, etc. A technology follower (Figure 2) is a company that invests in advance manufacturing in order to keep up with the technology leaders. The technology follower may experience a time lag before the benefits of advanced manufacturing are realized. This is attributed to the technology leader continuing to capture market share from other companies. A technologically inactive company (Figure 3) will lose market share.

A company can, also, be affected by market growth pattern. There are two types of market patterns considered in this paper. In the case of the growing market, the demand for the product grows at relatively high rate. In a saturated market, the demand for the product remains more-or-less unchanged.

For various combinations of technology leadership and market growth, the equations shown in Table 1 are proposed. These equations may be modified to incorporate quality and flexibility indices in addition to the price index. As shown in Figure 4, the gross incomes (G_j) to be used in a particular analysis should be equal to the difference between the enhanced revenues that result of investment in advanced manufacturing and the reduced revenues that result from no investment.

Table 1
Expected Gross Incomes

1. Company is technology leader in a saturated market.

$$G_j = M_0 [(p_j)^\alpha][b-ce^{-mj}]-M_0 S_0 (1 + r)^{-j}; \quad j = 1,..,m$$

$$= M_0 [(p_j)^\alpha][b-ce^{-mj}]; \quad j = m+1,..,n$$

2. Company is technology follower in a saturated market.

$$G_j = 0; \quad j = 1,...,d$$

$$= M_0 [(p_j)^\alpha][b-ce^{-mj}]-M_0 S_0 (1 + r)^{-j}; \quad j = d+1,..,m$$

$$= M_0 [(p_j)^\alpha][b-ce^{-mj}]; \quad j = m,...,n$$

3. Company is technology leader in a growing market.

$$G_j = M_0 [(p_j)^\alpha][b-ce^{-mj}][b_1 -c_1 e^{m_1 j}]-M_0 S_0 (1 + r)^{-j};$$
$$j = 1,..,m$$

$$= M_0 [(p_j)^\alpha][b-ce^{-mj}][b_1 -c_1 e^{m_1 j}]; \quad j = m+1,..,n$$

4. Company is technology follower in a growing market.

$$G_j = 0; \quad j = 1,..,d$$

$$= M_0 [(p_j)^\alpha][b-ce^{-mj}][b_1 -c_1 e^{m_1 j}]-M_0 S_0 (1 + r)^{-j};$$
$$j = d+1,..,m$$

$$= M_0 [(p_j)^\alpha][b-ce^{-mj}][b_1 -c_1 e^{m_1 j}]; \quad j = m+1,..,n$$

where,

G_j = Expected gross income in year j
M_0 = Market size at time zero
j = End of year (j = 1,2,....)
n = Capital recovery period
e = Natural base of logarithms
S_0 = Market share at time zero
P_j = Price Index (defined as price/the least price in the market for year j)
d = Year when the market share starts growing for the technology follower company
m = Year when the company will be forced out of business if it does not adopt advanced technology
b, c, m, b_1, c_1, m_1, r, and α are empirical constants

It is also recommended that a reasonable rate of return and a reasonable capital recovery period be used for determining the R_J's values.

DECISION CRITERION

The R_J's and G_J's, calculated in the fashion discussed earlier, are compared directly for decision making purposes.

EOY	R_J	G_J	
1	R_1	G_1	Accept the project if
2	R_2	G_2	$\Sigma(G_J - R_J)(P/F\ k_x, j) \geq 0$ (2)
.	.	.	
N	R_N	G_N	

If $G_J \geq R_J$ (for all $j = 1, \ldots, N$) then the project is desirable by inspection. Likewise, if $G_J < R_J$ (for all $j = 1, \ldots, N$) then the project is not desirable. If some of the R_J's are less then the corresponding G_J's and the others are greater, then equation (2) must be used to determine the project's economic desirability.

SUMMARY

This paper has presented an economic decision model for the evaluation of investments in advanced manufacturing. The decision model compares a series of revenue requirements to a corresponding series of expected gross incomes. Revenue requirements are generated in the fashion suggested by Stevens [4]. Gross incomes are generated as a function of market share. Market share, in turn, is defined as a function of technological leadership and of the market growth pattern.

REFERENCES

1. Buffa Elwood S., 1985, Meeting the competitive challenge with manufacturing strategy, National Productivity Review, Spring, pp. 155-169.

2. Hayes, Robert H. and Clark, Kim B., 1985, Explaining observer productivity differentials between plants: Implications for operations research, Interfaces, Vol. 15(6), pp. 3-14.

3. Kaplan, R. S., 1986, Must CIM be justified by faith alone?, CIMTECH, Conference Proceeding, March 10-13, Chicago, Illinois, CASA/SME.

4. Stevens, G. T., Jr., 1983, Engineering Economy (Reston Publishing Company, Inc.).

FIGURE 1

TECHNOLOGICAL LEADER

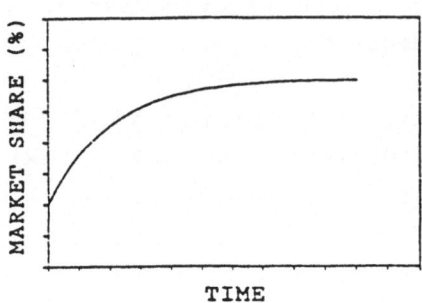

FIGURE 2

TECHNOLOGICAL FOLLOWER

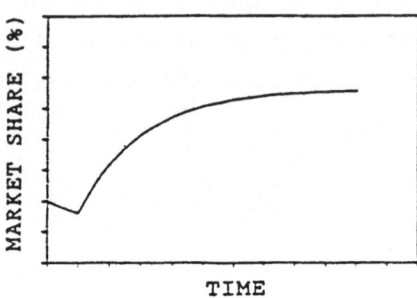

FIGURE 3

TECHNOLOGICALLY INACTIVE

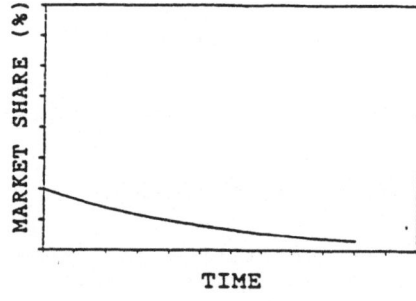

FIGURE 4

GROSS INCOME

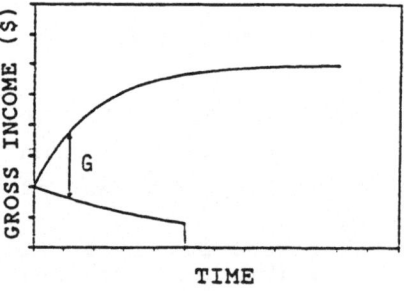

Computer-aided Analysis of the Manufacturing Process of Job Shop Production by a New Input-Output-Method

by Friedhelm Nyhuis and Wolfgang H. Dumke

tetra Gesellschaft fuer Technologie Transfer mbH,
Guentherstrasse 3, 3000 Hannover 81, West Germany

Introduction

Permanent control of the manufacturing process goes without saying in well-managed factories of today. However, this control concentrates mainly on processes that can be evaluated in terms of money. For example, it is common practise to sum up the monthly order entry, sales, investments, etc. and to compare these figures with the annual plan.

If we look at the production targets of a modern manufacturing company (figure 1), we have to realize that currently there is a lack of appropriate instruments to check the achievement of these objectives although quantified results of the manufacturing process can easily be derived from operational data (eg. ready messages).

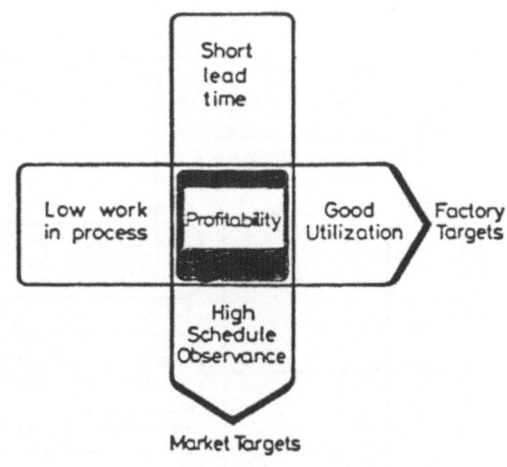

Fig. 1: Production targets /1/

In manufacturing we compare and evaluate planned and actually achieved production hours according to the resulting costs. But still we do not recognize the importance of a continuous monitoring of inventories, delays, and lead times. Without such a monitoring system an efficient control of production is not possible. Consider, for example, a Jumbo jet that is operated without the pilot having adequate indicators: He can only handle it in visual flight.

Looking at the main targets of production, additional information about its various influencing factors (changes of
quantities and due dates, batch sizing, sequencing, flexibility, breaks in production, etc.) is needed in order to
recognize and evaluate reasons for problems and their effects
in the manufacturing process. These influencing factors must
also be recorded if a comprehensive control of the manufacturing process is to be established.

According to the hierarchical structure of business scheduling
systems, which can be seen as linked control loops, the control
system of the manufacturing process must also show a hierarchical structure. Evaluations for the various control purposes
should help to make strategic decisions. The function of a
control system of the manufacturing process within production
planning and control is illustrated in figure 2.

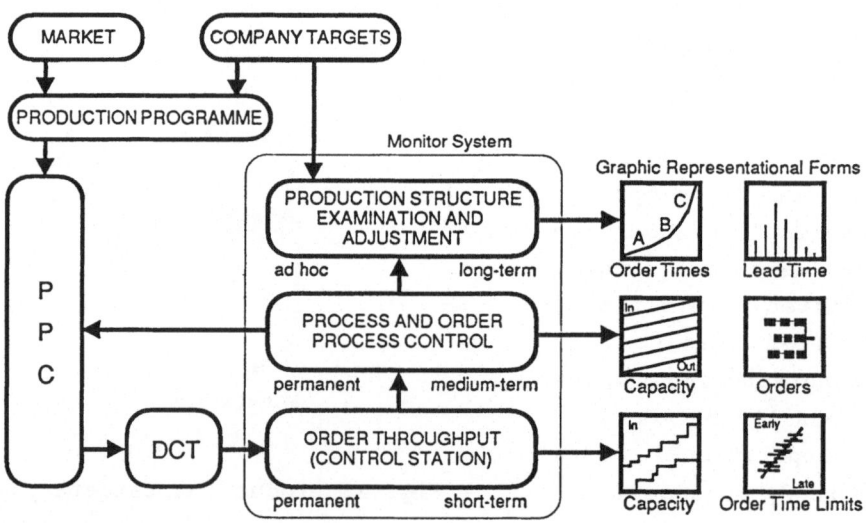

DCT=Data Collection Terminals
PPC=Production Planning and Control

Fig. 2: Functions of a Production Control System /1/

The module "Production Structure Examination and Adjustment" is applied especially to long term strategic decisions. Mid term control is mainly concerned with order control and due time observance whereas short term control supports work distribution. The modules mainly correspond with the constituents of the DUBAF system, a computer program for lead time and inventory analysis designed at the Institut fuer Fabrikanlagen (Factory Equipment Institute) of the University of Hanover, West Germany. This system has been widely extended and reshaped by the tetra company to a PC-based computer program called FAST (Graphic and Statistical Analysis of the Manufacturing Process). The program, which has been tested with 150,000 ready messages of about 500 different work centers. It has already been applied successfully in numerous production companies. The program's structure is shown in figure 3. It applies a new Input-Out-Method which is briefly described below.

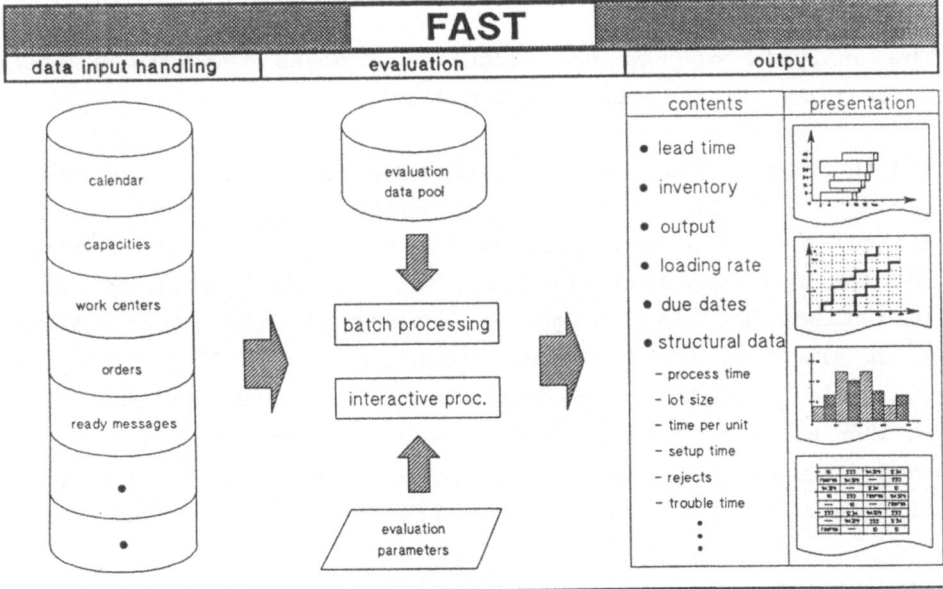

Fig 3: FAST - program structure

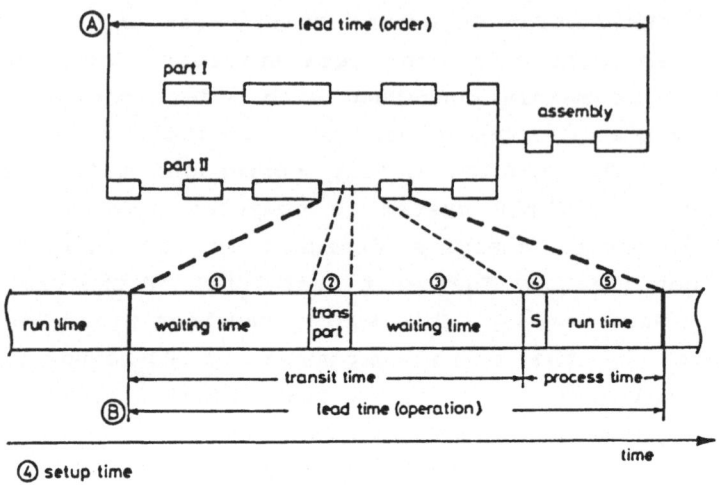

Fig. 4: Definition of lead time /1/

Analysis of the Manufacturing Process

The analysis of the manufacturing process requires a clear definition of lead time (figure 4).

The definition applied in this case is based on the "ready messages" of the work centers which are recorded in most plants. The time span between the completion of an operation and the completion of its predecessor is defined as operation lead time. In case of gateway operations the order release dates are used instead since there is no preceding operation. The sum of operation lead times equals the lead time of orders. It can also be expressed as the time span between the last ready message and the order release. The transit time is the difference between lead time and process time of the operation. The process time is the sum of set up and run time. Since data acquisition mostly does not supply the real set up and run times the standard time of the production plans must be used.

In general the deviation of real set up and run times from standard times does not make much difference because the process time rarely amounts to more than twenty per cent of the lead time.

There are two main possibilities of evaluating the data. One can analyse either the material flow which is the flow of orders through the job shop, or the operating and job sequencing at the work centers. To show the relationship between lead time and work-in-process (WIP) inventory we will first look into the operating at the work centers by using an Input-Output-Chart (figure 5).

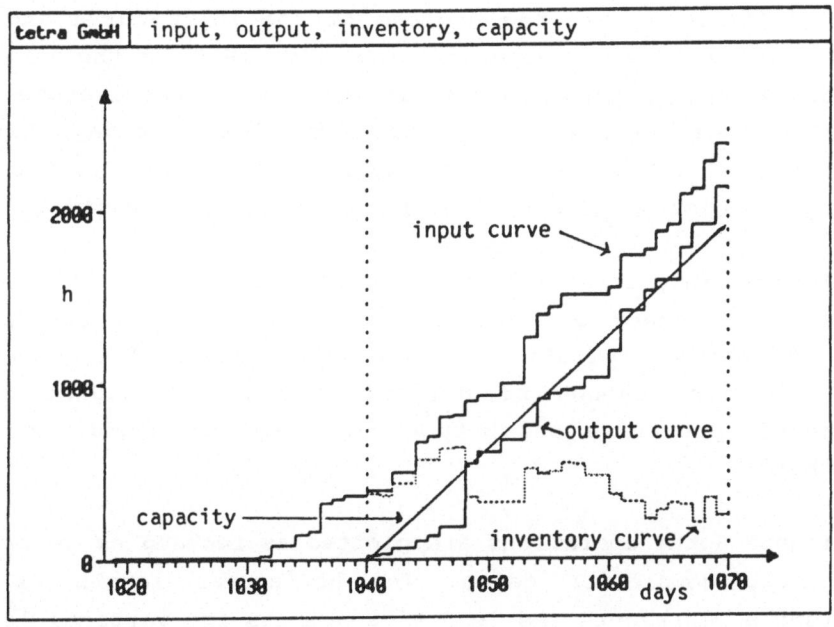

Fig. 5:   Input-Output-Chart

This chart consists of an input curve, an output curve and an inventory curve. The input curve describes the arrivals of orders processed at the work center whereas the output curve shows their completions. To generate the output curve the processed operations (orders) must be sorted according to their completion date. The values - in this case the orders' process times - are then accumulated and plotted. In this diagram a vertical segment of the output curve represents the work quantity of one or more orders completed at the same time. A horizontal segment represents the time span between two consecutive ready messages. The input curve analogously represents the arrival of the orders. The units to measure are usually hours for work quantity and process time and days for lead time.

The Input-Output-Chart clearly shows the relationship between work-in-process inventory and lead time. If one wants to reduce the lead time one has to reduce the inventory (WIP). The average lead time can be obtained by measuring the horizontal distance between the average of the input curve and the output curve. The corresponding inventory (WIP) is the vertical distance between these curves. In our example the average lead time of these work centers amounts to about six days with a standard deviation of about eight days. These results are among the best we have ever found in job shop production.

While the inventory-oriented Input-Output-Chart in figure 5 shows the process of inventory, the next figure illustrates the sequence of operations at the work centers. The sequences of operations cannot be derived from the preceding chart because input curve and output curve are independent of one another.

In figure 6 the operations are plotted as rectangles according to their completion dates. The horizontal length of the rectangle represents the lead time whereas the vertical length represents the process time at the considered work center. Therefore the left side of the rectangle corresponds with the the order's input at the work center; analogously the right side stands for the time of completion.

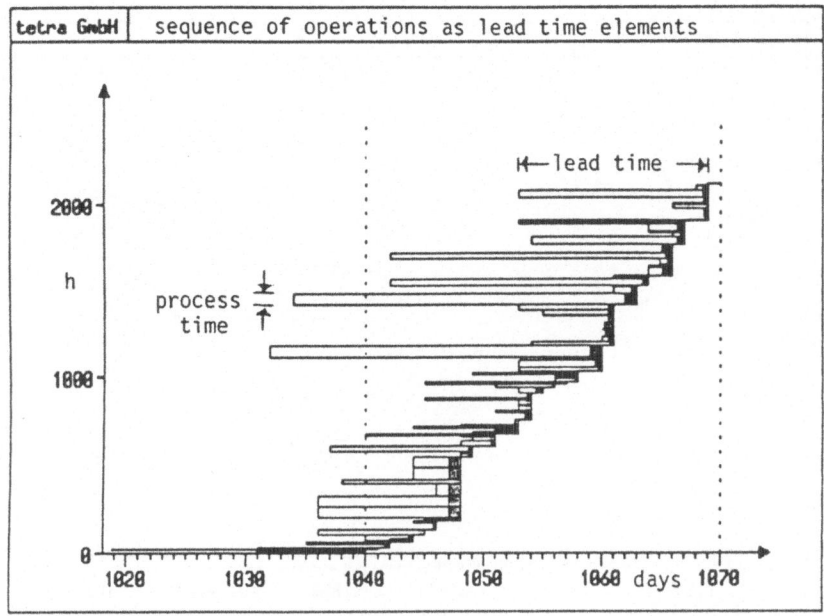

Fig. 6: Sequence of operations as lead time elements

Figure 7 shows the relative schedule deviation, i.e. a specific work center's contribution to the discrepancy between planned and actual lead time. This discrepancy is represented in the diagram by the lenghts of each bar. Those orders left of the output curve were processed slower, those right of the output curve were processed faster than planned. However, this representation does not explain the orders' absolute schedule deviation which is shown for the same work center in figure 8. Altogether 105 orders were processed within thirty work days (2119.40 work hours). The schedule deviation was minus eighteen days, i.e. on average the orders were completed eighteen days later than planned.

The summmation curve of the schedule deviation in figure 9 shows that fifty per cent of the orders were completed more than thirty days late. On the other hand about twenty per cent were completed more than ten days too early. Evidently the work center's operator was sequencing according to his personal aims, in mots cases to save set up time.

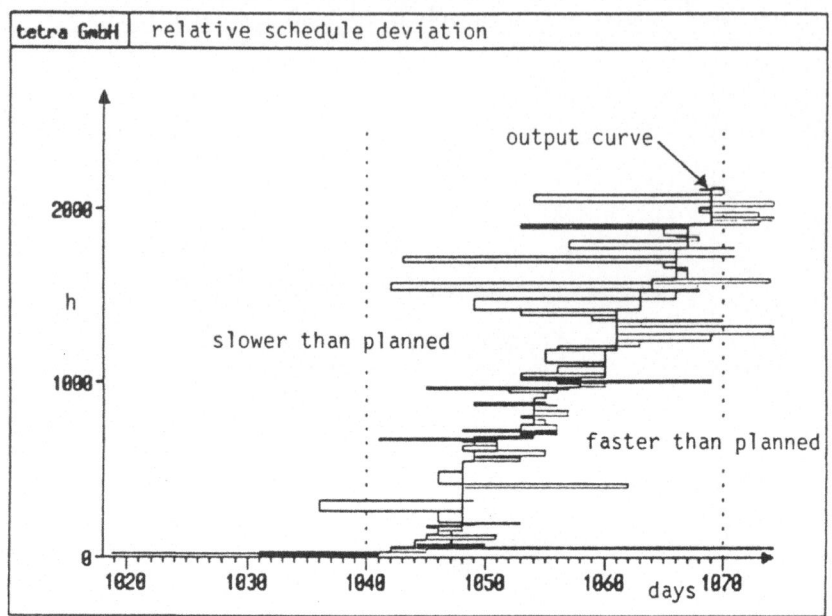

Fig. 7: Relative schedule deviation

However, the reason for this schedule deviation is produced only partly by that specific work center. The average planned lead time is about three days whereas the average actual lead time is 6.16 days (see figure 10). So the work center only contributes three days to the total delay of eighteen days.

The above described graphic and statistical evaluations are supported by those of periodically summed up data with which allow an overall view of various data for every single work center. Here a so-called 'hit list' , i.e. a ranking of the 'worst' work centers, supports the well-aimed search for weak points by determining and weighing plan deviations.

Conclusions

An efficient control of the manufacturing process is not possible without monitoring and analysis. The necessary data can be obtained without great difficulties from ready messages. A computer-aided analysis (as realized with FAST) illustrates the course of production, supports trouble-shooting, and thus creates transparency and efficiency of manufacturing.

References

1 Wiendahl, Hans-Peter: Load-oriented production control - Hanser-Verlag, Munich / Vienna 1987.

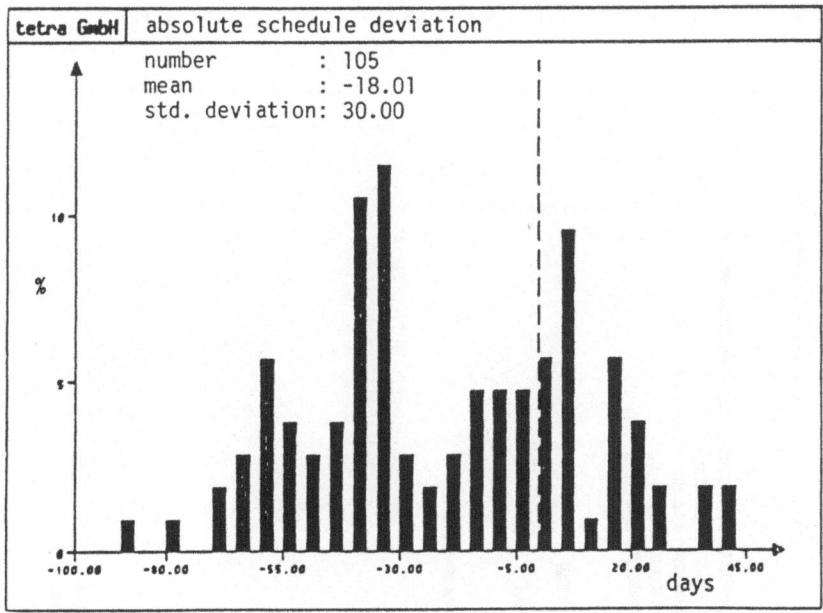

Fig. 8: Absolute schedule deviation

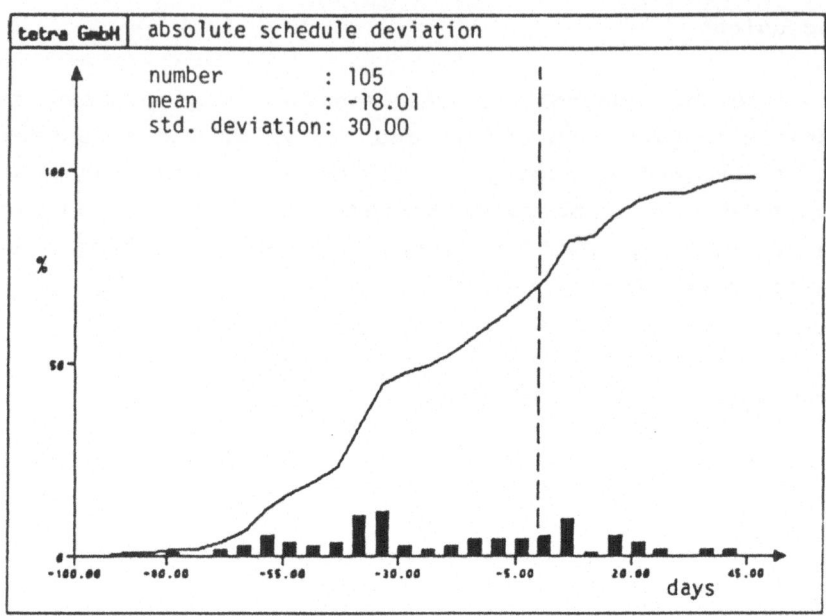

Fig. 9: Absolute schedule deviation, summation curve

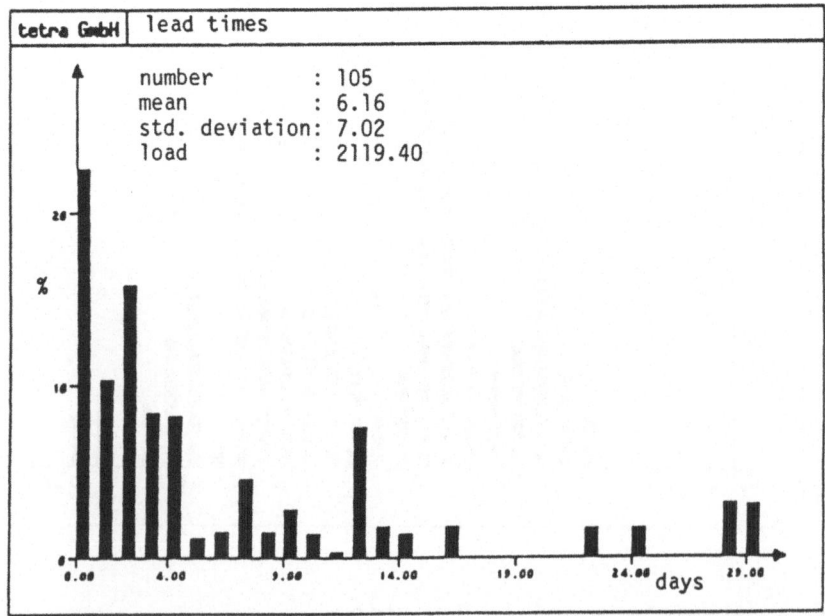

Fig. 10: unweighed lead times

Flexible Manufacturing Systems

Information System for Flexible Manufacturing System: A Conceptual Framework

DEVINDER K BANWET AND SUSHIL

Centre for Management Studies
Indian Institute of Technology
New Delhi-110016(INDIA)

Abstract

The present paper is an attempt to provide a conceptual framework for analysis, design and development of an information system to cater to the
needs of decision makers at various levels, be it strategic, tactical or
operational, in the context of flexible manufacturing systems(FMS). An
integration of the horizontal functional dimensions and the vertical hierarchical levels has been attempted in conceptualising an information system
architecture so as to make FMS more effective and productive.

Introduction to FMS

A proper corporate strategy needs to be devised to cater to the dynamic
changes that take place in the environment. With increasing competition,
rapid technological changes, customer quality awareness and shorter delivery
periods being demanded, there is a growing realisation that it is better
to conceptualise and design production systems that are 'flexible' or adpat
quickly and economically to a variety of changes that take place. An FMS
could be made more effective and productive, if it has an appropriate infor
mation system to aid the decision makers.

An FMS is an electic outcome of reconciliation of the conventional job
shop system and the line/mass production system, especialy for the mid
range volume situation in a P-Q chart. Bessant and Haywood [1] quote an
FMS as "...a system which combines micro-electronics and mechanical engineering to bring economies of scale to batch work. An outline computer controls
the machine tools and other work stations and the transfer of components
and tooling". Flexibility is the key word in an FMS. Various authors [2,4]
have conceptualised a large variety of flexibilities such as action, state,
job, machine, volume, routing,operation,process,short and long term flexibility. However, it is important to realise that the large spectrum of
flexibilities are difficult to achieve simultaneously. The mid volume variety
ushered in the need for group technology and further research brought in
the applicability of CNC,Flexible Manufacturing Cell, FMS, Flexible and
Dedicated Transfer Lines.

Decision Structure in an FMS

FMS is the provision of the production system so as to enable prompt delivery of goods by providing all relevant resources in the right quantity, quality and time by controlling and supervision of 'flexible' process of equipment, machines and tooling, methods and processes. In FMS, which is usually part of a larger system, certain decisions would be taken periodically, some continuosly in various horizontal functional dimensions, where as decisions are also taken by decision makers at the top strategic, middle tactical or bottom operational level of management. In order to improve the quality of decisions, it is important to devise an appropriate information system.

Information System Design Conceptual Proposal

An overall conceptual design of the information and control systems for an FMS should define the information sources, sinks, requirements, channels, information flow, major storage and processing stages etc. A horizontal functional dimension based information system architecture is depicted in Figure 1 in an FMS context. Data bases exist for design, product, CAD/CAM, Production Control, Tooling, Tool control and Machinability, Tool Inventory etc. to aid the FMS Decision Makers to carry out the computer Aided Process Planning, MRP, Capacity and Resource Planning, Scheduling and Routing. In the proposed architecture, due importance being given to Tool Management aspects of requirements, performance evaluation, transportation control, inventory management, inspection, reconditioning and disposal of tools, which can be catered to by interacting with the relevant data bases. There is also the vertical heirarchical dimension to be considered as made out by Elfving [3]. Suri and Whitney [4] discuss the type of decisions, hardware and software in proposing a decision support system for flexible manufacturing. The top strategic level decision makers deal with highly unstructured, non-programmable decisions situations interacting with the external environment of customers, markets, suppliers, technological R & D laboratories, governmental agencies etc. It is here where the top should be able to forecast and foresee the future changing events so as to feed this information to chart up appropriate plans, policies, budgets etc. which form the instructions to act as guidelines by the middle level decision makers. It should have the capability of being 'just-in-time' for whatever and whenever relevant is demanded by your customers, shareholders, governmental agencies etc. Tactical middle level FMS supervisors accordingly go about planning schedules, materials, tooling on one hand while assessing the performance and feedback from the lower operational bottom level so as to transmit feedback reports in an appropriate format to the higher level. Thus at this level MRP and MRP-II could be utilized alongwith Tooling and Material Handling Equipment availabilities so as to optimise/satisfice decisions pertaining to the FMS under consideration.The lowest operational level decisions pertaining to the shortest time horizon deal with work order scheduling and despatching, tool management, more specifically tool availabilities. The lowest level perhaps makes the maximum use of decentralised micro-computers to cater to the status of material handling equipment, machine tool status, robotics, AGV's, tooling etc.

In order to simplify the structure of the information system, a particular functional decision center node need to communicate with a maximum of two other functional decision center node, one above and one below the existing level. Essentially two sets of data bases, one for instructions/orders and the other for feedback reports need to be conceptualised. These are created appropriately and these ought to be maintained, updated and retrieve data at different levels for different uses at different times. A DSS,CAD/CAM interface is required for the FMS. An information system architecture taking the horizontal and vertical dimensions in an integrative framework in the context of an FMS is shown in Figure 2. A typical channel is illustrated.

Concluding Remarks

An FMS is a good proposition for the mid volume variety of batch manufacturing. However,an FMS environment,has to cater to a large number of uncertain and fuzzy situations.An FMS might by itself be technologically sound. But to make it more effective and productive, the FMS needs to be superimposed with a relevant and good information system architecture.Some suggestions for an integrative framework of an information system have been highlighted which could hopefully be suitably modified and adopted to specific situations so as to be able to derive much greater benefits from the specific FMS.

References

1. Bessant,J.and Haywood,B,"Flexibility in Manufacturing Systems",Omega,Vol. 14,No.6,1986.

2. Buzacott,J.A.,"The Fundamental Principles of Flexibility in Manufacturing Systems,Proc.of First Int.Conf.on FMS,1982,pp.13-22.

3. Elfving,J.E.,"Choice of Control Systems for FMS Cells with aspect of Economy and Reliability",Proc.4th Int.Conf.on FMS,1985,Ed.by Lindholm,R.pp. 15-17.

4. Suri,R.and Whitney,C.K.,"Decision Support Requirements in Flexible Manufac- turing,J. of Manufacturing Systems,1984,Vol.3,No.1.

302

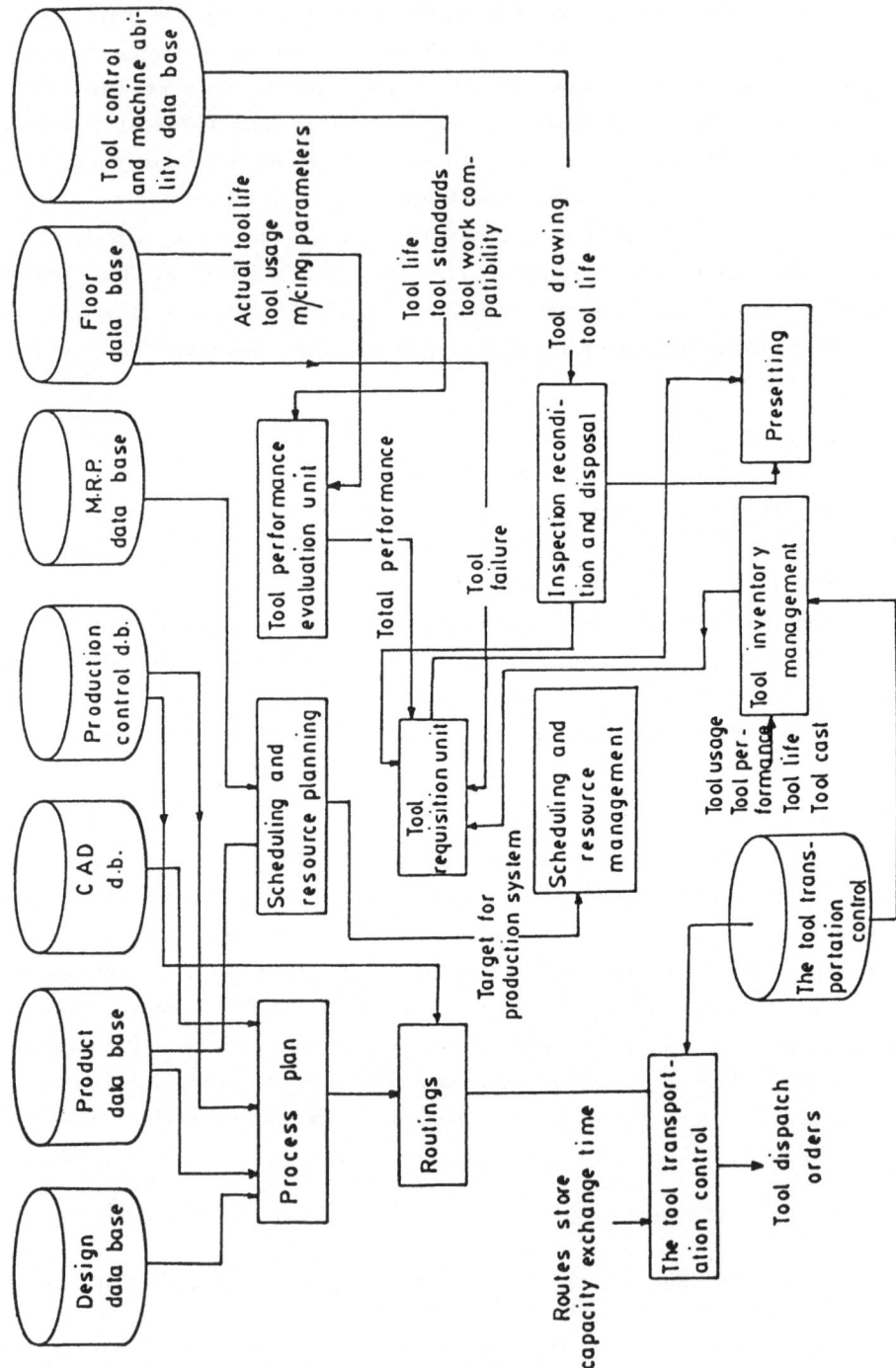

FIG.1 MANAGEMENT INFORMATION SYSTEM IN F.M.S. (Horizontal integration)

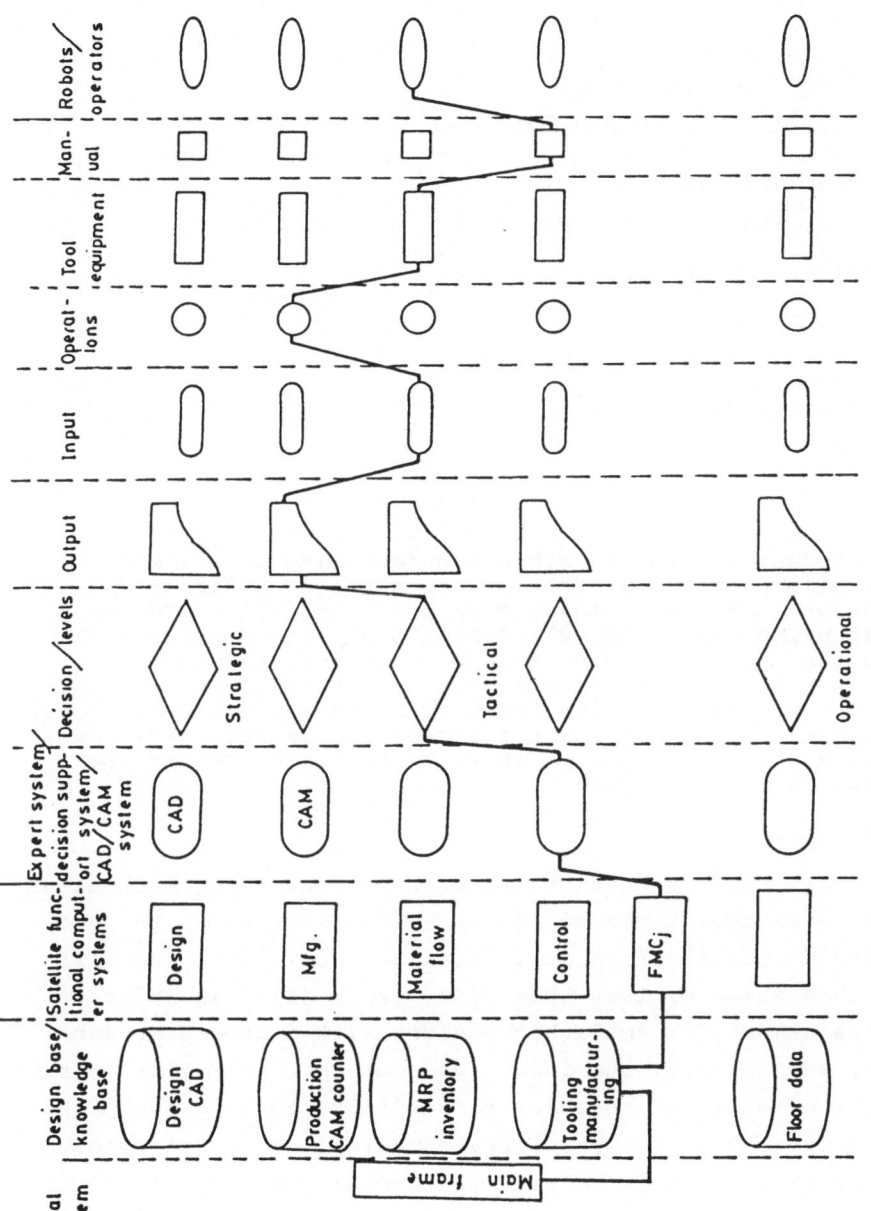

FIG. 2 DIMENSIONS OF MIS DESIGN IN FMS ENVIRONMENT

—— A typical channel

Modeling and Analysis
of a Flexible Manufacturing Cell

B. P. BANDYOPADHYAY

Mechanical Engineering Department
University of North Dakota
Grand Forks, North Dakota 58202

T. V. NGUYEN

Mechanical Engineering Department
Woodward Governor Company
Fort Collins, Colorado 80525

Summary

The introduction of a flexible manufacturing cell (FMC)
within a company represents a major advance in factory
automation. The paper focuses on the design of a FMC for
manufacturing family of cylindrical parts by computer
simulation modeling, using AUTOMOD software, and on the
economic justification of the cell. Costs and benefits of
the FMC were compared with those of stand-alone CNC machines.
It has been found that the net income in the next five years,
utilizing the FMC, would be about three times that of the
stand-alone CNC equipment.

Introduction

Substantial changes have occurred in manufacturing methods and
processes in recent years. It is estimated that by the end of
this decade seventy five percent of the parts produced will
have batch sizes of less than fifty parts [1]. In order to
meet the demand for small batch sizes with improved productivity, and quality, the industry needs FMC. A FMC is a group
of machine tools and associated materials handling equipment
that is managed by a supervisory computer. The supervisory
computer is utilized to direct the handling equipment to
transport parts from machine to machine and to control the
manufacturing operation of parts. A FMC is an independent unit
but may be tied together with other cells by a common material
handling system to form a flexible manufacturing system (FMS).
This paper addresses the design of a FMC for manufacturing a
family of cylindrical parts by simulation modeling and the
economic justification of the cell. The paper is divided into
three sections. The first section briefly concentrates on the
necessary hardware to create the FMC. The second section of

this paper discusses the modeling of the FMC using the AUTOMOD
simulation software in order to study the dynamic behavior of
the cell. The third section carries out the evaluation of the
overall system performance economically and functionally, for
ensuring the original production planning.

Cell Design

The first stage in the design of a FMC involves concentration
on the necessary hardware to create the cell: computers,
machine tools, forms of material handling equipment, fixture
and tooling. For a complex FMC, a more detailed design is
needed that takes into account the machine sensors and fixtures for efficient operation of the cell. During the design
process the following criteria were taken into consideration:
i) cell flexibility, ii) cell economic justification, and iii)
cell modularity. In order to accomplish the design of the
cell the following three steps were followed: i) selecting the
machines, ii) designing alternative cell configurations, and
iii) evaluating the design and its variations on technical and
economic grounds. The cell includes three computer numerical
controlled (CNC) lathes, one CNC milling machine and a robot
for handling the parts. Two system configurations were considered in the design of the cell layout, i) in-line type, and
ii) robot centered type. These two types of layouts were
thought to be appropriate because all the parts in the family
have a welldefined routing process, they are first processed
by the lathe and then by the mill with no back-flow routing.
The problem of selecting a specific layout among the two will
depend on i) shape of the available space and ii) the future
production plans. The in-line layout has the definite
capability of providing the space for future addition of
several more machine tools. However, it is not suitable when
back flow routing of any part is required. The robot-
centered layout has the capability of processing the parts in
any sequence, it allows back flow routing of parts with ease.
However, future addition of machine tools will be difficult
because of the space restriction. To achieve better control
of the robot joint movement, the cell robot should have

servo-driven DC motors for every joint. The maximum operating
speed of the robot is an important factor in the increase of
the production rate of the cell. A faster robot would minimize
the material handling time, and therefore, will increase the
total cell productivity. However, a fast robot with the
required accuracy for loading and unloading tasks will
definitely be much more expensive. The cell host computer will
be required to have the speed and power to service communication with all the machine controllers. The UNIX operating
system and the DEC VMS operating system were thought to be
appropriate as the cell host operating system because of their
capability to interface well with the programmer and machine
controller. A 16-bit or 32-bit processor is usually required
to provide sufficient speed and memory for the cell operations
[3].

COMPUTER MODELING AND SIMULATION

The design of the FMC was completed by computer simulation
modeling. For any given potential FMC application the number
of feasible FMC configurations is enormous. It will be too
costly to actually try out potential FMC solutions on the shop
floor and then select the "best" one. Computer simulation
permits us to study the dynamic behavior of the FMC. It
allows one to describe each FMC design and predict its
performance without actually building the system. The
analysis of the FMC was accomplished by using the AUTOMOD
simulation software. The AUTOMOD simulation model was
utilized to represent the cell characteristics such as
machine behavior, workflow, job routing, control and sequence
rules. AUTOMOD provided the results of the simulation in
statistics reports [4]. The simulation model of the cell
design was developed with two objectives in mind: i) performance prediction: to predict maximum possible throughput and
machine utilization of the cell and ii) optimization: given the
cell throughput and machine utilization, addition of in-process
storage with varied capacity was considered to observe the
effect of the in-process storage on the cell performance. The
model was restricted to one set of rules which represents the
cell characteristics and the controlling algorithm of the host

computer in an actual cell operation. The cell model was
implemented on a VAX 11/785 minicomputer at the University of
North Dakota (UND). The model was initially simulated for an
8-hour operation to obtain a steady state. After the steady
state has been reached, the model was simulated for another
24-hour operation. The major results of the computer
simulation are given in Table I.

TABLE I

Production Throughput of the Cell in 24 Hours

Part No	Buffer Storage 0	Buffer Storage of 5	Buffer Storage of 20	Buffer Storage of 50
I	104	127	118	113
II	61	54	62	58
III	71	69	63	64
IV	74	61	62	58
V	25	32	35	33
VI	30	36	36	40
Total Number	365	379	376	366

The computer simulation results indicate that addition of
in-process storage would increase the cell production rate and
machine utilization. The optimum capacity for the storage was
found to be 5 to 20 parts. This is due to the fact that the
parts that are completed on the lathes did not have to wait for
the milling machine but can go directly to the buffer storage,
thus making room at the lathes for incoming parts.

Economic Analysis

The purpose of the economic analysis of the cell design was
first to examine the economic feasibility of the cell based on
the performance measures obtained from the simulation study,
and secondly to compare the costs and benefits of the cell with
those of stand-alone CNC machines to determine which of the two
is a better alternative. The cost analysis was performed with
the assumption that both of the alternatives have the same

annual incomes; each was to produce the same batch of parts
annually [5]. It was found that in order to achieve the same
production rate as the FMC, with stand-alone CNC equipment, at
least 3 CNC lathes and 2 CNC milling machines would be
required. The extra milling machine is required because the
manual fixturing and handling of the parts at the mill take
more time than those performed by the robot.

The cost analysis shows that the FMC would require higher
capital investment than the stand-alone CNC machines, but would
drastically reduce the labor costs, and therefore the total
annual operation costs [2]. The economic comparison of the two
systems was based on the assumption that both systems have the
same production requirements per year and therefore the same
annual income. It has been found that the net income in the
next five years, utilizing the FMC, would be about three times
that of the stand-alone CNC equipment. This large gap in the
future net income is caused mainly by the large difference in
the annual operating costs of the two systems. Therefore it
can be concluded that the investment on the FMC can definitely
be justified in developed countries such as the U.S. where the
cost for labor is high.

Acknowledgement

The authors wish to thank AUTOSIM Inc., of Bountiful, Utah for
donating the AUTOMOD Software to UND.

References

1. Hartley, J., FMS at work, IFS Publications, Ltd. 1984.
2. T. V. Nguyen, The preliminary design and evaluation of
    a flexible manufacturing cell, M.E. Thesis, University of
    North Dakota, July 1987.
3. Cutkosky, Mark R., et al., Precision flexible machining
    cells within a manufacturing systems, The Robotics
    Institute, Carnegie-Mellon University, June 1984.
4. AUTOMOD User's Manual, Autosimulation Inc., Bountiful,
    Utah, Editiion 1.3.EB6, Feb. 27, 1986.
5. Sloggy, John E., "How to justify the cost of an FMS,"
    Tooling and Production, December 1984, pp. 72-75.

A Prototype Flexible Manufacturing System

M. MEHDIAN[*] and R. SAGAR[+]

[*]Lecturer. Manufacturing and Robotics Centre,
Department of Mechanical Engineering.
Design and Manufacture.
South Bank Polytechnic, Borough Road,
London SE1 0AA
[+]Assistant Professor, Department of Mechanical Engineering
I.I.T. Delhi. India.

Abstract

A computer integration of part processing and material/tool handling system with the intelligent communication and data processing nodes is presented as the Proto-type Flexible Manufacturing System (PFMS). The PFMS was designed on the basis that both the hardware (Mechanical and Electrical) and the software. the part programming, the Data Base Management System (DBMS), etc. had to be modular and compatible.

The system is developed to integrate many processing (material and data) stations with their real-time decentralised microprocessor/microcomputer structure which is the basis of the system's flexibility and adaptability. A real time-fault analysis and automatic error recovery in the case of breakdowns is provided by the Sensory Based Diagnostic Feedback (SBDF) organisation with an expert system linked to the System History Data Base (SHDB).

The main components of the PFMS are; Machining Centre, Robot, Gravity feed track, Automated Guided Vehicle (AGV) and Automatic Storage and Retrieval System (ASRS). The proposed PFMS also possesses special characteristics such as multi programming, multiuse, real-time parallel data transmission and processing, sensory feedback organisation incorporated with an expert system for part identification/recognition and task allocation, supervision and synchronisation.

INTRODUCTION

Dedicated automatic manufacturing systems are designed for manufacturing large quantities of one or a few products. Product changes often result in a new manufacturing sequence which necessitates conversion measures with regard to the existing system. Moreover it is required from the manufacturing system to be sensitive to everchanging market demands [1].

Designing the FMS projects may include the following sequence; problem definition at both sub-system and overall module levels, financial and technical abilities, system modelling and simulation [2]. For systems integration problem a connection solution based on a centralised control structure may be employed to coordinate the activities of the machines and provide a common programming language [3].

Alternatively a highly modular architecture which presents a distributed control and computing structure to relieve the higher levels from trivial tasks can be adopted [4]. Stochastic variables such as machine breakdown, tool breakages, material failure, etc., rely on "intelligent" machines which correct their own malfunctions, understand and plan their own requirements and communicate with each other in a network. The knowledge - based system of this intelligent network consist of a supervisor, global knowledge base, current system model, motion controllers and sensory subsystem modules [5].

Distributed real-time sensory feedback system makes the FMS faster, more intelligent, more reliable and more flexible. Non-contact sensing systems, i.e. vision systems, and tactile sensors [6] are used for a variety of applications such as; part recognition/orientation/transfer, assembly, inspection, condition monitoring, etc.

The Components of PFMS:
The Robot:
The prototype robot manipulator was designed and built with SCARA type kinematic structure, figure (1), according to an appropriate degree of compromise between mechanical rigidity, work volume, position accuracy, weight, energy consumption, manufacturing simplicity, maintenance and cost. The structural advantage of this design was in bringing the centre of gravity of the whole manipulator as close as possible to the axial line of the drive shaft of the base link. Therefore better controllability with lower actuator torques for heavier loads travelling with higher accelerations could be acheived.

Although many assumptions are common in robot control (e.g. piece wise constant acceleration and maximum velocity constraints), the maximum achievable accelerations and velocities can vary substantially with robot configuration and angular velocities, both because of the non linear manipulator dynamics and because the maximum torques that electric motors can produce depend on the angular velocities of the joints.

In theoretical control terms a robotic system may be described as multivariable, interacting, non-linear, time varying and partially modelled. For this robot the path (i.e. The Tool Centre Point Trajectory) is specified and the actuator torque limitations are known. Hence the time-optimal control technique is employed where the optimal open-loop torques are found, and a method is given for implementing these torques with a conventional linear feedback control system [7].

Automated Guided Vehicle (AGV):
The design for the AGV consisted of; load analysis, the AGV type, the number of pick and delivery stations, the FMS layout, the number of AGVs in the system , the power source and the control system structure.

The main parameters considered for the load analysis are the maximum load, the centre of gravity of the load and its dimensions, the type of pallets and fixtures used. In this design, the AGV is equipped with space for two pallets which are incorporated with an automatic pallet changer (APC) to speed up transfers by moving from the ASRS to two buffer stores without returning to base.

The light-guidance AGV designed for this PFMS is equipped with photosensors which detect the fluorescence (from the tape laid on the floor), AGV not only travels along straight tracks, but can turn round corners and spin (360°). The advantages of this type of AGV are; the routes can be changed easily, no extra sensors are needed for forward and reverse travel, the speed can be varied continuously and the installation cost is very low.

The number of pick and delivery stations is reduced to a minimum by adopting a mixture of functional and modular layouts for the FMS system. The number of AGVs required can be established by planning the routes through checking on utilisation and optimisation of times using computer simulation.

The theoretical calculation of the required number of vehicles (N_v) can be made by computing; the total transport time (T_t), the total operation time (O_t), the distance between stations (d), the AGV'S average speed (S_a) with and without load, the total load transfer (i.e. loading and unloading) time (L_t), the empty transport factor (E) and the maximum transport variations during the time of operation by the following equations:

$$T_t = \frac{d}{S_a} (1 + E) + L_t$$

$$N = \frac{T_t}{O_t}$$

$$N_v = NV_m$$

Where N is the number of AGVs and it is assumed that there are no transport variations.
A more accurate calculation of the number of AGVs needed can be obtained from the following equation [8]

$$Z = \frac{\sum_{i=1}^{n} t_i h_i}{3600\ \eta}$$

Where Z is the number of AGVs, t_i is the cycle time of a transport operation for the connection i from the set of all transport connections, i = 1, ...n, in seconds; h is the number of transport operations to be carried out for the connection i for the period of peak demand and η is the availability of the AGVs.

The Automatic Storage and Retrieval System (ASRS):

ASRS designed, figure (2), consists of a series of storage aisles that have arrays of storage locations (SL) for holding the materials to be stored. Every two aisles are serviced by one dedicated dual-masted storage and retrieval machine (SRM).
In this system the automated multiple command cycle storage - retrieval (MCCS) technique [9] with interleaving is adopted. The SRM is capable of transporting two pallets at the same time and these pallets can be loaded or unloaded independently of each other. Also the interleaving permits a store and a retrieve to be handled during the same storage/retrieval (SR) cycle.
The admissible cycles for the SRM could be divided into two modes:
1st mode: with interleaving mode: (S,R), (R,S) (S,S,R), (S,R,S), (S,R,R), (S,S,R,R) (S,R,S,R)
2nd mode: without interleaving: (S,S), (R,R)
The MCCS leads to decreased average travel time per transaction and hence higher throughput than any single command cycle storage-retrieval technique.
The ASRS needs to keep an accurate and updated model of the hardware and operations under its control. The model must include information about the operations and answers for the following questions:
Which SRMs are in service?
What commands are outstanding to each SRM?
Which SLs are currently unavailable?
What each SL's status is?
Which SLs or Input/Output(I/O) points the SRMs are servicing.
The real - time ASRS's computer system is an integration of the computer controls, the supporting information and record keeping system with the bi-directional communication links with other FMS modules.

Buffer Store

The parts delivery and storage system was designed with the Cart on the Gravity Track Conveyor System (CGTCS). The system configuration is shown in figure (3). The uniformly spaced discrete placement strategy for the pallets at the machining centre or at the assembly station was adopted.
The design of CGTCS was based on the following principles:
i) One direction of flow from AGV to the machining centre or to the assembly point.
ii) Single pallet per cart and multiple items per pallet.
iii)Unequal rates of loading and unloading.
iv) Single pick-up station and multiple drop-off stations.

v) Flow in multiple directions.
vi) Automated and computer controlled routing.
For the carts to reach the prespecified destination on the multiple workstation line, the input section of the track is operated by varying the angle of inclination. This angle is measured according to the total mass of conveying mechanism and the distance between the input section and the specified workstation.

Machining Centre (MC)

The term "machining" in the FMS context encompases a variety of tasks which in their entirety would account for automated expert part/tool selection, stable manipulation and feeding of components and tools with correct orientation. This involves supervising the entire process that is introduction of tool data and any compenstions and intervention required in an unstructured manufacturing environment.

The prototype MC incorporates an "intelligent" sensory feedback system to:
i) Position tools (work pieces) in specified locations with correct orientations from an unorganised tool pool.
ii) Have capability of different part processing in a flexible environment.
iii)Be able to carry out non-sequential routines if necessary.

Control

To maximise the utilisation rate of the facilities under production constraint a real time decentralised micro-processor/micro-computer control structure based upon interconnected computing elements is adopted.

A set of intelligent plant inferfaces (nodes) where each handles bi-directional communication link to the node, field peripherals and management functions lies at the lowest level.

All the nodes are connected to the main process bus of this master-slave configuration and all the process busses are connected to the line controller. The line controller supports the mass storage peripherals, manages the bus communication and provides high level of data manipulations.

The supervisor computer hosts the higher level modules and the Data Base.

The advantages of such a highly modular control architecture are:

i) Each node operates highly independent of the main data processing node.
ii) Diagnostic action can be taken locally without interrupting the whole system.
iii)Localised information processing results in transmission of decision taking data only.
iv) Distributed control and computing structure, relieves the higher levels from trivial tasks, thus increasing the overall throughput.

Sensory Based Diagnostic Feedback (SBDF)

The software architecture of the SBDF is modular and it is based on a real time knowledge base located in the main memory. The SBDF is designed to function on a multiprogrammed basis in a concurrent processing environment.

The knowledge base is continuously updated and contains, the system status information, input/output configurations, parameters for operaticns, control variables, communications management type definitions, display and report definitions. This knowledge base is linked to the System History Data Base (SHDB) through a expert system.

The task of the SBDF is to:
i)   Detect and correct error conditions.
ii) Real time fault analysis task suspension until recovery by and
     servicing all interrupts

(iii)Control all Input/output informaiton Transfers.

(iv)Schedule the running of tasks according to their priority.

(v) Allow concurrency and online program development.

(vi)Provide information management for real time remedial action.

Conclusions:

In building a fully computerized FMS there are two major problems; first is
the communications between the machines and the controllers. This can be
overcome by constructing multi level system of decentralised controllers
using line-by-line configuration language to control lower level
controllers.

The second problem is to adapt to, new products and different production
systems. In this case the control software may be designed for a real time
knowledge base and general control primitives to execute the controls
according to parameters in the knowledge base.

References

[1] H. J. Warneke, G. Vettin.
    "technical investment planning of flexible manufacturing systems"
    the application of practice oriented methods. Michigan, 1982. Journal
    of manufacturing systems, society of Engineers,

[2] P. Ranky
    The Design and operation of FMS
    IFS publications Ltd., 1983.

[3] R. Casins,
    "Hierarchial Control of integrated manufacturing systems"
    Technical papers, society of manufacturing Engineers, 1983

[4] L. Torri,
    "Flexible Manaufacturing System: A modern Approach"
    2nd internaitonal conference on FMS, 1982

[5] A.C. Kak, K.L. Boyer, C.H. Chen, R.J. Safranek, S.H. Yang
    "Knowledge based Robotic assembly cell"
    IEEE. Expert Sys. Vol.1. No.1 1986.

[6] M. Mehdian and H. Rahnejat
    "A Tactile Sensor with Automatic learning capability for industrial
    parts inspeciton"
    The international Journal of Advanced Manufacturing Technology,
    2(4), 11-26, 1987.

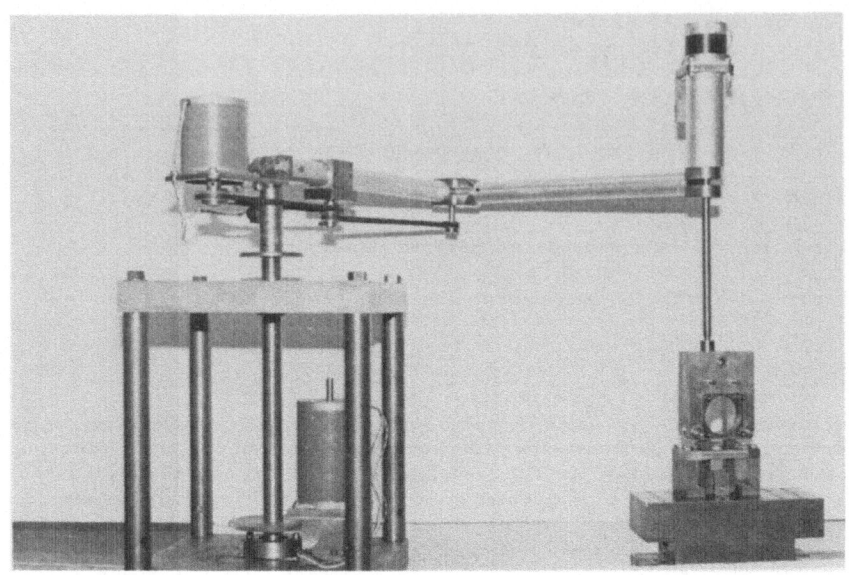

Figure 1: The Robot

Figure 2: The ASRS

[7] E. Bobrow, S. Dubowsky, S. Gibson,
"Time optimal control of Robotic manipulators along specified paths."
The international Journal of Robotics Research,
Vol.4, No.3, Autumn 1985.

[8] T. Muller
"Automated Guided Vehicles"
IFS publications Ltd., 1983

[9] R. Jaikumar, M.M. Solomon
"Real-Time Control of multiple command cycle storage-Retrieval
Warehousing systems."
7th international conference on Automation in warehousing, San
Francisco, Ca. USA. 1986

Figure 3: The Buffer Store.

Mathematical Modeling of the Machine Grouping Problem in Cellular Manufacturing Systems

K. RAJA GUNASINGH AND R.S. LASHKARI

Department of Industrial Engineering
University of Windsor
Windsor, Ontario, CANADA   N9B 3P4

Abstract

A methodology is proposed to group the machines in cellular manufacturing systems based on the tooling requirements of the parts, toolings available on the machines and the processing times. Two 0-1 integer programming formulations are proposed, assuming that the part families are known. The application of the formulations is illustrated using an example.

Introduction

This paper addresses one of the important planning problems in cellular manufacturing systems. The basis for implementation of cellular manufacturing systems is the allocation of parts or components with certain "similarities" into groups or families, and the formation of machining cells which are capable of manufacturing one or more part families.

The machine grouping problem has been studied by a number of researchers. Among them, McAuley [2] used single linkage cluster analysis to form machine groups based on jaccard similarity coefficients between machines. Rajagopalan and Batra [3] developed a graph theoretic method which uses cliques of the machine graph as a means of grouping. Steudel and Ballkur [5] suggested a two stage dynamic programming heuristic to form machine groups. Lemoine and Mutel [1] presented a dynamic clustering technique for machine grouping. This paper proposes a methodology to group the machines on the basis of their capabilities to process the parts under consideration using 0-1 integer programming models.

Proposed Methodology

The methodology proposed in this paper groups the machines on the basis of their "capabilities" to process a given set of parts. Most machine grouping methods reported in the literature assume that each operation of a part is restricted to one machine. In cellular manufacturing systems, however, each operation of a part may be performed on alternative machines. To account for this, the processing requirements of the parts are defined in terms of their tooling requirements, and the machine capabilities are expressed in terms of the tool availabilities on each machine. Indices are developed to define the "compatibility" of a machine with a part based on the tooling requirements of the part, toolings available on the machine, and the processing times. Subsequently, two 0-1 integer programming formulations of the machine grouping problem are developed. These formulations are explained in the following section.

Mathematical Modelling

  Notation:

i   - index of part $i=1,n$

j   - index of machine, $j=1,m$

k   - index of group, $k=1,p$

NP_i - number of tools required for part i

NM_j - number of tools available on machine j

NCT_{ij} - number of common tools between part i and machine j

CO_{ij} - set of operations of part i that need machine j

OP_i - set of operations required to be performed on part i

S_{ij} - compatibility index between part i and machine j

P_{il} - processing time for operation l of part i

q_i - annual production requirements of part i

f_j - annual fixed cost rate of machine j

L_i - size of unit load for part i

C   - average cost of an intercell movement

NMA_j - number of type j machines available

MNA_j - maximum number of type j machines that could be procured

NMG_k - limit on the number of machines in group k

a_{ik} = 1 if part i is in group k

         = 0 otherwise

The decision variables are defined as follows:

X_{jk} = 1 if machine j is in group k

         = 0 otherwise

Development of Compatibility Indices. The development of the indices to define the compatibility of a part with a machine is explained below.

(a)   The compatibility index based on tooling requirements is defined as:

$$S^1_{ij} = \frac{2.NCT_{ij}}{NM_j + NP_i} \qquad (1)$$

which implies that a large number of common tools between a part and a machine results in reduced intercell movement if the part and the machine are placed in the same group.

(b)   The compatibility index based on tooling needs and processing times is defined as:

$$S_{ij}^2 = \frac{\sum\limits_{\ell \,\epsilon\, CO_{ij}} P_{i\ell}}{\sum\limits_{r \,\epsilon\, OP_i} P_{ir}} \tag{2}$$

This index implies that the longer the processing times of a part on a machine, the higher the utilization of that machine, if the part and the machine are placed in the same group.

Formulation of the machine grouping model. The machine grouping model is formulated to represent the situation where one is interested in reorganising the existing manufacturing system into a cellular structure. The model is formulated under the following assumptions: a. part families are known; b. tooling requirements, processing times and production quantities for each part are available; c. toolings available on each machine, machine cost, and the cost of material movement are available; and d. the existing machines are all utilized.

Two possible objective functions are considered in formulating this model. The first objective function maximizes the sum of the compatibility indices of all machines and parts in all groups.

$$\text{Max} \quad Z_1 = \sum_{ijk} X_{jk}\, a_{ik}\, S_{ij} \tag{3}$$

where S_{ij} is defined by either eq. (1) or eq. (2).

The second objective function seeks a trade-off between the cost of allocating the machines and the cost of intercell movement. Intercell movement arises if a part in group k requires a tool which is available only on a machine in group m, m≠k. Thus, we have:

$$\text{Min} \quad Z_2 = \sum_{jk} f_j X_{jk} - \sum_{ijk} X_{jk} a_{ik} NCT_{ij} C\, q_i/L_i \tag{4}$$

The first term in the objective function represents the cost of allocating a machine to a group, and the second term represents the maximum savings achievable in the cost of intercell movement.

The constraints on the system are as follows:

$$\sum_{j} X_{jk} \leq NMG_k \quad \text{for all k} \tag{5}$$

$$\sum_{k} X_{jk} = NMA_j \quad \text{for all j} \tag{6}$$

$$X_{jk} \,\epsilon\, \{0,1\} \quad \text{for all j and k} \tag{7}$$

Constraint (5) imposes restrictions on the number of machines allowed in each group. Constraint (6) ensures that all the machines of each type are allocated to one or more cells. The last constraint meets the integrality requirements.

Application of the Formulations

The application of the machine grouping model is illustrated using an example problem . A manufacturing system with 25 parts and 10 machines is considered. The tooling and the production requirements of the parts are known; however, they are not presented here to save space. The toolings available on the machines, the annual fixed cost rate of the machines, and the number of machines of each type are given in Table 1. The average cost of an intercell movement is assumed to be $5 and the size of the unit load is assumed to be 5 for all the parts. The parts have been grouped into three families, based on their similarity in tooling requirements. Table 2 represents the composition of the three families.

Machine	Number available	Annual fixed cost rate ($)	Tools available
1	2	10,000	A01,A02,A03,A04,A05,A06
2	1	8,000	B01,B02,B03,B04,B05,B06
3	1	6,000	C01,C02,C03,C04,C05
4	1	7,000	D01,D02,D03,D04,D05,D06
5	1	8,500	E01,E02,E03,E04,E05,E06,E07
6	1	12,000	F01,F02,F03,F04,F05,F06,
7	1	11,000	G01,G02,G03,G04,G05,G06
8	1	9,000	H01,H02,H03,H04
9	2	10,500	M01,M02,M03,M04,M05
10	2	11,000	R01,R02,R03,R04,R05,R06

Table 1. Tools Available on the Machines

Part family	Part Numbers
1	3,5,8,9,13,18, 19,23,24, and 25
2	2,6,7,11,15,16, 21, and 22
3	1,4,10,12,14, 17, and 20

Table 2. Part Family Composition

Part family	Objective function-1	Objective function 2
1	1,2,5,6, and 10	2,5,6,7, and 10
2	1,3,7,9, and 10	1,3,9, and 10
3	4,8, and 9	1,4,8, and 9

Table 3. Results of Machine Allocation

The resulting machine grouping problem was solved on an IBM 4381 computer in 12 seconds of CPU time, using the SAS/OR [4] software. The resulting solution is given in Table 3.

Conclusions

Two integer programming formulations of the machine grouping problem are presented in this paper. Assuming that the part families are known, the grouping of machines is based on the tooling requirements of the parts in each family, and the toolings available on the machines. The application of these formulations is illustrated using an example problem. For small to medium size problems, commercially available integer programming codes may be used to solve the formulations. The authors are currently investigating the possibility of using this methodology to solve large size machine grouping problems.

Acknowledgement

Major funding for this research work was provided by the Natural Sciences and Engineering Research Council of Canada under grant no. A4187. Their financial assistance is gratefully acknowledged.

References

(1) Lemoine, Y.; Mutel, B., Automatic recognition of production cells and part families, in Advances in CAD/CAM, Ellis and Semenkov, O.I., (eds.) Amsterdam: North Holland Publishing Co., 1983.

(2) McAuley, J., Machine grouping for efficient production, Production Engineer, 51(1972), 53-57.

(3) Rajagopalan, R.; Batra, J.L., Design of Cellular production systems: A graph-theoretic approach, Int. J. Prod. Res. 13(1975), 567-579.

(4) SAS Institute Inc., SAS/OR User's Guide, Version 5 Edition, Cary, N.C.: SAS Institute Inc. 1985.

(5) Steudel, H.J.; Ballakur, A., A dynamic programming based heuristic for machine grouping in manufacturing cell formation, Computers and Ind. Engg., 12(1987), 215-222.

The Impact of Designing GT Cells on the Total Performance of Computer-Integrated Manufacturing

IBRAHIM AL-QATTAN and JAMES R. ROSE
Box 5002
Tennessee Technological University
Cookeville, TN 38505

SUMMARY

The total performance of Computer Integrated Manufacturing (TPCIM) represents the overall integration of productivity, quality and flexibility. Due to rapidly decreasing product life cycles, flexibility has become a major criteria in the evaluation of the total performance of manufacturing systems. This paper will discuss two group technology techniques which can be used to assist forming more flexible manufacturing systems. The impact of GT cells on the flexibility of products, machines, and processes will also be discussed.

INTRODUCTION TO TOTAL PERFORMANCE OF CIM

American Manufacturing productivity and quality have improved considerably over the past few years due to the application of a variety of CAD, CAM, and CAE tools within the manufacturing environment. These tools are referred to generically as CAX tools, and the application of these tools to manufacturing has given rise to another acronym, computer integrated manufacturing (CIM). It has been reported that, despite good productivity and quality, firms are encountering declining profitability because of a lack of flexibility in their manufacturing systems. Research has shown that the average life cycle for new products has declined in recent years [2]. The major reason for the shorter life cycles is related to the impact of technological innovation in materials and industrial techniques, thus necessitating rapid changes in manufacturing systems. Therefore, flexibility has arisen as an essential factor which is vital as the core strategy in planning for the optimization of total system performance. Hence, the achievement of total performance of computer integrated manufacturing systems (TPCIM) must include an integration flexibility, productivity, and quality criteria. Productivity, quality, and flexibility are critical measures of the total performance of any manufacturing system. CIM provides the potential to maximize overall manufacturing performance through the formation of an integrated network in which computers are used in all functions associated with a given enterprise.

Productivity is concerned with efficient utilization of resources (input)
in producing goods and/or services (output). Total productivity is the
ratio of total output to the sum of all input resources. Partial productivity is the ratio of output to class of input (labor, capital, material,
overhead, etc.). Quality is a measure of manufacturing performance which
represents the degree of perfection in making products. Two different
types of quality measures are considered. Process quality is the ability
of processes to make good products with a small prevention cost and product
quality is the degree of excellence of finished products [3]. Flexibility
is the measure of manufacturing performance which indicates a manufacturing
system's adaptability to change as new or modified products are demanded.
There are several types of flexibility which influence overall manufacturing
system flexibility [2]. These are:

. Machine flexibility: ease of making the changes required to produce
  a given set of part types.
. Product flexibility: ability to change over to produce a new set of
  products economically and quickly.
. Process flexibility: ability to produce a given set of part types in
  several ways.
. Operation flexibility: ability to interchange the ordering of certain
  operations for each part type.
. Scheduling flexibility: ability to handle different routes along which
  a given part can be produced.
. Volume flexibility: ability to operate the system profitably at different production volumes.
. Expansion flexibility: capability of expanding the system size as needed,
  easily and modularly.
. Material handling system (MHS) flexibility: ability to handle different parts in a number of different routes.

The overall flexibility therefore is the universe of part types that a
system can produce.

Group Technology Schemes

Group technology (GT) is a manufacturing philosophy used in production
planning in which similar products or parts are identified and grouped

together to take advantage of their similarities in design and manufacturing characteristics. GT does not include a classification and coding scheme only. It also contains a technique for developing methods for grouping parts into part families (PF) and machines into machine cells (MC). Therefore, a product family is a collection of products which are similar, either because of their geometric shape and/or size or because similar machining steps are required in their production. Grouping helps to decompose a complex manufacturing system into a series of smaller and simpler subsystems. Two GT schemes are proposed in this paper. The first, called Product Code (PC), is used for product design specifications. The second, called feature code (FC), is used for decomposing a product into features. Figures 1 and 2 illustrate the product and feature codes, respectively [1]. The product code contains general information such as geometric shape, material, dimensional ratio, number of internal/external features, the machines to be used, part number (which includes all features), and the batch size of the product to be produced. The product code provides a sort of horizontal level communication to facilitate access to the database at the function-

PRODUCT CODE

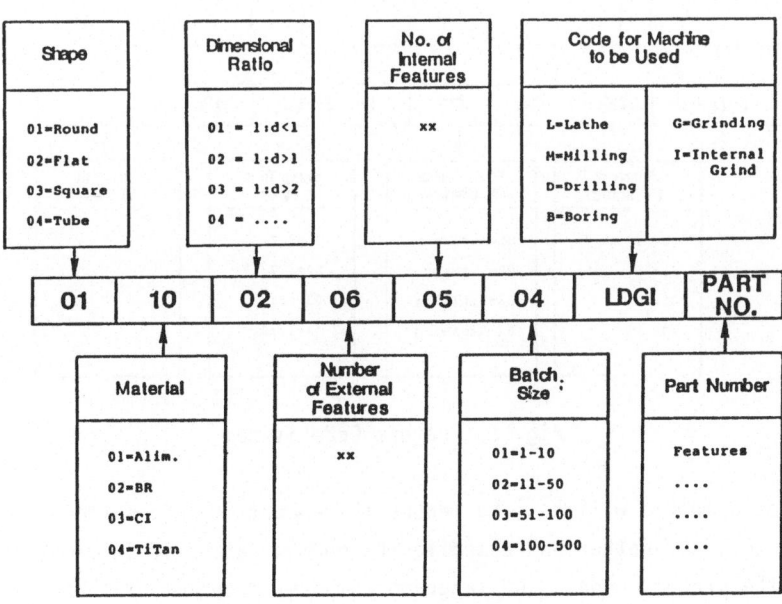

Fig. 1. Product Code System

al level, thereby enabling the transfer of data and information among CAX tools. The feature code information includes tolerance level, surface finish, machining operations, setup and processing time, and special fixtures required for feature extraction. The feature code provides, at the operational level, vertical communication which enables easy access to the database system in order to implement the search for optimization parameters [1]. The proposed method dissects the data and information into two parts. One part is used for generating the NC cutter path and contains information associated with the configuration of the solid model to be fabricated. The other part is used for production planning, such as forming GT cells, the sequence of machining operations, tooling and fixtures selection, and searching for optimization parameters.

FEATURE CODE: Fxx

Feature Name	Tolerance Level	Machine Used	Processing Time Minute
Cyln	01=±.0005"	L=Lathe	01=<1
Cone	02=±.001"	M=Milling	02=(1-3)
HoleT	03=±.005"	D=Drilling	03=3.-5
HoleB	04=±.01"	B=Boring	04=5-15
	05=±.10"	S=Shaper	05=16-30

| Hole-# | 03 | 04 | 02 | D | 02 | 03 | |

Surface Finishing	Machining Operation	Setup Time Hour	Special Fixtures
00: Rough	01=Facing	01=<.2	
01: V	02=Turning	02=(.2-.5)	
02: VV	03=Drilling	03=.51-1.	
03: VVV	04=Renaming	04=1.-2.	

Fig. 2. Feature Code System

The proposed method will help to reduce the amount of information transfer from one tool to another and expedite the communication of information through the product code. "C" language computer programs have been written to create an interactive environment to generate both the product and features codes.

THE IMPACT OF DESIGNING GT CELL ON TPCIM

The basic idea behind group technology is using coding and classification
methods to group together similar parts into part families and similar
machines into machine cells. Grouping allows for the decomposition of
a manufacturing system into a number of smaller subsystems. Dealing with
a number of subsystems will simplify the flow of the information and will
minimize the amount of data to be managed. The main problem in process
planning is the complexity of scheduling required for machining operations.

GT cells provide a tool to reduce the complexity of this problem because
parts with the similar process plans (routes) are grouped into part families
and corresponding machines are grouped into machine cells. The scheduling
of each part family is far easier than scheduling all parts together.
Designing GT cell will enhance the flexibility of machines, parts and
routing. After completing the design phase, the designer will interact
with two "C" computer programs, one to generate product codes and the
other to generate feature codes. Each coded feature is extracted from
the CAD tool and will include all necessary information about the geometrical
shape of the feature. This information will be assigned to a single feature
code. Data for obtaining the NC tool path can be accessed easily by calling
the feature code. Each encoded product includes its machining operations
which will be used to form GT cells. GT cells coupled with product codes
will reduce scheduling problems and will enable the generation of automated
process plans. Further, the use of feature codes will reduce the amount
of data to be managed and will accelerate the obtainment of NC toolpath
data.

References

1. Al-Qattan, I. and James Rose, "Using GT/CAPP to Enhance Product Data
   Data Exchange Standard-Key to CAD/CAM Integration", CG International
   Proceedings, May 24-27, 1988, Switzerland.

2. Gordon, H. McDougall and Hamid A. Noori, "Manufacturing-Marketing
   Strategic Interface: The Impact of FMS," Modeling and Design of FMS,
   Elsevier Science Publication, Amsterdam 1986, 189-205.

3. Young, Kuy Son, Chan S. Park, "Economic Measure of Productivity, Quality
   and Flexibility in Advanced Manufacturing Systems." Journal of Manufacturing Systems Vol. 6, No. 3, 1987.

Robotics and Automation

Project-Based Undergraduate Robotics Education

WAYNE W. WALTER

Mechanical Engineering Department
Rochester Institute of Technology
Rochester, New York 14623

Abstract

Robotics is a subject that cannot be taught directly from a
textbook. It can best be learned "by doing". This paper
describes the format of the present nontraditional project-
based undergraduate robotics course taught to our senior
mechanical engineering students. Student design projects,
both at local industries and done in-house, provide a means
to a hands-on learning experience for students that a
textbook approach cannot provide.

Introduction

The robotics program in the Mechanical Engineering
Department at RIT is being developed with virtually no funds
from the departmental budget [1]. Our program began in 1983
with a student working on an independent study project at a
local industry.

Today, we have a senior-level robotics course which is an
integral part of our program, and a robotics laboratory
which contains five industrial robots and three educational
robots. The major emphasis of our program is on hands-on
projects involving real-world problems. Such projects are
the focus of this paper.

Background

Our program got its start with an independent study project
at a major automaker here in Rochester. At the time, we had
no experience or hardware in the robotics area to allow him
to work on-campus. As a result of his success in
establishing a robotic workcell around an IBM 7535 robot,
many more projects were made available for subsequent
students. This also led to part-time work for me during the
academic year and full-time activities during the summers.
It was through these opportunities that I was able to
develop some meaningful background other than from
information available in texts.

Labwork

The Mechanical Engineering Robotics Laboratory currently contains four Unimation PUMA five axis industrial robots, an IBM 7535 four axis scara industrial robot, three educational robots, two Gould programmable controllers, and items constructed by students as part of their project work. Some of these items include a conveyor system used to pass parts between two robots during an assembly process, some hand-made robots controlled by personal computers, interfaces which allow various pieces of equipment to talk together, and tooling and fixtures for the robots.

All five industrial robots were provided on long term loan by a local automaker here in Rochester. Although the automaker initially sought to donate these units to us, it quickly became clear that the required paperwork on his part was prohibitive. Although loans greatly accelerated the delivery of the robots to RIT, they are not without some unique problems. Expenditures for maintenance are sometimes difficult since the University does not own the robots. As a result, students will experience some frustration using the robots during their project work when downtime occurs. Close attention to the robots by a technican is normally required to prevent this from becoming a major problem. There is also, of course, always the possibility that the robots may need to be returned at some point, although this seems rather remote at present.

The lab also currently contains a small industrial robot used for light assembly tasks which belongs to a local robot supplier with whom we have established a working relation-ship. The lab provides a convenient place for him to demonstrate the robot to potential customers. While it is in the lab, it is also available for use by students for programming exercises and as part of their project work. This is a new robot only recently released for marketing. The supplier has encountered new applications from customers which require that development work be done on the robot. The lab has recently completed a development project for him, and will shortly begin a second. It is our hope that more funded industrial development projects like this will follow, and that the lab will become an applications development center for local industry.

Design Projects

The essence of the educational experience in our robotics course is a applied project in which students learn "by doing". These projects, both at local industries and done in-house, provide a means to a hands-on learning experience that a textbook approach cannot provide.

In the recent past, almost all of the projects were done on-site at local industies [1]. This was particularly true in 1983 when the course was first taught [2]. At that time we had no equipment on-campus, as we do now, to permit in-house projects. By going to the industrial site, students were able to work with high-tech facilities and equipment that RIT cannot afford on real-world problems. These problems were generally of great interest to the industries involved, and this provided great motivation to students. Some of our projects continue to be of this type.

Industrial projects are arranged through the contacts I have developed at local industries in Rochester. After matching the student's interests with the available projects, I will visit the site with the students and introduce them to an engineer who will be their key industrial contact. He/she is usually the engineer responsible for the project at the site. They also facilitate students' entry to the plant, and help students locate needed resources and equipment.

Although industrial projects have been a great success, and extremely well-received by students, they have not been without problems. On occasion, business trips and production crises have prevented students from meeting with their industrial contact. Many of the projects, by their nature, require students to work entirely on the industrial site with consequent travel way from campus each week. It has sometimes been difficult for students to work at the industrial site on weekends when access to the plant may be restricted. In a few instances, resources that were needed by students were not made available as expected.

During the present academic year, in contrast to previous years, almost all projects have been done in-house in our robotics lab. Some reasons for this were mentioned above, but I believe the major factors are convenience to the students, and a recognition on their part that these projects greatly develop the state-of-the-art of our lab. Students can work in the lab between classes when they an hour or two of free time, and they can work in the lab on the weekends by signing out a key from the Department secretary. As a general rule, students must work in pairs when in the lab at night or on the weekend, and follow all prescribed safety procedures. A short description of projects done during the current academic year is given in Table I.

Students are required to document their progress on their project by keeping a day-day log of their activities. Weekly progress reports are prepared by the group leader. Each project team submits a set of design drawings at the end of the design phase. Recent design drawings have done on RIT's new Intergraph CAD facilities. The Mechanical Engineering Department has held a series of Intergraph seminars which many of the robotics students have attended to become proficient with the software. Although most of the robotics students have been mechanical engineering students, more industrial and electrical engineering students are becoming involved with mechanical engineering students on joint projects. This allows a technical interaction which is so typical of what students will encounter in industry. At the conclusion of the ten week quarter, students submit a final written report, and make a formal presentation to the class and invited quests from industry, if any have been involved in the projects. To date, all projects have been well received by students. As we continue to build the facilities in the lab, the quality and effectiveness of the projects will continue to advance.

TABLE I: Short Description of Student Projects Done During the 1987/88 Academic Year

Project #   Description

1.* Design, build, and test (D/B/T) a singulator to present a stack of sheet-metal parts to an ASEA robot one at a time. Parts may be of arbitrary shape and stick together.

2. D/B/T an odd-form electronic component pick-up device. Must be able to pick up a variety of shapes. Must lock in place.

3. D/B/T an eight position electronic component assembly turret. Want min. weight, max. no.of signals, and exellent positioning acuracy.

4. D/B/T a tool-changer to be used with the turret of Project #1. Want min. weight, max. no. of channels for electric and pneumatic signals.

5. Show feasability of robotic assembly of a paper baffle with 12 parts. Redesign and debug of existing tooling. Design-for-assembly (DFA) analysis required.

6. D/B/T a three-axis stepper driven jointed-arm robot, and control it by a pc. Want min. cost, fast response, and easy programming.

7. D/B/T a dual track parts feeder for a PUMA robot. Each track to have photo-sensors for part presence. Monitor sensors through the I/O module. (See Fig. 1)

8.* D/B/T a gripper to pick up a stack of payroll cards. Initial rectangular crossection must become a parallelogram before stack is presented to a card reader.

9. D/B/T a conveyor to pass parts between two robots. Want sensors for part presence, and min. cost. Control through robot I/O.

10. D/B/T a compliance device for a PUMA robot. Want min. weight.

11. D/B/T spot-welding tooling for a PUMA robot. Robot to have a tool-changer to pick up the welding unit with electrodes. Part clamping and transformer control via solenoids interfaced to the robot I/O. (See Fig. 2)

12. Use a programmable controller to interface the track feeder of Project #7 to a PUMA robot. Robot to receive instructions from the controller.

* Project done at the request of a local industry.

References

1. Wayne W. Walter, "Undergraduate Robotics Education on a Shoestring", Second International Conference on Robotics and Factories of the Future, San Diego, CA, July 1987.

2. Wayne W. Walter, "Initiating University Robotics Education Through Industrial Student Projects", Robotics and Factories of the Future, Springer-Verlag, 1984, pg.677.

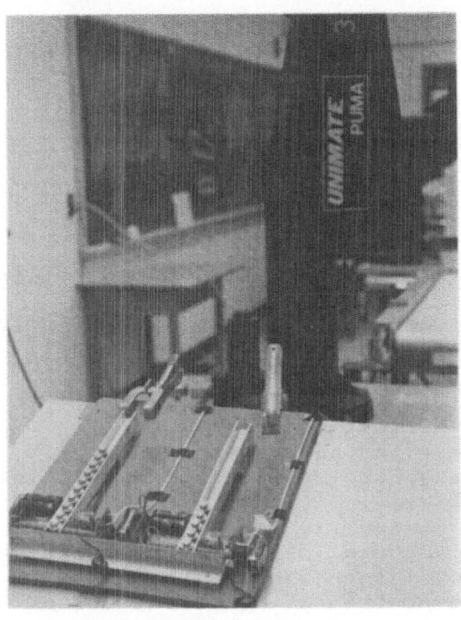

Fig. 1 Dual Track
parts feeder

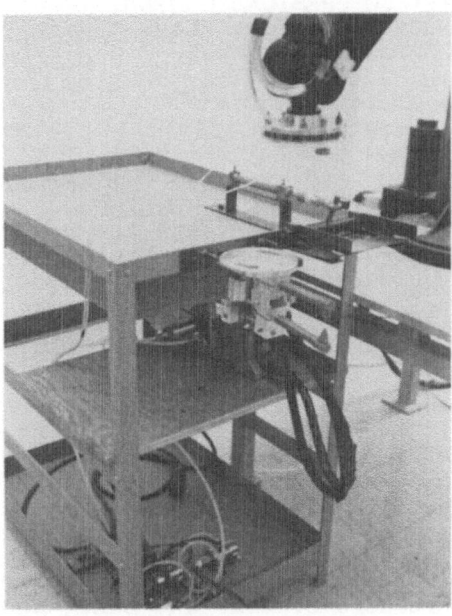

Fig. 2 Spot-welding Tooling
for a PUMA robot

Numerical Control and Robot Technology for the Process Industry Maintenance Function

P. R. HALE, P.E.

TEXACO CHEMICAL COMPANY
NECHES CHEMICAL PLANT
PORT NECHES, TEXAS

Summary
This report presents the need and application of numerical control and robotic technology to the repair of mechanical equipment and components, and to welding at a petrochemical plant or refinery. At present, machining and welding, two large classifications of the process industry maintenance function, are performed on manually controlled machinery and equipment.

Automated Welding
It is well known that welding is the most economical and efficient means for joining metals. In the process industry, pipe welding and to a lesser degree plate welding are extensivily employed to join metal pipe and plate.

The three basic welding processes are gas (oxyacetylene) welding, arc welding, and resistance welding. The high quality welds produced by GTAW and GMAW, and the continuous nature of these processes make them ideal candidates for automated welding.

With advances in the programming capabilities of robotic welding systems, their applicability is extending from long repetitive production runs to low volume production applications found in large chemical plants and refineries.

Automated welding can be applied to pipe welding by two means. First, orbital welders attached to stationary pipe crawl around the circumference of the pipe welding a joint. The other means utilizes a robot welder in conjunction with a workpiece positoner which can be used for pipe or plate welding.

In some applications it is be more practical to move the workpiece than rotate the weld gun about the workpiece. These applications require a robot welding system. The robot welding system, which would be applicable to the process industry, has the following five basic components which will be described in detail.

1. Operator Controls
2. Robot Controller
3. Robot or Manipulator
4. Automated Welding Equipment
5. Workpiece Positioner

Operator Controls -
The operator controls are usually a microprocessor with a
portable control pendant capable of off-line programming and
programming by the leadthrough method. Microprocessor software
can be used to pick and control the welding process. One model
contains ninety preprogrammed arc welding schedules for mild
steel applications. The operator inputs the weld joint
configuration, shield gas and metal plate thickness and the
microprocessor automatically sets the optimum arc voltage and
current, filler wire diameter and welding gun travel speed.
[Reference 6]

The microprocessor software can incorporate statistical process
control programs to monitor welding variables, such as, voltage,
wire feed speed, shielding gas consumption, and number of welds
performed, providing management with a valuable quality and
production control tool.

The microprocessor software can take feedback information from
the robot or positioner and stop their motion when it detects
that either one has hit something. This helps protect the human
operator and the equipment. The software and feedback can also
be used to stop motion on arc failure which can occur when the
wire feeder reel runs out or when welding arc initiation fails.
The software can make provision for periodic robot weld nozzle
cleaning at an automatic cleaning station.

Robot Controller -
The robot controller translates and relays information and
commands from the operator controller to the robot and it
coordinates the robot and workpiece positioner movements.

Robot or Manipulator -
The robot or manipulator consists of an articulate or rectilinear
robot arm, depending on the application, with an arc welding gun
end effector. For the weaving motions and maneuvering required
in pipe welding and the relatively small scale plate welding
performed in the process industry weld shop, the articulate type
robot arm is more effective. Rectilinear type systems are more
applicable when the workpiece is not curved. Welding robots can
be powered electrically, hydraullically or pneumatically.

Automatic Welding Equipment -
The automatic welding equipment for GTAW or GMAW includes the
power source, wire feeder, welding gun and shielding gas. The
power source or welding machine is generally a motor generator or
rectifier driven by a gasoline or diesel engine or wired to an
electric power supply. The power source can supply and control
direct current. The wire feeder controls the rate of filler wire
feed and shielding gas flow. The weld gun performs the arc
welding and can prepare the workpiece for welding or auxiliary
grinding equipment can be utilized for edge preparation.

In a new process, called pulsed current arc welding, the welding
machine produces two levels of welding current. One level is
high where arc "spot" welds are formed and the other level is low
where the spot welds of the high current are allowed to cool and
partially solidify before the next adjacent overlapping high

current spot weld occurs. The controls are set to maintain a
continuous weld. This would be very advantageous for welding
pipe and other contoured surfaces because it allows for much more
misalignment without sacrificing weld quality and uniform weld
appearance. With other arc welding methods variation of
electrode position (electrical current) or travel speed is
required for manuevering around curves.

Workpiece Positioner -
The workpiece is loaded manually at a specific reference location
(for the microprocessor or programmable controller) into the jaws
or chucks of the positioner. The simplest positioner is the
lathe type which can only rotate the workpiece about the
horizontal axis. The more versatile models can tilt and rotate
(two axes). The positioner can be driven manually,
pneumatically, hydraulically or electrically.

There are two types of drive mechanisms. First, the indexing
type which utilizes a cam-indexing mechanism in combination with
multiposition drive cylinders. The motion is in non-adjustable
incremental steps which is a distinct disadvantage. The other
type is the gear type which utilizes high precision gears that
allow continuous motion and can be integrated readily with
sophisticated control features.

Switches, sensors and potentiometers can be integrated into the
drive system to send information back (feedback) to the robot
controller to sychronize and control the robot and positioner
movements. A more versatile way to control would be to have a
seperate programmable controller for the positioner. This would
make it easier to interchange positioners with different robot
welders and possibly other equipment, and it would reduce the
required number of robot controller output channels. The
programmable controllers would allow positioners to be programmed
by off-line programming and programming by the walkthrough and
leadthrough methods.

Computer Numerical Control Machining
The machine shop in the processing industry is responsible for
maintaining a wide variety of equipment and components. Most of
these machine shops have a number of conventional engine lathes,
a vertical variable speed metal cutting band saw, a bench model
sensitive drill press for light duty drilling, a single station
radial drilling machine for heavy duty drilling, a vertical
column and knee type milling machine, a vertical and a horizontal
boring mill, and a shaper.

By a wide margin, the engine lathes are of most value. This is
because the majority of the work done in the shop is on the
shafts and associated accessories, such as cylinderically shaped
sleeves, glands, journal bearings, and collars, of the process
rotating equipment. Machining on a conventional engine lathe is
time consuming, labor intensive and at times tedious. The other
machines are used mostly for preliminary and secondary machining
operations on pieces machined primarily in the engine lathes.

Original equipment manufacturers (OEMs) do not stock replacements
for the vast majority of the shafts and other accessories that

they supply as part of their equipment so upon receipt of order, delivery is usually measured in weeks. Even if the OEM does stock the item, unless it is stocked locally, delivery is minimum one or two days with extra freight charges for fast delivery. Generally when process rotating equipment comes down for repair it needs to be put back in service immediately, if not sooner, to prevent loss of production.

The time and labor required to machine the aforementioned components from roundstock on manually operated machines coupled with long deliveries from the OEMs makes it necessary for chemical plants and refineries to stock huge inventories of rotating equipment components so they will be on hand when needed. If these components could be machined quickly with the required accuracy and finish then this inventory of expensive finished parts could be dramatically reduced and turnaround time for equipment with non-stocked spare parts could also be reduced.

A continuous path control CNC, chucking, saddle-type horizontal turret lathe with compound cross slide would be ideal for machining the variety of components used in the process industry.

This would require the purchase of a new machine completely designed for versatile CNC operation. An existing engine lathe revamped with point to point NC or continuous path NC can still only hold one tool at a time which means the machine must be shut down for tool changes. In contrast, the turret on a turret lathe can hold a number of tools and a simple rotation of the turret puts another tool into cutting position. Revamping and engine lathe with point to point NC requires only controlling cutter tool movement. Revamping to continuous path NC requires controlling cutter tool movement and also entails controlling feed rates. Revamping to control cutter tool movement (point to point NC) would not be too difficult. However, the difficulties involved in revamping a manually operated engine lathe with NC to control feedrates, as is required for continuous path NC, would outweigh the benefits gained. Revamping an engine lathe with a point to point NC positioning system is feasible and would aid in general turning but would be of no benefit in the many applications which require taper turning, thread cutting and contour forming.

The CNC lathe should feature point to point positioning, continuous path linear and circular interpolation, internal and external thread cutting control which eliminates cumulative error from thread to thread, sequence number display and search, command and dimension display, part program storage and editing capability, off-line keyboard manual data input, extensive self diagnostics, tool selection control, spindle speed control, tool offset control to compensate for tool sharpening and position change, automatic chamfering and corner rounding, and integrated adaptive control: automatic spindle acceleration/deceleration with load change. No bar loader would be needed because it would be of inconsequential value for the low volume, fast turnaround components to be machined on a CNC lathe in this application.

To facilitate CNC machine production some type of microcomputer based CAD/CAM system is necessary. The system needs to be easy to use and be readily accessible to the machine shop. The system should consist of the following complete package.

1. A mechanical drafting program complete with geometric dimensioning and tolerancing, and finishing symbols;
2. a fixture design program; and
3. an NC part programmer that can generate an NC part program from parameters of the mechanical drafting program.

These systems are currently available.

References
1. Giachino and Weeks, WELDING SKILLS, American Technical Publishers, 1985.

2. P. T. Houldcroft, WELDING PROCESS TECHNOLOGY, Cambridge University Press, 1977, Chapter 11.

3. Edgar Graham, MAINTENANCE WELDING, Prentice-Hall, 1985.

4. Robert C. Rosaler, Editor in Chief and James O. Rice, Associate Editor, STANDARD HANDBOOK OF PLANT ENGINEERING, McGraw-Hill, 1983, Chapter 16-3 by Howard Cary, Hobart Brothers Company.

5. Mikell P. Groover and Emory W. Zimmers, Jr., CAD/CAM, Prentice-Hall, 1984, Chapters 10 and 11.

6. Robert C. Baldwin, "Maintenance Welding By Robot" , Plant Engineering, 38:44-50, October 25, 1984.

7. Milton F. Pierce, "A Guide To Robotic Arm Welding Positioners", Welding Journal, 64:28-30, November 1985.

8. J. Hanright, "Selecting Your First Welding Robot", Welding Journal, 63:41-5, November 1984.

9. Victor E. Repp and William J. McCarthy, MACHINE TOOL TECHNOLOGY, McKnight Publishing Company, 1984.

10. C. Thomas Olivo, ADVANCED MACHINE TECHNOLOGY, Brenton Publishers, 1982.

11. Ray Milton, "Single-Setup Turning Eliminates Secondary Operations", Modern Machine Shop, 60:74-78, August 1987.

12. Michael Page, "Mill/Turning Gains in Europe", Modern Machine Shop, 60:78-88, September 1987.

13. Edward G. Hoffman, "Microcomputer CAD/CAM For All Shops, Part I", Modern Machine Shop, 60:82-90, October 1987.

14. Edward G. Hoffman, "Microcomputer CAD/CAM For All Shops, Part II", Modern Machine Shop, 60:102-107, November 1987.

A Configuration of Parallel Pipelined Processing with DSP-Based Controller for High-Order Linear Systems

MYOUNGHO SUNWOO

General Motors Research Laboratories
Warren, Michigan 48090

SUBRAMANIAM GANESAN and KA C. CHEOK

Oakland University
Rochester Hills, Michigan 48063

Abstract

A configuration of parallel pipelined processing for an explicit adaptive
controller with multi-Digital Signal Processors (DSP) is investigated.
Employed is the pole-placement Explicit Adaptive Control (EAC) or Self-Tuning
Regulator (STR) algorithm which consists of on-line parameter estimation,
compensator design, and manipulation of the control parameter with the DSP
system. For higher-order digital control systems, the throughput of the
controller can be maximized by parallel processing with multi-DSPs.

1. Introduction

The introduction of DSPs by NEC and AT&T in 1980 became a major impetus for
the implementation of real-time digital signal processing for digital
filters, speech analysis, image processing, and telecommunications [1]. The
DSP's pipelining technique for instruction can be applied effectively to the
intensive computation processing required. The throughput of the pipelined
system increases as the degree of pipelining is increased. The fast
execution cycle of the dual-bus Harvard architectured DSP is very attractive
for many digital control areas. The DSP's can also be pipelined to provide
systolic array processing.

However, the potential of this pipelining technique for digital control
application has not been fully exploited in either research or practice.
This paper introduces a configuration of parallel pipelined processing for a
high-order controller system such as an adaptive controller. The parallel
architecture enhances the throughput of the processor with fast processing
time and reduces the calculation algorithm with multi-DSP's. These DSP's
share duties such as parameter estimates and convolution calculation of a
linear control system by a load balancing strategy. An EAC [4,5] is applied
and implemented in the DC servo laser scanner which scans the position within
a prespecified boundary. The implemented results are shown in this paper.

2. DSP-based Controller Architecture

A general equation for a linear system in the Z-domain is as follows:

$$H(z) = \frac{Y(z)}{U(z)} = \frac{\sum_{k=1}^{N} b_k z^{-k}}{1 + \sum_{k=1}^{N} a_k z^{-k}}$$

A systolic pipelined realization of the above linear system is the modular controllable canonical form shown in Figure 1 [2,3]. This synchronous data flow can be fitted for the pipelined DSP architecture.

TMS 320 Digital Signal Processor

The TMS320, which employs a modified dual-bus Harvard architecture, enables simultaneous fetching of data and instructions. The architecture allows the pipelining of instructions which use addressing modes such as auto-increment, auto-decrement modes for arrays, and indexed addressing modes for data processing. It allows efficient access to relevant data structures and the transfer of data from program memory to data memory and vice versa. The major hardware and software key features of the TMS320 DSP are specified in references [1,7].

Most instructions and arithmetic operations are executed in a single cycle. The following subset of instructions illustrates the unique operations available in DSP: RPTK, MAC, LTD, BLKD, DMOV, MACD and so on are instructions which implement several operations in one instruction cycle. This type of processing enhances the DSP-based digital control system. It can be applied and implemented to complex digital control algorithms in real time.

Parallel Architecture

Figure 2 is a block diagram of a dual DSP parallel processing configuration. The flexibility of the second generation of Texas Instruments (TI) TMS320 architecture DSP allows the operation of a multi-DSP system. As shown in the figure, the configuration uses global memory space for common data memory storage. Two DSPs share the global memory while executing the local program memory.

The software arbitration of global memory is facilitated by Bus Request (BR) and the external flag output (XF). The combined logic READY notices whether the DSP is ready or not. There are two cases when the bus is used: 1) to access the global memory area, and 2) to acquire the input data and store it in the global memory. For instance, when DSP#1 wants to access the global memory, the BR of DSP#1 is asserted. Then the instruction BIOZ of DSP#2 polls the BIO pin. When DSP#1 reads the data from the input channels, the sample clock changes the state of BIO, so the BIO pin can be polled by a branch instruction BIOZ. The action of BIOZ instruction depends on the state of BIO. At the same time, since the requested data address is in global memory, the BR of DSP#1 will be asserted. The XF of DSP#1 must be low for smooth data transfer.

For synchronizing the multi-DSP system, a special SYNC pin of the TI DSP is used. This signal allows the internal clocks of two or more DSPs to be synchronized. Since the processors operate on the same internal clock signal, all external signals will be synchronized. By synchronizing the inter processor signals, the need for external arbitration logic is eliminated.

The software of each processor can be balanced by the length of the program for each DSP. As long as the control algorithms do not interfere with each other, the DSPs can share their duties. Figure 3 is the flow chart of the system which shows the duties of each processor for real-time implementation of the explicit adaptive control algorithm.

The amount of processing duty each DSP will provide is determined by the scheme of the control algorithm, the number of the I/O channels, the order of the system, and the bandwidth of the system. In high-order linear systems, equal distribution of the convolution calculations among the processors provides optimization of software and hardware resources.

3. Explicit Adaptive Control

The three steps of the EAC or STR are:

- On-line Unknown Plant Parameter Identification
- On-line Design of controller parameters
- On-line Realization of the Controller.

The STR, based upon pole-placement, is selected in this application because of its versatility and the ease implementation. A lowest-order convergent Recursive Least Squares (RLS) is used for on-line parameter estimation schemes. The forgetting factor λ is applied for giving more weight to the most recent data and for ensuring global convergence of the process. A closed-form solution is solved using Macsyma and used for on-line calculation of controller coefficients.

Figures 4 and 5 depict the test setup and the block diagram of the STR laser scanner. Figure 6 shows the result of the implementation. The detailed description of DC servo system is not included in this paper in view of space limitation.

4. Conclusion

A parallel pipelined processing with multi-DSPs approach to STR was initiated for time-invariant high-order linear system control. The pilot framework for this case was designed and successfully implemented for the STR DC servo motor controlled laser scanner which focused the position within the predetermined overshoot and settling time constraints with a general-purpose microprocessor. These experimental results indicate feasibility and flexibility for the proposed parallel processing architecture with multi-DSPs and different applications.

References

[1] Lin, Kun-Shan, Frantz, Gene A., and Simar, Ray, "The TMS320 Family of Digital Signal Processors," Proceedings of the IEEE, Vol. 75, No. 9, September 1987.

[2] Rao, Sailesh K. and Kailath, Thomas, "VLSI Arrays for Digital Signal Processing Part I-A Model Identification Approach to Digital Filter Realizations," IEEE Transactions on Circuits and Systems, Vol. CAS-32, No. 11, November 1985.

[3] Hung, H. T., "Why Systolic Architectures?" IEEE Computer, pp. 37-46, January 1982.

[4] Chalam, V. V., "Adaptive Control Systems," Marcel Dekker, Inc., 1987.

[5] Astrom, Karl J. and Wittenmark, Bjorn, "Computer Controlled Systems," Prentice Hall, 1984.

342

[6] Sunwoo, M., Cheok, Ka C. and Ganesan, Subramaniam, "An Implementation of Explicit Adaptive Control with DSP-based Control System," Proceedings of 19th Annual Pittsburgh Conference on Modelling and Simulation, May 1988.

[7] "TMS320C25 User's Guide", Texas Instruments, 1986.

z^{-1} : Delay element (storage)

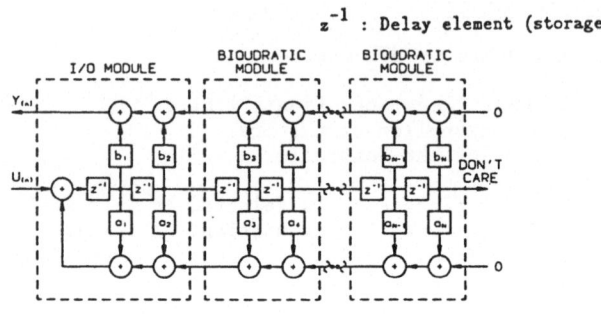

Figure 1. Systolic Processing of Controllable Canonical form
Nth-order System

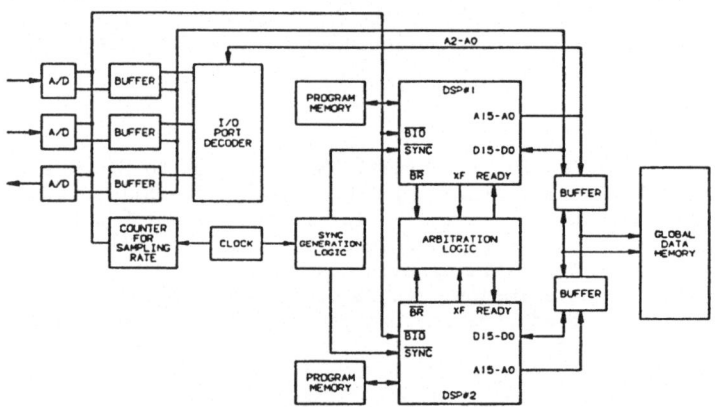

Figure 2. Parallel Processing Architecture

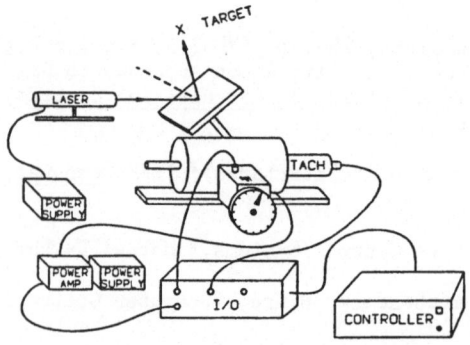

Figure 4. DC-servo Laser Scanner

343

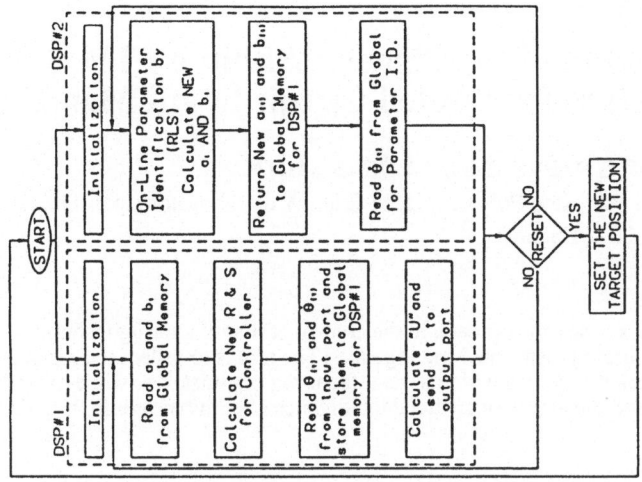

Figure 3. Flow chart of Parallel Processing of STR

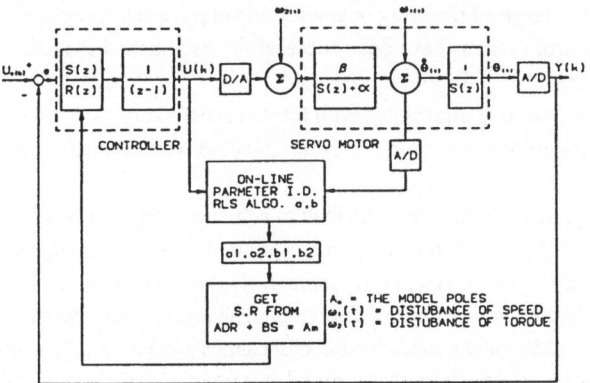

Figure 5. Block diagram of Self-tuning Adaptive Controller

Parameter		Input	Output
a	b	U	Y
-0.8060	1.4455	1.2	40
-0.8065	1.4450	1.0	35
-0.8070	1.4445	0.8	30
-0.8075	1.4440	0.6	25
-0.8080	1.4435	0.4	20
-0.8085	1.4430	0.2	15
-0.8090	1.4425	0.0	10
-0.8095	1.4420	-0.2	5
-0.8100	1.4415	-0.4	0

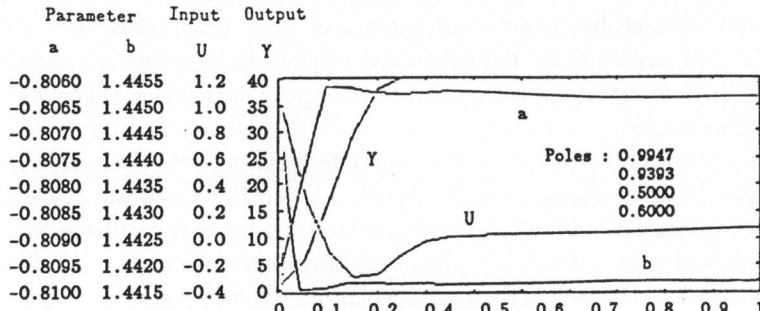

Figure 6. Implementation Result

Distribution of Intelligence Within a CIM Architecture for Printed Circuit Board Manufacture

PROFESSOR G RZEVSKI and A LUCAS-SMITH
CIM Centre, Kingston Polytechnic, 21 Eden Walk, Kingston upon Thames,
Surrey KT1 1BL, England

Summary
This paper describes an approach to the control of manufacturing technology by computer systems possessing distributed intelligence. The aim is to reflect the manner in which humans attempt to make decisions within a manufacturing organisation, while exploiting the speed of automation and the ability of artificial intelligence to eliminate conflicts in decision making.

Introduction
A vast body of research and development in manufacturing systems has mirrored the evolution of the mainframe computer by producing monolithic scheduling and control systems to run on them. MRP, MRP2, large engineering databases and application programs are typical results. Although realtime control systems have been successfully established as essential for continuous flow production the search for all purpose manufacturing schedulers of much wider application has resulted in scheduling algorithms with limited satisfactory performance. Usually they perform well when manufacturing conditions are subject to well defined parameters and predictable variations.

The fully optimised schedule is difficult to achieve within the large solution space available. Requirements, priorities, tradeoffs and implicit assumptions are human concepts of complexity and ambiguity. The fuzziness of manufacturing, described in [1], suggests that very often a heuristic schedule will be more effective than an algorithmic solution. Current algorithms require considerable data, run in unacceptable timeframes and can still be unresponsive to the many unexpected events of a manufacturing environment.

The need for expertise to be encapsulated within a manufacturing system described in [2] and [3], has prompted the demand for artificial intelligence techniques in the achievement of CIM. The distributed nodal intelligence described in this paper is considered applicable to a wide range of manufacturing environments. It is an attempt to capture the many facts, concepts, rules and methods that humans employ in manufacturing and place them where, and only where they are required in practice.

A similar, but distinct architecture [4] described by Parunak proposes 'actors', equivalent to nodes, capable of negotiating amongst themselves. Ultimately this would enable behaviour like that of cooperating humans under critical conditions which require conflict resolution.

The concept of neural networks, [5] also has similarities with the nodal approach. It can be viewed as the special case of a large number of nodes holding identical rules solely concerned with the GO/NO GO state of machines.

Rzevski [6] has proposed the concept of distributing machine intelligence within a manufacturing system by providing knowledge systems for cell scheduling and control. The method and its underlying mechanisms are applied here to a typical printed circuit board factory.

Fundamental Concepts
Manufacturing knowledge may be related to a specific process, or to a group of processes at a

higher level. The aim is to produce an architecture which will support an appropriate distribution of intelligence and enable processing and updating of knowledge in order to schedule and control events, in a rational manner. The ultimate measure of success will be the ability of the system to perform under conditions of uncertainty at least as well as human experts.

Associated with manufacturing processes and resources are the human activities of making decisions, initiating and controlling processes, recording results and communicating. The activities of moving inventory or money, and of providing services are to be associated with **Transaction Nodes,** the locations where **Business Transactions** are carried out. It is proposed that in all cases a **Transaction Account** be maintained by crediting or debiting movements into and from the Transaction Accounts. This is directly parallel to the use of ledgers and files in any organisation, but in this case each logically associated with a particular Transaction Node.

Similarly, decision making activities are to be associated with intelligent **Decision Nodes,** located at whatever location and level the decisions are made. It is emphasized that the function requires knowledge only of how decisions are to be made and not the details of the resulting transaction, which are the concern of the Transaction Nodes.

Finally, all aspects of communications are to be the concern of intelligent **Communication Nodes,** required to pass simple messages between nodes as well as issue longer reports. Communications may be amongst single and multiple nodes, and externally to and from humans and machines.

It is proposed to employ three types of knowledge-based sub-systems or nodes in order to separate the concerns of transaction handling, decision making and communicating. Each is to contain an appropriate inference engine and a knowledge base comprising both rules and facts (or data).

fig (i) Transaction Node Architecture	TRANSACTION INFERENCE ENGINE	COMMUNICATION NODE INTERFACE
	TRANSACTION KNOWLEDGE	

Capability: Possesses knowledge about how transactions are carried out and is able to update Transaction Accounts.

Transaction Knowledge: Machine and process capabilities; Process initialisation; Transaction accounts; Process success and failure.

Example: Capability and availability of machines M1, M2, M3 to carry out processes P1 - P8

fig (ii) Decision Node Architecture	DECISION INFERENCE ENGINE	COMMUNICATION NODE INTERFACE
	DECISION KNOWLEDGE	

Capability: Posseses knowledge about the decision making process.

Decision Knowledge: Rules and criteria for decision making.

Example: Rule 1. If operation X is to be performed on Part A check availability of machines M1, M2, M3 in that order of preference.

Rule 2. M3 to be used only if no other equivalent is available.

fig (iii) Communication
Node
Architecture

COMMUNICATION INFERENCE ENGINE	COMMUNICATION NODE INTERFACE
COMMUNICATION KNOWLEDGE	

Capability: Posseses knowledge about what to communicate and the destination of messages and reports.

Communication Knowledge: Contents and formats of messages and reports: Destination of nodal communications; How to communicate with nodes, machines and humans.

Example: If machines within group M1 - M9 become out of action, inform Transaction Nodes T1 - T9.

Intelligence Architectures

Sample configurations include the following with the three node types shown as (D) (T) (C)

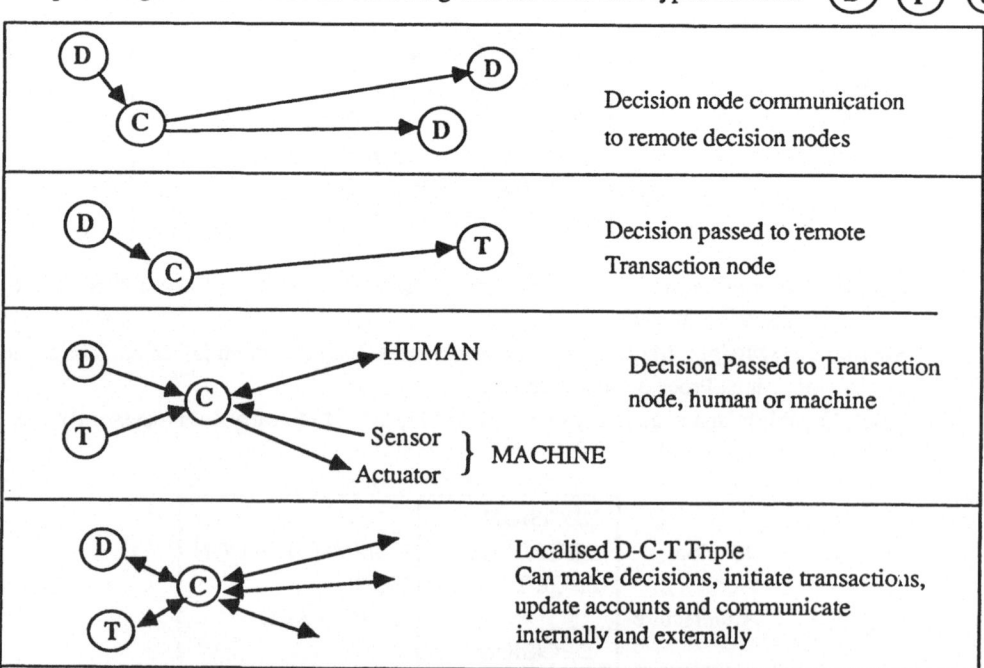

fig (iv) Examples of node configurations.

Printed Circuit Board Manufacture, A Case Study

This case study applies in outline the concept of distributed intelligence to a typical factory manufacturing printed circuit boards (PCBs) and in the process of implementing CIM.

Products

Around 800 types of PCB with approximately 20% change of products per annum. Around 10,000 boards produced per week, typically in batches of 20-80. Principal board types are double-sided (30% by number), multi-layer with 3-14 layers (60%) and multi-layer platters (10%)

Principal Production Activities

- Scheduling production
- Materials purchasing and supply
- Manufacturing layers and testing them
- Assembly and bonding of layers into boards, testing
- Component insertion and final test

Functions Required

A number of issues currently require resolution by means of human intelligence in scheduling and controlling workflow. The following principal functions are considered suitable for handling by intelligent nodes:

Decision	- Priority scheduling at job level
Nodes	- Material usage scheduling
	- Material purchasing schedule
	- Identification of late running jobs
	- Identification of jobs with potential for delay
	- Synchronisation of layer production for bonding
	- Decisions to expedite
Transaction	- Jobs completed on time
Accounts to	-Late running jobs
be maintained	- Scrapped layers or boards
	- Process yields
	- Test failures and successes-
Communication	-Pass along production line the details of job progress
Nodes	ie. if behind, up to, or ahead of schedule
	- Warnings of approaching materials shortage
	- Regular reporting on transaction accounts to appropriate nodes.

Location of Nodes (N)   ie D-C-T Triples

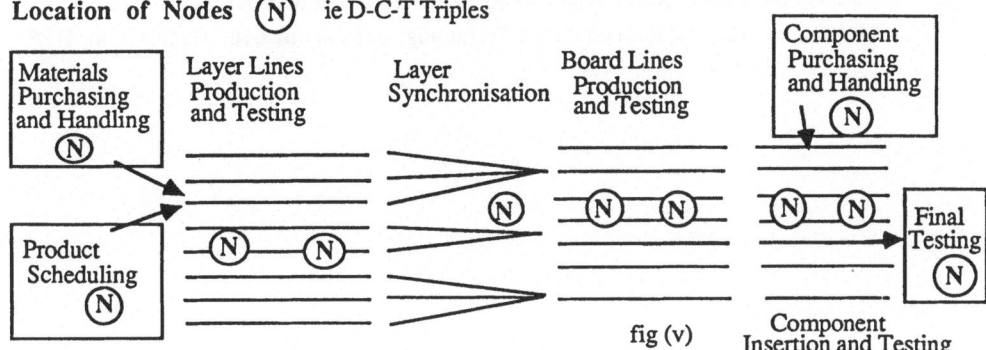

fig (v)

Component Insertion and Testing

Conclusions

In the context of comparison with centralised control systems, the principal benefits and advantages anticipated are that distributed nodal intelligence:

(1) Emulates human decision making, transaction initiating and communicating at an appropriate level, thereby assisting comprehensibility. Centralised monolithic control at many levels is more difficult to implement and maintain.

(2) Eliminates contradiction by choice of rules,if applied consistently and at the appropriate level. The distinction between the three nodal types is intended to assist the separation of concerns.

(3) Limits complexity by making decisions and transactions localised to where relevant.

(4) Defines and limits the boundaries within which change is propagated. For decisions and transactions the boundaries would normally encompass one node. In the case of communication nodes complementary updating is likely since for every transmitter there is at least one receiving node.

(5) Makes full use of specific local knowledge. Where it is well defined it can be exploited.

(6) Similarly, as a corollary to (5), development and tuning can be carried out at a local level without affecting the performance of other parts of the system.

References

[1] Wing, M. and Rzevski, G. The Use of AI Techniques in Automated Guided Vehicle Systems, 5th International Conference on Automatic Guided Vehicle Systems, Tokyo, 1987.

[2] Parunak, H.V.D. Distributed Artificial Intelligence Systems, Artificial Intelligence Implications for CIM, (ed. Kusiak,A) IFS (Publications) Ltd, New York, 1988.

[3] Buchanan, J.T., Burke, P., Costello, J., and Prosser, P. An Intelligent Knowledge-Based Scheduler for Heavy Manufacturing, Proc. Institute of Mechanical Engineers, May, 1988.

[4] Parunak, H.V.D., Irish, B.W., Kindrick, J., and Lozo, P.W., 1985, Fractal Actors for Distributed Manufacturing Control, in Proc. 2nd IEEE Conference on Artificial Intelligence Applications, pp 653-660.

[5] Smith, A.W. and Rzevski, G., Applying Neural Networks to Manufacture, CIM Centre, Kingston Polytechnic, Technical Report 9, 1988.

[6] Rzevski, G. Distributed Knowledge-Based Systems for Cell Scheduling and Control. 19th International Symposium on Allied Technology and Automation, Monte Carlo, 1988.

Production Processes and Control

A Scheme for Translating Control Flow in the C Programming Language to Grafcet

Bruce Hunter Thomas
National Bureau of Standards
Gaithersburg, MD

Abstract:

The purpose of this paper is to illustrate a translation scheme from control flow in the C programming language to the Grafcet language. Grafcet is a graphical language for expressing control flow. Grafcet is used to design parallel systems such as in a manufacturing environment. The control constructs covered in this paper are: conditional statement, while, do, for, switch, break, continue, goto, label, and null. The Grafcet used in this paper is the language, as augmented by Savoir. The C programming language is the one described by Kernighan and Ritchie. This translation is to be used as a reference for programmers to translate existing C source code into Grafcet.

Introduction:

Grafcet is a powerful graphical language for expressing control flow. Savoir* has implemented a version of Grafcet[1]. Grafcet is an excellent tool for designing, documenting, and demonstrating the control flow of a system. A system which is already written in C can be translated into a set of Grafcet programs. This paper illustrates a translation scheme from the control flow in a C program into a set of Grafcet programs. The C source code constructs are those defined by Kernighan and Ritchie [2].

A basic introduction to the portion of the Grafcet language needed for the translations is given in the paper. A number of the extensions to Grafcet by Savoir used for this translation are described. The translation of these C source code control constructs are described: conditional statement, **while, do, for, switch, break, continue, goto, label,** and null.

Background:

For the purposes of this paper we will define our problem as translating a set of C source code control constructs into a set of Grafcet program fragments. The Grafcet language has three major primitives: regular steps, macro steps, and transitions. Regular steps allow the user to represent an arbitrary control action in the form of embedded C source code. The Grafcet places a dot in the regular step's square to indicate that C source code is embedded in it, see the example below. Macro steps allow the user to name embedded Grafcet macro programs as control actions. Transitions act as gates on the flow of control through the Grafcet program. Associated with each transition is a C expression which determines if control can pass through that point. If the expression evaluates to **False**, the gate stops the control flow. If the expression is **True**, the gate allows the control flow to continue. These primitives can be connected to form a control flow path by attaching a link between two primitives. A link can only be attached between a step primitive and a transition primitive; that is to say a transition primitive cannot be linked to another transition primitive. These links are represented by a line from one primitive to another; sometimes an arrowhead is used to indicate the direction of the control flow. Examples of what these primitives look like are shown below. In these examples 0 and M1 are labels supplied by the Grafcet programming environment, and **label** is a user supplied name for a step.

* The Grafcet used in this paper is the language defined by Savoir, Oakland, CA, which has some very useful extensions. Savoir provides an interactive development environment FLEXIS DESIGNER™ to help design control systems with Grafcet. There is a research agreement between NBS, Savoir, and Xerox Corporation for the use of FLEXIS DESIGNER™ and two Xerox 1186 workstations. The FLEXIS DESIGNER™ runs on a Xerox 1186 artificial intelligence development workstation.

352

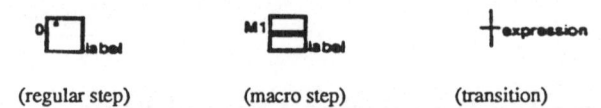

(regular step)　　　　　(macro step)　　　　　(transition)

It has been shown that all the other C control constructs can be formed from just conditional and **goto** statements [3],[4]. Therefore by translating the C conditional and **goto** statements first, the rest of the C control constructs will follow.

Savoir has extended the Grafcet language in many important ways. This paper uses three of these extensions: transitions with C expressions, embedded C source code, and an **otherwise** clause. Savoir has extended Grafcet to support C source code embedded in their "regular step." The Grafcet transitions use C expressions for the condition that determines whether or not control can pass through the transition. In an asynchronous branch a special transition condition of **otherwise** was added. An asynchronous branch is a single step followed by two or more transitions. Control can flow through one or more of the transitions whose conditions are true. **Otherwise** is defined as true if the conditions of the transitions to its left in a set of asynchronous branches are false.

Definitions:

In the examples of Grafcet programs, T is defined as **True** or: T != 0

Description of Translation:

The translation of these C source code control constructs are described in this section: null, label, **goto**, conditional statement, **while, do, for, switch, break,** and **continue.** The translated Grafcet statements are fragments which may be linked together to form a complete Grafcet program.

Null statement: ;
A null statement translates to a regular step with no C source code associated with it and has the label **null.**

Labeled statement: *identifier* :
The label statement in C is used for a destination location for the **goto** statement. There is no explicit translation into Grafcet, but the **goto** statement is translated.

Goto statement: **goto** *identifier* ;
As described above, control follows the links in a Grafcet program. Each link acts similar to a **goto** from point to point in a Grafcet program. A link must be contained in one Grafcet macro program. An example of a goto statement is shown in Figure 1.

Conditional statement: **if** (*expression*) *statement*
　　　　　　　　　　if (*expression*) *statement1* **else** *statement2*
The conditional statement is made with two asynchronous branches. One branch is a transition with the condition *expression*, and the other transition has the condition **otherwise.** The translation of the **if then else** statement is shown in Figure 2. The **if then** statement is translated the same, but *statement2* is a null statement.

While statement and **Do** statement : **while** (*expression*) *statement*
　　　　　　　　　　　　　　　　do *statement* **while** (*expression*) ;
The translation of the **while** statement is shown in Figure 3, and the **do** statement is shown in Figure 4.

For statement : **for** (*expression1* ; *expression2* ; *expression3*) *statement*

The translation of the **for** statement is shown in Figure 5. Each of the three expressions in the **for** statement are optional. If the expressions *expression1* or *expression3* are not used in a **for** statement, the unused expression is replaced by a null statement in the Grafcet program. If the expression *expression2* is not used in a **for** statement, the expression in transition is replaced by **T** in the Grafcet program.

Switch statement : **switch** (*expression*) *statement*
                       case *constant-expression* :
                       default :

The switch statement can easily be expressed with only conditional and goto statements [3]. The **case** label is replaced by an equivalence comparison between the constant-expression and the expression. The **default** case is handled by the **otherwise** condition in the Grafcet. The translation of the **switch** statement in Grafcet always produces a **default** statement to allow the control flow to continue. If there is no **default** in the C source code, a null statement must be inserted after the **otherwise** condition. An example of the **switch** statement is shown in Figure 6.

Break statement : **break** ;

The **break** statement is equivalent to a **goto** statement which passes control to next statement after the smallest enclosing **while, do, for,** or **switch** statement. An example of the **break** statement in a **while** statement is shown in Figure 7. The link in the example which was added for the **break** statement is shown as a dashed line.

Continue statement : **continue** ;

The **continue** statement causes control to pass to the loop-continuation portion of the smallest enclosing **while, do,** or **for** statement. An example of the **continue** statement in a **while** statement is shown in Figure 8. The link in the example which was added for the **continue** statement is shown as a dashed line.

Conclusion:

The paper provides a scheme for translating control flow in C source code into Grafcet. The control constructs are easily translated from C source code to Grafcet, but control flow through the use of C functions is difficult to translate[5]. Grafcet is being used for designing and documenting Cell Controllers in the Automated Manufacturing Research Facility (AMRF) at the National Bureau of Standards. A major language used for writing these controllers is C and this translation scheme was used to translate the high level logic of the Cell Controller [6]. Grafcet has provided an excellent means of visualizing the control structure of these large systems.

References:

[1] Savoir [1986] "Savoir Grafcet" Savoir, Oakland, CA.

[2] Kernighan, B. W., and Ritchie D. M. [1978] *The C Programming Language*, Prentice-Hall, Inc.

[3] Knuth, D. E. [1974] "Structured Programming with *go to* Statements", *Computer Surveys*, 6, 4, 261-277.

[4] Wirth N. [1974] "On the Composition of Well-Structured Programs", *Computer Surveys*, 6, 4, 247-259.

[5] Thomas B. H. [1988] "A Scheme for Translating Control Flow in the C Programming Language to Grafcet with Examples" NBS Internal Report NBSIR 88-3741

[6] Thomas B. H., and McLean C. [1988] "Using Grafcet to Design Generic Controllers" *First International Conference on Computer Intergrated Manufacturing* (to be presented).

The NBS Automated Manufacturing Facility is partially supported by the Navy Manufacturing Technology Program. Certain commercial equipment, instruments, or materials are identified in this paper in order to adequately specify the experimental procedure. Such identification does not imply recommendation or endorsement by the National Bureau of Standards, nor does it imply that the materials or equipment identified are necessarily the best available for the

354

Example goto statement :
           statement1;
           if *expression* goto NEXT;
           statement2;
       NEXT: *statement3*;

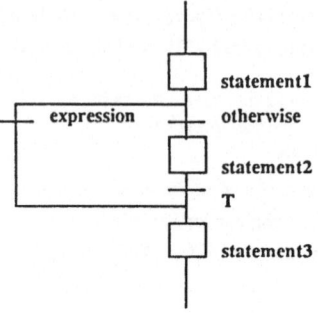

Figure 1 goto statement in Grafcet

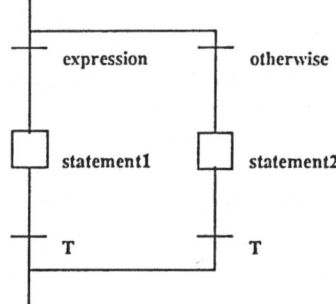

Figure 2 if then else statement in Grafcet

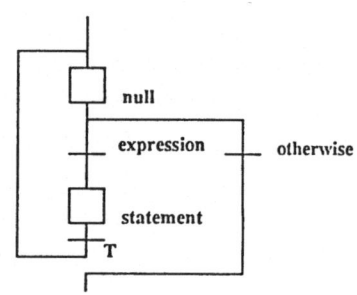

Figure 3 while statement in Grafcet

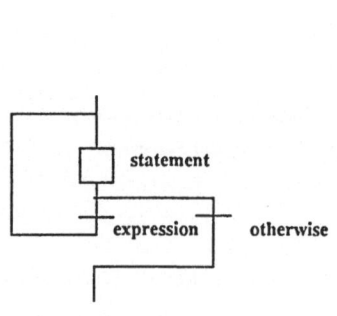

Figure 4 do statement in Grafcet

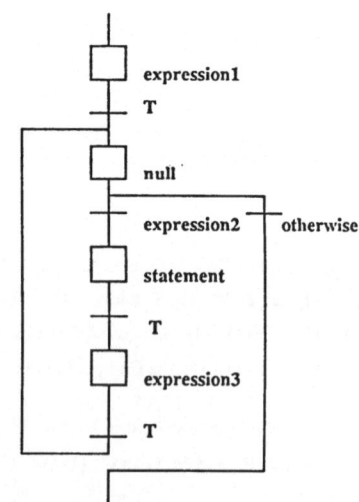

Figure 5 for statement in Grafcet

Example **switch** statement:
    switch (*constant- expression*) {
        case *case1* : *statement1*;
        case *case2* : *statement2*;
        case *case3* : *statement3*;
        default : *statementd*;
        }

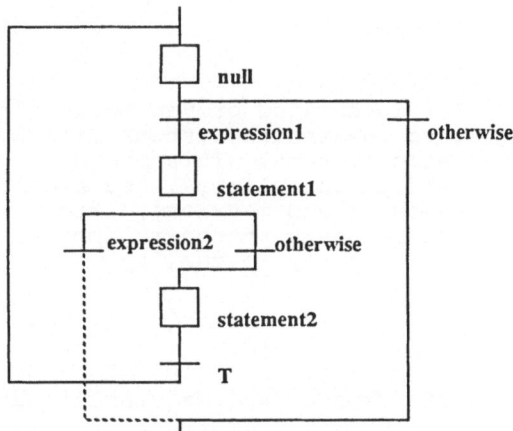

Figure 6 **switch** statement in Grafcet

Examples :   **while** (*expression1*) {
                statement1;
                if (*expression2*) **break**;
                statement2;
                 }

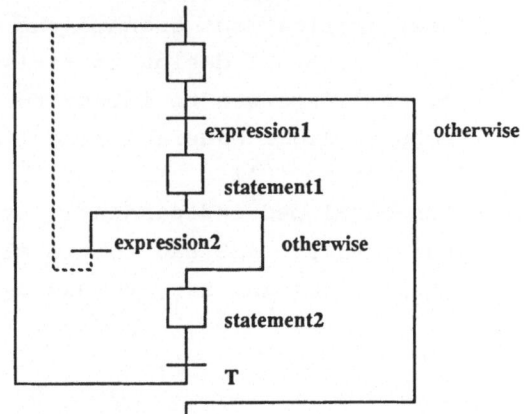

Figure 7   **break** in a **while** statement in Grafcet

Examples :   **while** (*expression1*) {
                statement1;
                if (*expression2*) **continue**;
                statement2;
                 }

Figure 8   **continue** in a **while** statement in Grafcet

The Challenge of Developing Large-Scale Software Systems

Charles A. Zonca, Jr., CQA
John F. Watkins, Jr., CQA

Electronic Data Systems Corporation (EDS) ·
Troy, Michigan

Summary

EDS integrated proven methodologies and quality assurance
into the entire Systems Life Cycle for the development of
large systems. This included human factors (training and
uniform guidelines), technical solutions (automated tools,
measures and metrics), strong management commitment, and an
evolutionary phased-in approach. This led to dramatic
improvements in quality.

Overview

EDS' Product and Manufacturing Engineering Systems Division
(PME) was faced with the challenge of integrating numerous
engineering applications into a fully integrated system.
These applications spanned the entire engineering process,
from conceptual design to finished manufactured product.
The challenge was to integrate Computer Aided Design (CAD),
Computer Aided Manufacturing (CAM), and Computer Aided
Engineering (CAE). Many of the pieces of the system had
been developed independently by small teams or by
individuals. EDS had to tie all of the pieces together into
a harmonious and easy-to-use system.

To face this challenge, EDS' PME Division selected a Systems Life Cycle and a methodology so that our developers would have a uniform approach and a proven method for software development. The incredible complexity of the system forced us to document more thoroughly, increase our communication with other department teams, and perform more testing across applications. EDS also had to ensure high quality and reliability.

This paper describes the evolutionary process that took place to bring cultural cohesiveness to a large, diverse group of software developers. It discusses various successful approaches for removing defects from software products to improve quality and customer satisfaction.

Management Commitment

In May of 1984, management recognized the need for uniform methods. Uniform methods became especially important in enabling movement of personnel onto hot projects without retraining, to allow for a systematic and coordinated approach to training, and to allow for the use of automated tools to increase productivity. Management made a commitment to improve the development of large systems. They selected activities and deliverables appropriate for a CAD/CAM/CAE Systems Life Cycle. They also selected the Structured Analysis and Design Methodology. They recommended the use of automated tools and training to support the life cycle and methodology.

Systems Life Cycle and Methodologies

A Quality Assurance Department was chartered to set uniform guidelines for the Systems Life Cycle and promote acceptance among the project development teams. The guidelines emphasized the use of structured methods.

A manual was developed which described each phase of the
Systems Life Cycle. Each phase had its own set of activities
and deliverables. All deliverables are reviewed and approved
prior to sign-off of a particular phase.

Training

Overview training to management personnel was provided.
Senior personnel were trained in the front-end portions of
the life cycle, and all personnel were trained on structured
techniques for the testing and maintenance phases. Workshops
for structured programming, real-time techniques, and
walkthroughs were provided. In-house consulting is now
provided as a follow-up to the training effort to supply
guidance on new projects and help all project personnel to
gain confidence in their ability to apply what they have
learned.

Testing Methods

The concept of cyclomatic complexity (the number of basic
paths through a module) was introduced and a recommendation
was made that each program module's complexity should be kept
to ten or less. Test cases can be written to test each of
the basic paths through the module. Once the code and unit
test cases are written, a walkthrough is scheduled to ensure
that the code is correct and the test cases will adequately
test the code. This became a key element of our quality
assurance success.

Walkthroughs

Quality can be defined as CONFORMANCE TO REQUIREMENTS. A
Quality Assurance function is to make management aware of the
cost of NON-CONFORMANCE to requirements.

Detection of errors early in the life cycle is essential for a quality product. The earlier a defect is found, the less expensive it is to correct.

One of the most effective methods for producing quality, defect-free software is through the use of walkthroughs.

By monitoring the walkthrough process, we demonstrated the high cost of non-conformance. We have introduced walkthroughs at each phase of the life cycle and we have found that they are critical to the success of a project.

PME WALKTHROUGH QUALITY IMPROVEMENTS

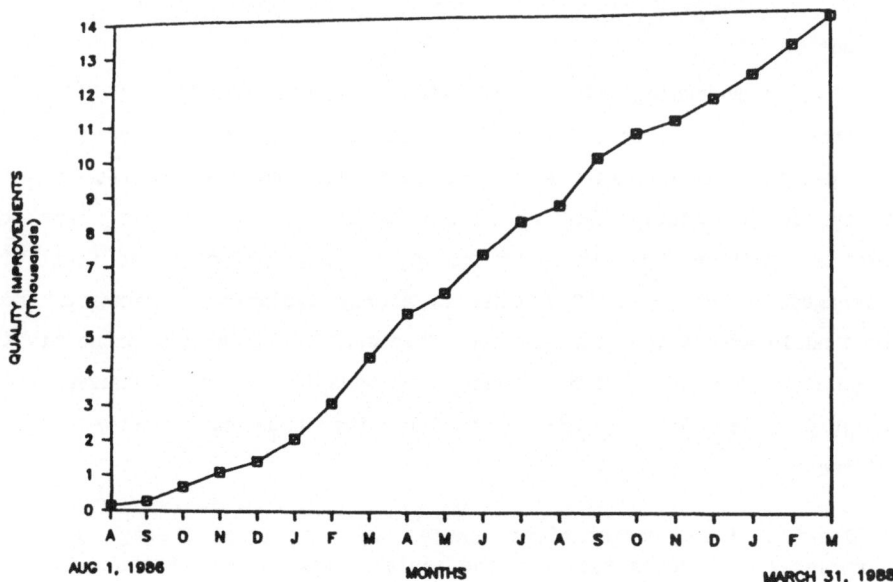

Design of a Decision Support System for Computer Integrated Manufacturing

Behnam Pourbabai
Department of Statistics and Operations Research
New York University
New York, New York 10006, U.S.A.

Abstract

In this paper the necessary components for design of a Decision Support System (DSS) for on line control of various activities of a Computer Integrated Manufacturing System (CIMS) will be delineated.

Introduction

Different technological and operational factors directly or indirectly affect the performance of a CIMS. The technological factors are related to those specific characteristics of the equipment and the facilities which directly or indirectly influence the manufacturing process, and the operational factors are related to the way people, equipment and facilities are managed. In general, if a CIMS is poorly designed or managed, those factors could adversely influence the efficiency, utilization, productivity, and profitability of a CIMS. Among the undesirable circumstances which frequently arise in a CIMS, the following problems deserve special considerations.

1. Blocking of the workstations, caused when the local storage of a workstation becomes full and the arrival work pieces at that work station are blocked;

2. Blocking of the material handling system, caused when the number of workpieces (to be transported) exceed the capacity of the material handling system;

3. Formation of excessive inventory of ready for delivery products, caused when direct communication links between the marketing department and the manufacturing department are non-existent;

4. Tardiness problem, caused when the completion time of a finished product exceeds its due date;

5. Under-utilization of the workstations and the material handling system, caused when the available production capacity exceeds the demand;

6. Sequencing/dispatching problem, related to the process of prioritizing the earliest start times of different stages of operation of each component of a product at each machine at the floor of the factory;

7. Manpower management problem, related to managing the work force, including the machine operators, shop floor supervisor, engineers, department managers, etc.;

8. Unreliability of the components of the system which influences both the production capability and the quality of the manufactured products.

Necessary Components

To design, plan, and control a CIMS such that the occurrence of the above undesirable operational circumstances are considered, and also to improve the overall performance, efficiency, productivity, and profitability of a CIMS, a capable computer support system is required for on line control of the system. For this purpose, a DSS should be used as the central component of the computer control system of a CIMS. For a review of the related literature, see De et. al. [1].

In Figure 1, the interrelationships among the central decision making components (e.g., modules) of a DSS for a CIMS are demonstrated. Those decision making modules are i) strategic planning, ii) organizational planning, iii) financial planning, iv) capacity planning, v) manpower planning, vi) marketing planning, vii) process planning, and viii) manufacturing planning modules.

In the strategic planning module, the specific long, intermediate, and short term objectives of the company should be specified. At this stage the marketing and manufacturing strategies should be evaluated. That is, this module should assist in generating the fundamental policies for governing a CIMS.

In the organizational planning module, the hierarchy of the administrative system should be evaluated and designed. More specifically, because the arhitecture of the flow of information directly depends on the organizational structure of a CIMS, at this stage the responsibilities and limitations of different departments, their operators, and their managers should be evaluated and be identified.

In the financial planning module, the specific financial limitations and capabilities of a CIMS during the planning time horizon should be evaluated. At this stage, the financial limitations and capabilities of every single department in a CIMS should be identified. Furthermore, capital budgeting decisions regarding replacing the existing technologies with updated technologies should be made based on a comprehensive cost-benefit analysis.

In the capacity planning module, the primary function should be identification of the production limitations and capabilities of a CIMS such as the type of product which can be manufactured, the processing limitations of each machine at its operators at each shift, the storage limitations, the material handling limitations, the marketing and the distribution limitations, quality control limitations, etc..

In the manpower planning module, the number of operators, engineers, and managers and their job responsibilities should be identified. Furthermore, the limitations and capabilities of each employee with regard to working different shifts should be explicitly considered.

In the marketing planning module, the demands at various regional, national and international markets for various products should be forecasted. Furthermore, the marketing strategies for both introducing and selling different products should be developed.

In the process planning module, the ideal mechanism for manufacturing a product should be developed. At this stage, the components of each product should be identified and the engineering drawings for each component of each product should be provided. For this purpose, CAD/CAM design facilities are utilized.

The manufacturing planning module consists of several other modules including master scheduling, inventory planning, material requirement planning, lot sizing, sequencing/dispatching, tooling, loading, balancing, reliability control, performance evaluation, and quality control modules. More specifically, the master scheduling module should provide the bill of material, lead times, batch sizes, etc.. The inventory planning module should consider the storage limitation and capabilities of the raw materials, semi-finished parts, finished parts, and completed products at every stage of manufacturing. The material requirement planning module should be responsible for identifying the number of units of each part to be manufactured, its due date, the specific tooling requirements, specific handling requirements, raw material requirements, etc. The lot-sizing module should be used to identify the ideal batch sizes which have to be processed at various work stations. The sequencing/dispatching module should identify the priorities of different jobs to be processed at different work stations according to the process planning specifications and other manufacturing data. The tooling module should allocate different jobs among various work stations which contain the compatible processing tools. The loading module should identify the size of each batch of each job to be processed at each work station and the size and the type of the material handling mechanism. The balancing module should attempt to allocate jobs among various workstations such that all the work stations have compatible work contents. Hence, underutilization of the work stations and their operators are eliminated. The reliability control module should analyze the possibility of both machine breakdown time periods and repair time periods. The performance evaluation module should verify feasibility, implementability, and profitability of a recommended operation plan before its implementation at the floor of the factory. The quality control module should investigate the performance of the processing units by sampling and analyzing the quality of the manufactured parts and products.

Concluding Remarks

We conclude this paper by recommending that after generting an operational plan and before implementing the plan at the floor of the factory, the feasibility of an operational plan be verified by a performance model (e.g. either a simulation or a queueing network model), see Figure 1. The role of the performance model is to quantify the resulted congestion along the material handling system and at the workstations, and also to evaluate how efficiently the material handling system and the workstsations are utilized. That is, if the performance model indicated that a recommended operational plan results in excessive congestion inside the CIMS, then a new operaional plan should be generated.

Reference

De, S., Nof, S. Y., and Whinston, A. B., Decision Support in Computer-Integrated Manufacturing, Decision Support Systems, (1985), Vol. 1, pp. 37-56.

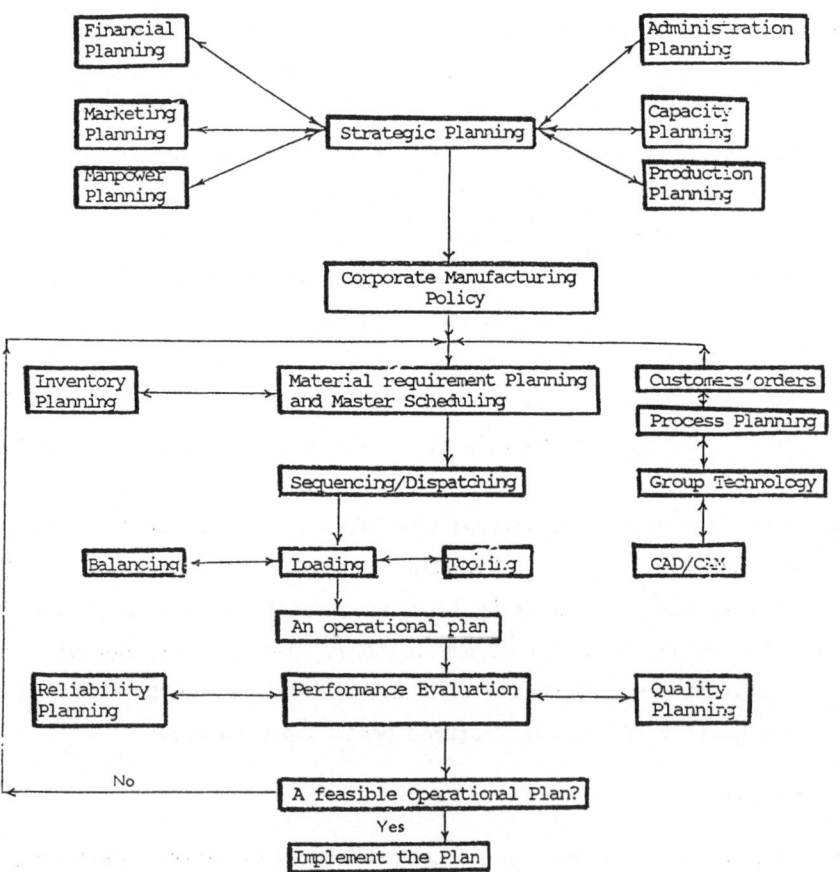

Figure 1 - Components of a decision support system for an integrated manufacturing system.

Preventive Maintenance Scheduling in Automated Factories

Cheickna Sylla

Engineering Management
Drexel University
Philadelphia, PA 19104

ABSTRACT

The research paper examines the rationale for the development of generalized procedures for the establishment of Preventive Maintenance (PM) schedules in the automated factories and the ongoing management of such schedules at a minimum cost. The approach of the generalized procedures proposed here is centered on a technique that will require that each scheduled task in a PM program be generated by an evaluation of failure consequences followed by an examination of the explicit relationships between that task and the reliability characteristics of the machine and/or equipment to determine whether the task is either:

1) Essential from a safety viewpoint, or
2) Desirable from a cost-benefit viewpoint.

The resultant PM schedules should result in a lower cost program, with the information available at any given time, that will ensure realization of inherent safety and reliability capabilities. While the local operating conditions are not considered here, these procedures can be easily extended to the remaining aspects of most local operating conditions.

INTRODUCTION

Preventive Maintenance (PM) is any pre-scheduled task or activity that is performed on an operational system or facility with one of the three objectives in mind: 1) to prevent equipment deterioration and failure, 2) to detect incipient failures, and 3) to discover hidden failures in off-line systems before an operating demand is made. Effective PM involves all three objectives. PM tasks tend to be derived on the basis of what can be done, and not necessarily what should be done and why. Current factory practices focus immediately on an item of equipment without a conscious concern for its function and the priorities in allocating resources. Such a process can only be suboptimal at best as no organized rationale or structure for selecting PM tasks is ignored, and therefore no way of really knowing whether the selected tasks represent a wise allocation of resources or, in some cases, if they are technically correct. While such practices although suboptimal may be acceptable in the regular factory context, they may turn out to be financially disastrous and even unsafe in an automated factory environment. Hence, the need for generalized procedures that can increase the effectiveness of development of the PM schedules in the automated factory environment clearly exists.

The generalized procedures that will be considered here are extended from PM scheduling technical context developed in the airline industry (Ref. 5). This technical concept requires that each scheduled task in a PM program be generated by an evaluation of failure consequences followed by an examination of the explicit relationship between that task and the reliability characteristics of the equipment to determine whether the task is either:

(1). Essential from a safety viewpoint, or
(2). Desirable from a cost-benefit viewpoint (we will consider the degradation of operational capability and non-availability for operations as costs)

In general, there are only four basic types of PM tasks that can be used in a maintenance program (Ref. 1,2,3):

- Items of equipment can be inspected at appropriate intervals to determine whether potential failures exists.
- Items can be reworked before a maximum permissible age is exceeded.
- Items or parts in them can be discarded before a maximum permissible age is exceeded.
- Items with hidden functions can be inspected at appropriate intervals to determine whether a functional failure has already occured.

The first three tasks reduce the likelihood that a single item will fail, whereas the fourth reduces the likelihood of a multiple failure involving an item with a hidden function. Thus all four types of tasks are preventive. The selection of an appropriate task requires that the conditions under which each type of task is applicable be identified and the term task-effectiveness be identified. The term "item" is applied to the object being anaylzed. An engine or a system may be an item as may also be a small part of a component. The first of the types of task) is called on-condition inspection. Three criteria must be satisfied for it to be applicable to an item:

- It must be possible to detect reduced resistance to failure for a specified failure mode.
- It must be possible to explicitly define a condition that represents a potential failure (it must precede a functional failure).
- It must be possible to approximate the time required for a potential failure to progress to a functional failure.

The second type of task is called schedule-rework. Three criteria must be satisfied for it to be applicable to an item:

- There must be an identifiable age after which the probability of failure increases rapidly.
- A large proportion of the items must survive to that age.
- Rework must restore the failure resistance lost due to the aging process.

The third type of task is called schedule-discard. Here four criteria must be satisfied for it to be applicable to an item:

- There must be a failure mode that has a direct and adverse effect on operating safety.
- It must be either impossible or impractical to observe reduced resistance to failure.
- Failure must have operational (but not safety) consequences.
- The local probability of failure must be strongly age-dependent (wearout).

The applicability of the preceeding tasks (which include both the economic-life discard, tasks and the safety-life discard tasks) is a direct function of the reliability characteristics of the item being studied.

The fourth type of task is called failure-finding inspection (including tests). These tasks are applicable to items with hidden functions where failures are not obvious to the user or operating crew, thus they have no immediate consequences. However, a first failure if it is followed by a second independent failure, might have adverse consequenced perhaps even on safety. It is always necessary, therefore, to schedule inspection tasks to ensure that the mechanics find and correct failed hidden-function items.

Components of the Generalized Procedures

The components of the generalized procedures for setting up optimal PM schedules in an automated factory reduce to the following major activities (which are explained below).

(1) defining the criteria for a PM task effectiveness;
(2) defining the required PM task activities and deriving the task identification matrix showing the tasks and their related PM benefits,
(3) establishing maintenance intervals for the tasks identified above,
(4) dividing the maintenance tasks and their corresponding intervals into a sequence of maintenance packages for implementation.

Effectiveness

Very often on-condition inspections or failure-finding tasks can be performed effectively while the item is still installed on equipment, whereas, a scheduled-rework or a scheduled-discard task requires the item to be removed and sent out to a major maintenance base to be effectively carried out. When a task is applicable to an item whose failure directly and adversely affects the operating safety of the equipment, the task must reduced the hazards to an acceptable level before it can be considered effective. Failure-finding tasks are always effective if the interval between repeated inspections is short enough. In all other cases the benefits of improved reliability and availability must be greater than the costs of performing scheduled maintenance before a task can be considered effective.

Task Identification

The first activity involves partitioning the equipment into major divisions (e.g., engine, functional systems, structure, etc.). Each division is then partitioned into subdivisions and these subdivisions may again be partitioned until a desirable level of simplicity is reached. Ideally, at this end an item, say a component, has been identified as one whose failure consequence is significant (we can use constraints to force the items with hidden-functions to become sifnificant items). Once the significant items have been identified the functions of each one must be carefully defined. The definition of the functions lead to the identification of functional failures and their consequences. These can be set up into a matrix format, called the task identification matrix.

Maintenance Intervals

After the types of tasks to support various items have been identified, the intervals between successive performance of each task must be identified. These intervals will be centered around the basic task types defined above. Here, safety, failure prevention, and basic cost-effectiveness will be major considerations.

Maintenance Packages

A maintenance package consists of a small group of tasks with comparable intervals. Therefore each maintenance package will have its own aggregate intervals. It is much easier to schedule and monitor the accomplishment of such packages than it is to control much larger number of individual tasks (Ref. 4).

Development of PM schedules from GP

For each major equipment (e.g., machines and tool units) a decision diagram should be set up for reliability and availability decisions relative to whether there is a need to perform a PM task with critical consideration about failure consequences (i.e., critical safety, economic operation, economic non-operation, exposure to multiple failures, etc.). When this decision diagram is established, the PM analyst (or scheduler) must carry out the following major activities for identifying the tasks to be included in the PM schedule for each equipment:

(1). Partition the equipment into significant and non-significant items.
(2). Define the functions of all significant items.
(3). Determine whether functional failures of the significant items are evident to the conductor or to the operating crew.
(4). Identify the failure modes that causes a loss of function.
(5). Identify the consequences of functional failures (and modes) and partition them into four categories.
- Safety.
- Other consequences (economic-operational).
- Cost of repair (economic non-operational).
- Exposure to multiple failures.

(6). Identify candidate maintenance tasks directed at the modes and partition them into four categories:
- On condition inspection
- Scheduled-rework.
- Scheduled-discard.
- Failure-finding inspection.

(7). Assign applicable tasks to items whose functional failures have safety-consequences to ensure that the residual hazards are reduced.
(8). Asssign applicable tasks to item whose functional failures have economic consequences; provided that tasks are cost-effective.
(9). Assign tasks to items with hidden functions to ensure that the likelihood of multiple failures is reduced to an acceptable level.
(10). Assign intervals to each tasks in the program, and group tasks into a small number of maintenance packages.

The analyst must continue with the following steps after he derives the maintenance packages for every piece of equipment of interest (at the end of step10).

(11). Define the PM planning horizon and set up priorities among the maintenance packages.
(12). Establish optimal PM schedules applicable in the automated factory with the use of a simple optimization tool (which ever appropriate), taking into consideration the local operating constraints

References

1. Corder, A.s., "Maintenance Management Techniques", (McGraw-Hill, New York) 1976.
2. Sherif, Y.S., and Smith M.L., "Optimal Maintenance Models for Systems Subject to Failure -- A Review", Naval Research Logistics Quarterly, Vol 28, No. 1, pp. 47-74, 1978.
3. Sylla, C. "Microcomputer Based System For Scheduling Preventive Maintenance", a manuscript report in preparation, Dept. of Engineering Management, Drexel University, 1986.
4. Lawrence, M.Jr., "Maintenance Management" (Lexington Books, D.C. Heath and Co., Lexington, Mass.) 1976.
5. Nowlan, S.F. and Heap, H.F., "Reliability-centered Maintenance", in Proceedings of the Annual Reliability and Maintainability Symposium, pp 38-44, 1978.

Monitoring of End Milling Operations in Unmanned Machining

Andrzej A. Markowski

Manufacturing Engineering Technology Department
Mankato State University
Mankato, Minnesota

Summary
In end milling, on line measurements of a milling torque have been employed as a model of an operation. Other signals like cutting force components, temperature, mechanical vibrations, and acoutic emission were investigated as a potential source of information about end mill state of wear.

Functioning of a Monitory System

The functioning of monitoring systems can be supported on a operations model, which has to be created when production is prepared, and then stored in memory to serve as a reference. It is obvious that the model's structure and it's accuracy are essential for functioning of an entire system. The attempt of creating the theoretical model of the end milling operations in the form of mathematical input-output structure, are not successful up to now [1]. The experimental models are more practical for monitoring. They are considered as records of processed physical signals generated by the cutting process. If during the manufacturing of successive parts from the same series of products, generated signals will differ from those registrered as a model more than assumed tolerance, monitoring systems send information to machine tool control centers. As it was found out [2], the experimental model of the end milling, based on cutting torque record, can be used for on-line identification of the workpiece location on machine tool table, or to distinguish too big or too small allowances of raw material, etc. (Fig. 1). This method, however, is not sensitive enough to detect changes in end mill due to its wear. Additional analysis of other cutting signals are required to obtain the information about the mill state when cutting.

Signals in End Milling

Experiments presented were conducted on middle carbon alloy steel 4040. Two kinds of end mills were used: HSS 5/8 inch four teeth and 1.0 inch in diameter tipped with two carbide SECO S25S type inserts.

Cutting Forces

Three component table dynamometers were used. Analysis of results show that near exact linear relation exists between unit tangential force q_v (N/mm, on the cutting edge length) and with wear land VB. This character remains the same over the entire tool's life up to the mill collapse without any additional warnings in measured signal. Small chippings of cutting edges cause similar increases of force, as regular wear land VB does (Fig. 2).

Cutting Temperature

Cutting temperature or thermoelectrical force from a monitoring point of view is interesting so far, as they can reflect important changes in tools shapes, due to its wearing. Natural thermocouple method of temperature measurements (tool material - work material) were used because it seems to be better applicable for end milling operations the other temperature measurement methods. Experiments which were carried on, using specially designed experimental tools, have provided much evidence that TEF (thermoelectrical force) is not related to the mill wear. TEF remains constant when mill wears and when VB increases to the maximum allowed value. Other experiments show that the value of TEF for a given operation represents tool wear rate (dVB/dt) (Fig. 3). This is a very important fact which opens new potential applications of TEF signals in monitoring or controlling of automated machining.

Mechanical Vibrations

To get answers to the questions if the vibration signal can be used for monitoring, several series of experiments were carried on regular and CNC BRIDGEPORT milling machines. Workpiece fixturing devices were chosen in both cases as a place of location of the vibrational gauges. Measured signal was processed by the HP Spectrum Analyzer
(0-25 kHz) to get RMS spectrum. It was found that:

- within the frequency range of 0-7, kHz vibrations are produced by a machine tool driving system rather then by a cutting process. The shape of a frequency spectrum depends on the rigidity of a workpiece, a tool and machine tool, but no spectrum changes related to the end mill wear have been found.

- the most interesting band of frequencies was found around frequencies 20kHz for both machine tools (conventional and CNC). RMS spectrum amplitudes in that range increased significantly only above the noise level during cutting. No define relation between tool war and RMS spectrum ampliltude was stated, but by simple processing of a vibration signal in that range of frequencies, important information for unmanned machining can be obtained. If a narrow range spectrum signal is keeping certain value, it might mean that cutting is on width preset parameters, (Fig. 4.)

Acoustic Emission (AE) Measurements

To retrieve the information concerning the state, an end mill wear from the AE signal rather simple amplitude spectrum analysis method was chosen. For the cutting conditions identical to the conditions of vibration experiments, AE signal was measured by PCB Dg203A, sensor attached to the workpiece vise. Then AE signal was filtered and analyzed on a Digital Processing Oscilloscope, NORLAND IQ 400. By summarizing results of several series of experiments made on two different machine tools, it can be said:

- the frequency spectrum of the AE signal shows that for a sharp mill, the energy of acoustic waves is concentrated within a relatively narrow frequency range: 60-130 kHz. This kind of frequency distribution dominates throughout the entire useful lifetime of a mill.
- the broadening of the frequency spectra to 350 kHz was observed for a substantial mill wear. The end of cutting properties of the end mill is associated with the rapid decrease of the AE signal just to level of the measurement noise within the entire investigated frequency band.

Conclusions

1. An automated monitoring system installation is one of the major requirements in an unmanned end milling to achieve maximum productivity and safety of the operation.

2. The cutting force or torque signal, as it was tested, can be used in monitoring mostly as a source of information about geometry, dimensions, and location of a workpiece.

3. The temperature on TEF in end milling are not related to the amount of mill wear. However, functional relations between dVB/dt and TEF is highly promising for futher applications in automation of metal cutting.

4. Vibrational and acoustic emission signals show good relation to rapid changes in end mill geometry due to its wear. In both cases, however, signals processing in frequency domain are necessary.

References

1. Abakumow, A.M.; Vidmanow, J., "Modeling of the Turning Process," Machine and Tools, September 1972. (in Russian)

2. Markowski, A., "Recorn of Static and Dynamic Milling Models," Proceedings of International Conference organized by Technical University of Wroclaw, 1985.

3. Colwel, L., "Cutting Temperature Versus Tool Wear," Annals of the CIRP January 1975.

4. Iwata, K.; Moriwaki, T., "An Application of Acoustic Emission Measurements to In-Process Sensing of Tool Wear," Annals of the CIRP, January 1977.

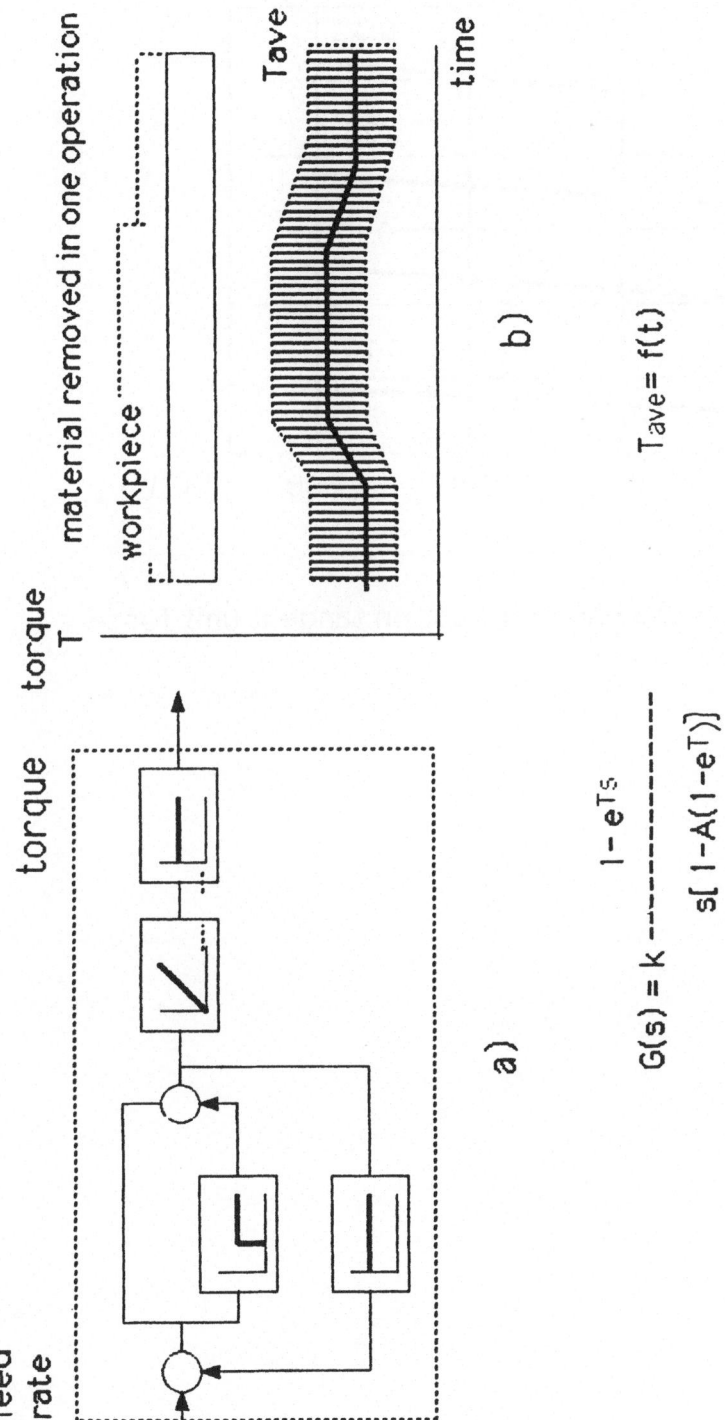

$$G(s) = k \frac{1 - e^{Ts}}{s[\,1 - A(1 - e^{T})\,]}$$

$$T_{ave} = f(t)$$

a)   b)

Fig 1. Models of the end milling operations a) theoretical,
b) experimental

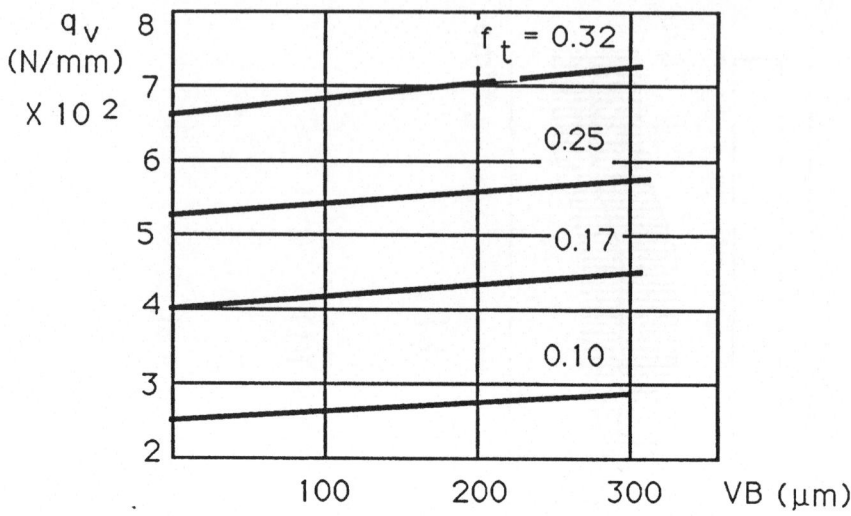

Fig 2. Effect of the end mill wear on tangent unit force

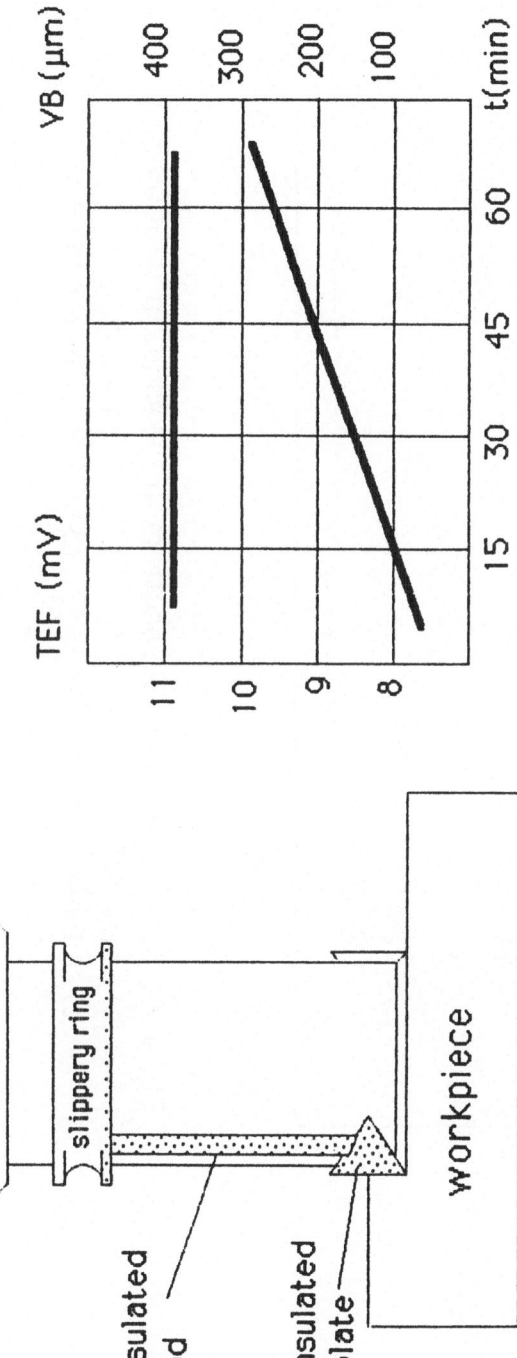

Fig 3. Experimental tool a), Thermoelectrical force and mill's wear land VB, versus time of cutting b) (v=50 m/min, ft=0.1mm).

378

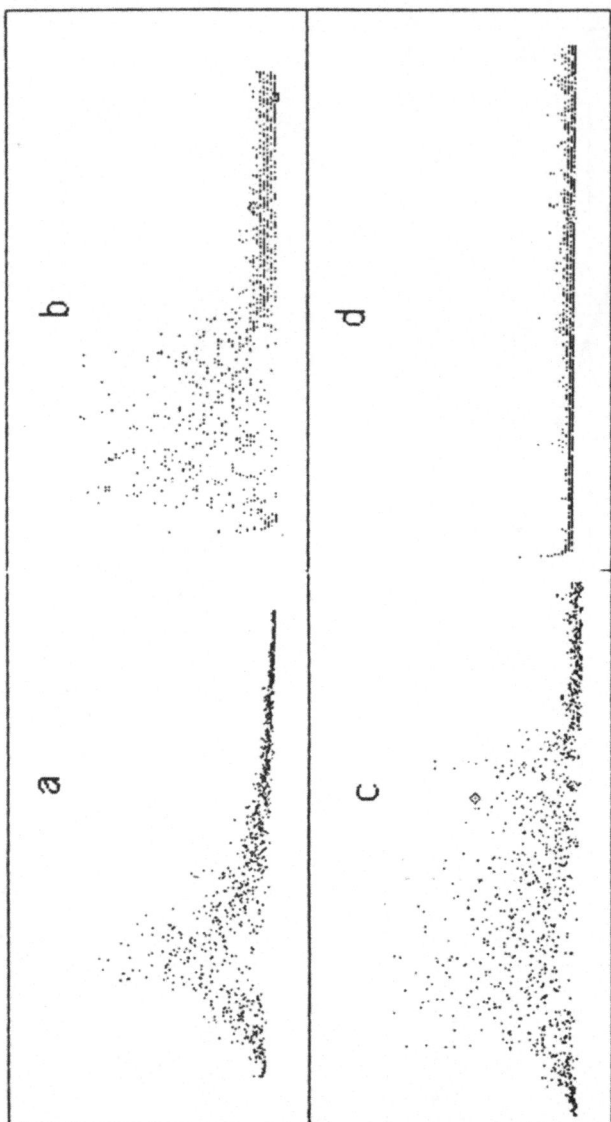

Fig 4. Acoustic emission (AE) spectrum in the range of 0–600 Hz.
a) a new mill b) medium wear
c) criterial wear d) mill totalIly worn down

A Total Production Information Service System for the Factory

M. SHIOJIMA, K. YAMAGUCHI, T. YOSHIHARA, M. GAU, T. TORII[**]

[*]Otsuki Plant, Transmission Division, NEC Corporation, Yamanashi, Japan

[**]Production Engineering Lab., NEC Corporation, Kanagawa, Japan

SUMMARY

Companies cannot expect to survive in today's cutthroat business market without the timely control and utilization of increasing amounts of information. This paper addresses this problem and explains the total information management system which we refer to as "ISS" (Information Service System). The system was developed to solve the problem of coordination and utilization of information. "ISS" consist of three subsystems they are: Planning Control, Manufacturing Control, and Factory Management.

Harmonization of material requirements planning "MRP" and shop floor control "SFC" and unification of information concerned with manufacturing activities enables personnel to know at all times, the status of the franticly changing manufacturing process. While this is commonly known, until now there has not been a cost effective way of managing this information. However this can now be done with "ISS", which has been successfully utilized in an NEC factory, in production areas that manufacture small quantities of various devices, as a tool to improve product quality and manufacturing efficiency.

INTRODUCTION

An important part of company strategy is to adequately understand the needs of a changing market and develop suitable new products. Today's modern factories are required to produce a multitude of high-quality products, that change frequently, and have a short life cycle. Given this production situation it is difficult, without an adequate information management system, to efficiently produce these items.

Errors that often occur in information passed from person to person create delays and impair judgement. For these reasons, we believe that the underlying production strategy for a successful enterprise, is to obtain accurate information on the constantly changing production activities and distribute this information in a timely manner. Upon analysis of this information, improvements can be initiated to bring cost down and improve quality.

ISS OBJECTIVE

The objective of "ISS" is to provide applicable production information in a package that is flexible and can be expanded as requirements change, but, is user friendly. The results of this will be improved efficiency, a decrease in costs, and an increase in quality.

ISS CONFIGURATION

"ISS" consists of three subsystems: Planning Control, Manufacturing Control , and Factory Management. The following paragraphs explain the function of each subsystem.

1. Planning Control Subsystem

   This subsystem plans production schedules on the basis of material requirements, material bills, and production lead time. The "ISS" processes this data and provides real-time delivery predictions.

2. Manufacturing Control Subsystem

   This subsystem stores data on quality, cost, the state of machine work, and etc...into the database, in a unified form. The information is indispensable when determining what improvements need to be made to increase work efficiency.

3. Factory Management Subsystem

   This subsystem provides strategic informations which is necessary for factory management.

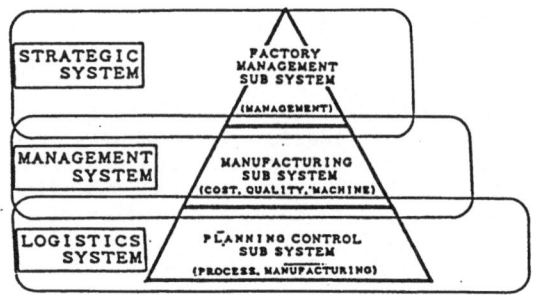

Fig. 1 'ISS Configuration

ISS NETWORK STRUCTURE

The "ISS" hardware configuration is shown in Fig. 2. The hardware configuration can be separated into two systems, one for text information and

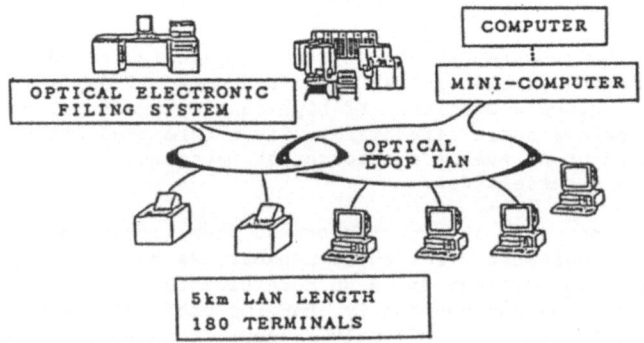

Fig. 2 "ISS" Hardware Configuration

one for graphic information. A typical system consists of a general use
computer, a mini computer', and 160 personal computers. The general use
computer is primarily used for managing the planning control subsystem. The
Mini computer is used to manage the manufacturing control subsystem. Personal
computers are used as data input and output terminals; one terminal for every
two employees. System hardware is integrated using an optical loop LAN.
Actual manufacturing data is inputted into the "ISS", which processes the data
and supplies real-time information.

SYSTEM CHARACTERISTICS
Listed and explained below are the characteristics of "ISS".

(1) Data Unification
    We have standardized and unified production data supply and collection
    methods, which operates without regard to the type of product or
    manufacturing process. This enabled us to develop the software used in
    the "ISS" equipment at a relatively low cost and in a shorter period.
(2) Gathering and Disseminating a Variety of Real-time Production Data
    Actual production data is inputted as it is generated. The inputted
    information is displayed by "ISS" in graphic form on the "ISS" display.
    The information is arranged and processed by the "ISS" equipment to allow
    easy understanding of the present production situation.
(3) Use of outer-OS for Mini Computer
    Use of outer-OS software to perform common functions in a mini computer,
    which alleviate the need to develop software that performs the same
    functions. The software writer is then only required to write the
    software for the different operations,
(4) Image Information Service
    "ISS" provides drawing information services. The drawing information is
    stored on optical disks and can be accessed from the electronic filing
    system of the "ISS". Additionally, "ISS" improves speed and the delivery
    accuracy by tracking information on materials throughout the parts vendor
    manufacturing process using VAN.

IMAGE INFORMATION SERVICE USAGE EXAMPLES
Image display services are used to provide manufacturing and management
improvement information. Typical images that can be produced by the "ISS"
equipment are listed in Table 1.

Table 1. "ISS" Display Image

VARIETY OF INFORMATION	NUMBER
ADMINISTRATION	12
COST	32
QUALITY	24
PRODUCTION CONTROL	168
MACHINE WORK	10
MANUFACTURING CONTROL	185
ENVIRONMENT CONTROL	5
TECHNICAL DOCUMENT	72
TOTAL	518

382

Listed below are some common types of image displays used for manufacturing
and management improvements.

1. When used for planning manufacturing schedules.
   Depicted in Fig. 3 is detailed manufacturing schedule for each process and
   the order number.

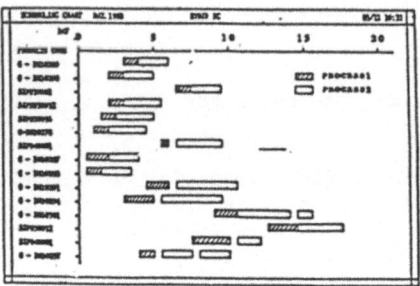

Fig. 3 Process Schedule

All employees are indicated in this planning schedule, also contained on
this display are the actual manufacturing results. This allows production
planners to easily forecast an accurate completion date, in order to
prepare the completed goods for shipment to the customer.

2. When used for quality improvements.
   Figure 4 shows the daily changing data for the defect ratio of each
   process. Figure 5 shows the defect ratio per each cause of the defect,
   using the pareto chart method. The charts can be used to highlight areas
   where numerous defects occur, corrections can be made which will rapidly
   improve quality.

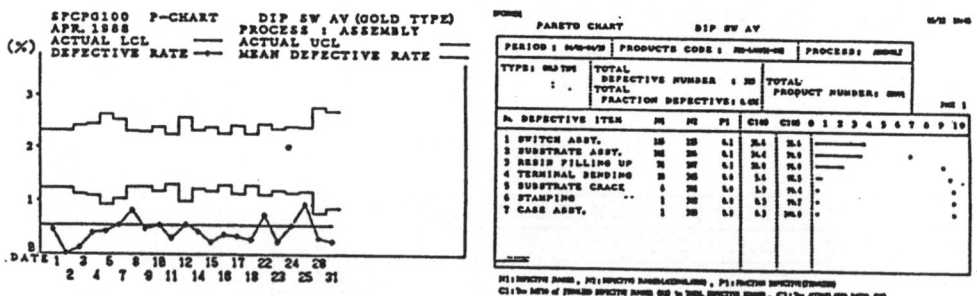

Fig.4 Daily Defect Ratio   Fig. 5 Defect Causes

3. When used for factory management.
   Figure 6 shows the sales per each business unit, which allows immediate
   comprehension on sales per unit to manage daily sales activity.

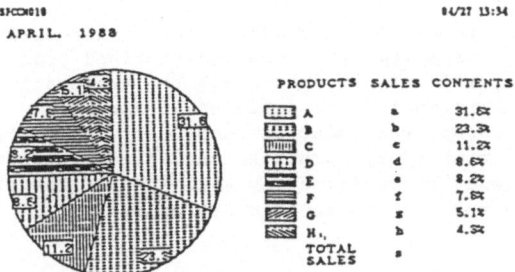

Fig. 6 Sales of each Business Unit

ISS Application

The information system increases the efficiency of a production area by quickly pointing out deficiencies. Depicted in Figure 7 is a typical model of the "ISS" process. It is referred to as TOP Activity, which means tactics for the Otsuki Plant. The TOP activity is deployed in all tasks. For example, "TOP-PLANS" activity stimulates innovations in production management, "TOP-QUP" activity helps to improve the product's quality, "TOP-DIET" activity helps decrease cost. The purpose of all these activities is to provide job innovations quickly and correctly by using information supplied by "ISS". A rough estimate of production gains from the use of "ISS", is that the time necessary to collect and arrange the data decreases one hundred times, and the time necessary to perform the task decreases ten times. As a result of these activities, in a short amount of time, we can improve quality by ten times and double efficiency. The use of this system will create new requirements that will help to improve "ISS".

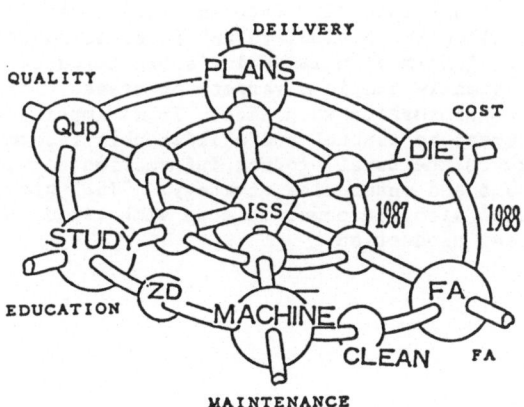

Fig 7 ISS Application

384

Development of ISS

Presently, we strive to develop the ideal factory automation system. This system unites the machine that produces the finished product with the distribution system, effectively decreasing the tedious functions that must be performed by man. This concept is depicted in Figure 8. The outer circle is the size of the business, the inner circles (Carrying, Information, Machine) are the functions performed by machines, the spaces between the circles are the functions that must be performed by man. As we integrate these circles man plays a smaller role in routine factory operations improving automation.

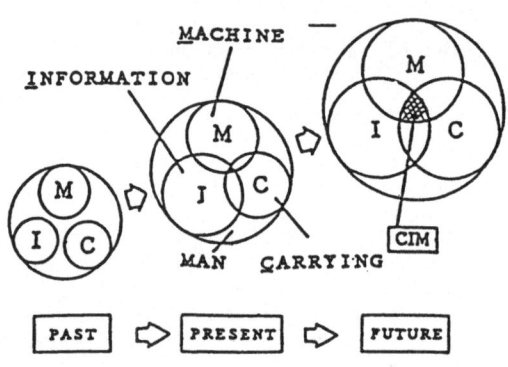

Fig. 8 Shows the conception

CONCLUSION

"ISS" is now strengthening the decision support functions of the information services, and is extending into the "Intelligence Support System". The next step of our plan is to develop "ISS" into an "Integrated Strategic System", with functions that allow it to simulate and forecast as the hub system for management strategy. Information is an invisible resource for management, which can be used instantly and in a variety of areas. The more information we use, the more new information we create. It is important for the strategy of enterprise management to control and utilize this information. In our opinion, "ISS" which is the total product information service system, is the foundation for product and enterprise strategy. "ISS" provides the manufacturing location with the power to cope with rapid changes and realize flexible and efficient production.

References

1. M. M. Barash "Computer Integrated Manufacturing Systems", ASME Winter Annual Meetings, Special Volume: The Factory of the Future, Chicago, Illinois, November 1980.
2. K. E. Stecke and J. J. Solberg, "Scheduling of Operation in a Computerized Manufacturing System", The Optimal Planning of Computerized Manufacturing Systems, Springfield, Virginia, Nationa Techinical Information Service#PB-300599.

Effects of Components' Reliability on the Transient Performance of an Integrated Manufacturing System

Behnam Pourbabai

Department of Statistics and Operations Research

New York University

New York, NY 10006, U.S.A.

Abstract

The effects of reliability of the components of a computer integrated manufacturing system consisting of a set of workstations, an inventory system, a loading station and an unloading station, linked by a closed loop material handling system on the transient performance of the system are quantified.

Introduction

Commonly, to design a Computer Integrated Manufacturing System (CIMS), a performance model is develpoed to study and address the effects of various operational control strategies on the performance of the system. For this purpose, appropriate performance measures are introduced and are evaluated to investigate either the transient (e.g., short term) or the asymptotic (e.g., long term) performance of the system. Such an investigation often will lead to operational control strategies and policies which ultimately will result in more efficient utilization of the system.

For a review of related literature of simulation of different job shop, see Baker and Bertrand [1], Kiran and Smith [2], Ragatz and Mabert [3], and their references.

The CIMS which is considered in this paper consists of 4 workstations, a loading station and an unloading stations, an inventory system, linked by a closed loop material handling system, see Figure 1. Each workstation consists of one flexible machine, capable of processing different workpieces belonging to the same family of parts, a preprocessed inventory (e.g., local storage), and a postprocessed inventory. Each workstation, contains a robot for dismounting workpieces from appropriate fixtures, transporting workpieces from the preprocessed inventory to the machine, transporting workpieces from the machine to the postprocessed inventory, and mounting workpieces on the appropriate fixtures. The material handling system consists of 25 line conveyors, each with a beginning and an end, a specified length, velocity, and capacity. Each workpiece is mounted on a fixture and is transported by the appropriate line conveyors from one location to another.

Our objective in this paper is first to develop a performance model for the above CIMS, and then to use the model to quantify the effects of two operational control strategies on the transient performance of the system. For this purpose, two scenarios are introduced, and a simulation model is developed, using MAP/1 simulation language, see reference [5]. In one scenario the system is assumed to be reliable and in the other scenario the system is considered to be unreliable. That is, each machine may independently breakdown.

Components of the Model

In this section the characteristics of the CIMS and the components of the model are described. Consider a CIMS consisting of four workstations (i.e.: A, B, C, and D), a local inventory system, a loading station and an unloading station linked by a closed loop material handling system, see Figure 1.

More specifically, two class of workpieces, each belonging to the same family of parts are considered. The first class of workpiece has to be processed first at workstation A, then at workstation D, and finally at workstation B. The second class of workpiece has to be processed first at workstation D, then at workstation B, then at workstation C, and finally at workstation A. Each workstation has a robot, one machine, a preprocessed inventory (e.g., local storage) and a post processed inventory. At each workstation the robot is used to dismount workpieces from the appropriate fixtures, to transport them from the preprocessed inventory to the machine, to transport them from the machine to the post processed inventory, and to mount them on the appropriate fixtures. The processing times for both workpieces at each workstation is assumed to be identical.

In Figure 2, the processing time values for an arbitrary workpiece at each work station is provided. It is noted that, each processing time value includes .25 minutes for dismounting each workpiece from the fixture, .5 minutes for transporting each workpiece from the preprocessed inventory to the machine, the actual processing time, .5 minutes for transporting each workpiece from the machine to the post processed inventory, .25 minutes for mounting the workpiece on the fixture. In Figure 2, the available number of fixtures for each type of workpiece are also specified. Furthermore, the possibility of the machine breakdowns at each workstation and the required repair times are also considered. The interbreakdown time for each machine and the required repair time for each machine are assumed to be exponentially and normally distributed random variables with Expon. (860.) and Norm. (30. and 5.) parameters, respectively. In Figure 3, other input parameters are presented. It is also noted, at each workstation, the longest waiting time dispatching rule for processing those workpieces waiting at each preprocessed inventory is used. That is, among the workpieces waiting to be processed, the one which has waited for the longest time is selected for processing, and in the case of existing a tie between two heterogeneous workpieces, the workpiece with a higher priority is selected for processing. We also point out that, the loading process is deterministic and is performed by a robot, see Figures 2 and 3 for the loading parameters. Furthermore, batches of the workpieces 1 and 2 are alternately loaded in to the system.

The material handling system consists of 25 lines conveyors, see Figures 2 and 3. It is assumed that if a workpiece is blocked at any workstation, the corresponding conveyor leading to the that workstation is also blocked. Furthermore, because we are interested in the transient performance of the system, we arbitrarily chose to simulate the performance of the system for only one working week. For this purpose, the simulation results for six working days were obtained, but to warm up the system the data from the first day was deleted. It is also assumed that each working day lasts 480 minutes. The reported simulation outcomes are given in Figures 4 to 6. It is noted that each reported outcome for the scenario number 12 is its approximate 90 percent confidence interval, and is based on ten independent runs.

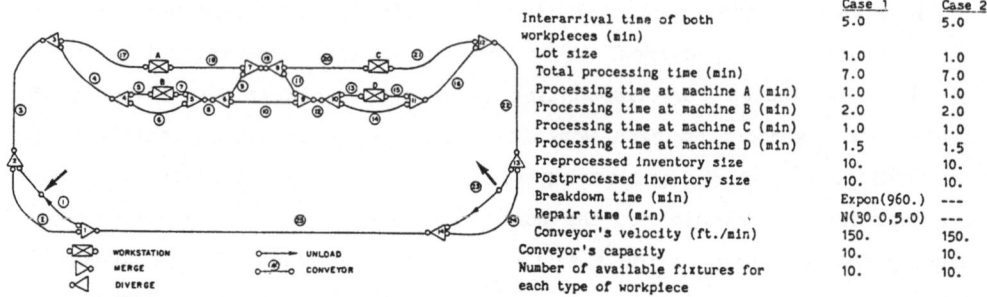

	Case 1	Case 2
Interarrival time of both workpieces (min)	5.0	5.0
Lot size	1.0	1.0
Total processing time (min)	7.0	7.0
Processing time at machine A (min)	1.0	1.0
Processing time at machine B (min)	2.0	2.0
Processing time at machine C (min)	1.0	1.0
Processing time at machine D (min)	1.5	1.5
Preprocessed inventory size	10.	10.
Postprocessed inventory size	10.	10.
Breakdown time (min)	Expon(960.)	---
Repair time (min)	N(30.0,5.0)	---
Conveyor's velocity (ft./min)	150.	150.
Conveyor's capacity	10.	10.
Number of available fixtures for each type of workpiece	10.	10.

Figure 2 - Input parameters

Figure 1 -- Graphical representation of the CIMS

Conveyor Number	Length (ft.)	Begin Point	End Point
CONVEYOR 1	2	SWITCH 1	SWITCH 2
CONVEYOR 2	3	SWITCH 1	SWITCH 2
CONVEYOR 3	4	SWITCH 2	SWITCH 3
CONVEYOR 4	3	SWITCH 3	SWITCH 4
CONVEYOR 5	1	SWITCH 4	WorkStation B
CONVEYOR 6	2.5	SWITCH 4	SWITCH 5
CONVEYOR 7	1	WorkStation B	SWITCH 5
CONVEYOR 8	0.5	SWITCH 5	SWITCH 6
CONVEYOR 9	1	SWITCH 6	SWITCH 7
CONVEYOR 10	2	SWITCH 6	SWITCH 9
CONVEYOR 11	1	SWITCH 8	SWITCH 9
CONVEYOR 12	0.5	SWITCH 9	SWITCH 10
CONVEYOR 13	1	SWITCH 1	WorkStation D
CONVEYOR 14	2.5	SWITCH 10	SWITCH 11
CONVEYOR 15	1	WorkStation D	SWITCH 11
CONVEYOR 16	3	SWITCH 11	SWITCH 12
CONVEYOR 17	2.5	SWITCH 3	WorkStation A
CONVEYOR 18	3	WorkStation A	SWITCH 7
CONVEYOR 19	0.5	SWITCH 7	SWITCH 8
CONVEYOR 20	3	SWITCH 8	WorkStation C
CONVEYOR 21	2.5	WorkStation C	SWITCH 12
CONVEYOR 22	4	SWITCH 12	SWITCH 13
CONVEYOR 23	2	SWITCH 13	SWITCH 14
CONVEYOR 24	3	SWITCH 13	SWITCH 14
CONVEYOR 25	14	SWITCH 14	SWITCH 1

Figure 3 - Input parameters

Output Analysis

To analyze the transient performance of the CIMS, a simulation model was developed based on the input information in section 2 was developed, using the MAP/1 simulation language, and the following results are obtained.

Based on Figures 4, 5, and 6, the following observation are made. The unreliability of the system has resulted in: 1) increasing the expected flow (e.g., sojourn) time, 2) increasing the expected travel (e.g., transportation) time, 3) increasing the expected waiting time at all the appropriate workstations, and 4) increasing the average waiting time in the system. However, the other performance measures have not been significantly changed.

	Workpiece 1		Workpiece 2	
	Case 1	Case 2	Case 1	Case 2
Flow Time	10.13±0.39	8.44	16.1 ±1.73	9.61
Travel Time	0.64±0.11	0.44	0.93±0.16	0.71
Station Waiting Time	2.49±0.31	1.01	6.65±1.61	0.40
Transporter Waiting Time	0.00±0.00	0.00	0.20±0.01	0.00
Production Rate	0.20±0.00	0.20	0.20±0.01	0.20
Throughput	480.30±0.14	480.00	480.90±3.09	480.00

Figure 4 - Expected values of the performance measures

	Average number of parts waiting		Average number of blocked workpieces		Average waiting time	
	Case 1	Case 2	Case 1	Case 2	Case 1	Case 2
Station load	0.09±0.02	0.08	0.00±0.00	0.00	0.23±0.042	0.20
Workstation A	0.21±0.07	0.00	0.01±0.01	0.00	0.52±0.20	0.00
Workstation B	1.12±0.28	0.10	0.03±0.01	0.00	2.89±0.75	0.25
Workstation C	0.01±0.02	0.00	0.00±0.00	0.00	0.07±0.09	0.00
Workstation D	0.30±0.12	0.10	0.01±0.01	0.00	0.77±0.31	0.25
Station Unload	0.05± .03	0.00	0.00±0.00	0.00	0.12±0.08	0.00

Figure 5 - Expected values of the performance measures

	Busy		Idle		Down		Blocked	
	Case1	Case2	Case1	Case2	Case1	Case2	Case1	Case2
Station load	0.20±0.00	0.20	0.79±0.01	0.80	0.01±0.01	0.00	0.00±0.00	0.00
Work stationA	0.40±0.01	0.40	0.58±0.01	0.60	0.02±0.01	0.00	0.00±0.00	0.00
Work stationB	0.81±0.01	0.80	0.18±0.01	0.20	0.01±0.01	0.00	0.00±0.00	0.00
Work stationC	0.20±0.01	0.20	0.78±0.01	0.80	0.02±0.01	0.00	0.00±0.00	0.00
Work stationD	0.60±0.01	0.60	0.39±0.01	0.40	0.01±0.01	0.00	0.00±0.00	0.00
Station unload	0.20±0.01	0.20	0.79±0.01	0.80	0.01±0.01	0.00	0.00±0.00	0.00

Figure 6 - Expected values of the performance measures

References

1. Baker, K.R. and Betrand, J.W.M., A Dynamic Priority Rule for Scheduling Against Due-Dates, Journal of Operations Management, (1982), 3,37.

2. Kiran, A.S. and Smith, M.L., Simulation Studies In Job Shop Scheduling-I, A Survey, Computer and Industrial Engineering, (1984-a), 8, 87.

3. Kiran, A.S. and Smith, M.L., Simulation Studies In Job Shop Scheduling - II, Performance of Priority Rules, Computer and Industrial Engineering, (1984-b), 8, 95.

4. Ragatz, G.L. and Mabert, V.A., A framework for the Study of Due Date Management in Job Shops, International Journal of Production Research, (1984), 22, 685.

5. Map/1 Simulation Package, Pritsker and Associates, Lafayette, Indiana, (1984).

Statistical Analysis of System Performance in a FMS

MICHAEL J. HENNEKE and RICHARD H. CHOI

Department of Industrial and Management Engineering
Montana State University, Bozeman, MT 59717

Abstract

A study was undertaken to investigate the effects of production scheduling rules on the performance of a FMS. Data were generated by a digital simulation model and evaluated using both factorial analysis and the Waller-Duncan Bayes LSD (BLSD) method.

Problem Outlines

One of the many production control functions in a Flexible Manufacturing System supervised by the system's computer is the decision as to when to launch a specific part into the system from storage, and to which machine it will be routed to. This is accomplished by establishing priorities among the parts in storage and the machine centers, called scheduling rules. In general, these rules are based on some attribute of the part being selected, for example, the decision of which part will next enter the system may be based on that part's processing time or sequence, due-date, traveling time, etc. Because there was some question as to the extent machine center queue capacities would influence the results, three different queue capacities were included in this study. A computer model of the system was developed capable of simulating all of the system's operations. The model was developed using SIMAN, a high level language developed specifically to simulate manufacturing systems.

Once the model was built and data under the various rules were generated, care was taken to insure that the conclusions drawn were reliable. Because many of the operations carried on by the system were stochastic processes, the analysis was based on statistical inferences. Test statistics were calculated by using an interactive statistical analysis software package.

The FMS Simulator

The FMS adopted for this research had the following major components [1]: an AS/RS, a parallel storage structure including eight storage, a parallel machine center structure including six identical NC machining centers, one turning cell including a

robot, two vertical NC lathes, a washing station, two AGVS. The
model was capable of processing seven different part families,
each with unique processing times, due-dates and production
costs. The AS/RS cart had a capacity of one part. Whenever the
cart was idle, one of the seven raw materials or a semi-finished
part was selected for routing to one of the six machine centers.
The model included failures of the major components. The number
of failures was generated by an exponential distribution with
availability values for the AS/RS cart, the machine centers and
the robot obtained from the actual manufacturing system.

Performance Criteria

Because no single criterion accurately reflected the overall
performance of the system, six measures of performance were
included in this study [1].

- actual system effectiviy   (%)
- average traveling time of parts   (min.)
- actual production output   (units)
- average manufacturing throughput time   (min.)
- work-remaining (work-in-process inventory)   (min.)
- average production lateness   (min.)

Scheduling Rules

Traditional job shop scheduling rules [2] must be classified into
two groups on the basis of FMS characteristics: part dispatching
rules and machine center selection rules. For example, the SPT
(shortest processing time) rule was not appropriate for machine
center selection. Because a FMS normally does not require setups,
rules considering setup time were also inappropriate.

Parts were assigned a priority value for resolving part selection
conflicts. Any one of seven traditional job shop scheduling rules
[2] was employed: RANDOM, first storage first served (FSFS), the
shortest processing time (SPT), the earliest due-date (DDATE),
the least remaining slack time (SLACK), the smallest ratio of
slack time divided by remaining processing time (S/PT), and the
highest production cost (VALUE).

Machine centers were also assigned priority values. These were
RANDOM, first machine first served (FMFS), the lowest number of
parts in queue (NINQ), and the least work in queue (WINQ).
Whenever the queue was full, the machine center was assigned the
lowest priority no matter which machine center selection rule was
in effect.

Queue Capacity

Because of the speed at which the parts were dispatched, each
machine center contained its own waiting line to increase system

utilization. The capacity of the waiting lines were studied at 1, 2, and 4 parts.

System Performance Observations and Evaluation

The experiment was designed to determine whether or not the choice of a scheduling rule or machine center queue capacity significantly affected the means of the performance criteria. When significant differences were detected, further efforts to establish which part/machine selection rule or machine center queue capacity performed best under each of the performance criteria were undertaken.

Preliminary calculations using approximations of the maximum error of estimation and the maximum standard deviation indicated that five independent trials would produce reliable tests at 95% confidence. Each trial consisted of simulating the system's operation for one hundred forty hours, the length one week's operation of the actual FMS.

The method used to determine whether the performance criteria were significantly affected by the choice of scheduling rule or machine center queue capacity was a two-factor ANOVA analysis. Scheduling rules and queue capacity served as the factors with twenty-eight and three levels, respectively. Because this experiment could detect interaction effects, it was also determined if the effect of schedulings rule acted independently of machine center queue capacity, which was crucial to properly interpret the experimental results.

By comparing the resulting F ratios to F values corresponding to a 95% confidence level, it was determined whether or not the performance criteria means differed significantly at that level. The F ratios, summarized in Figures 1 and 2, show that the choice of part/machine selection rule and machine center queue capacity did significantly effect all of the system performance criteria. Interaction effects, while marginally significant, were small enough compared to the main effects to have no influence on the conclusions drawn by this experiment.

Once it was established that the choice of scheduling rules and queue capacity significantly effected system performance, the top treatment means were selected and are shown in Table 1. To test for significant differences between individual means, the authors chose the Waller-Duncan Bayes LSD (BLSD) method. The method utilized a minimum average risk, dependent upon the confidence level, F ratio, and degrees of freedom to calculate the least significant difference (LSD) between means. Table 1 shows BLSD values for each performance criterion. Finally, Table 2 ranks the best performing queue capacity for each of the performance criteria. Rule combinations or queue capacity were recommended

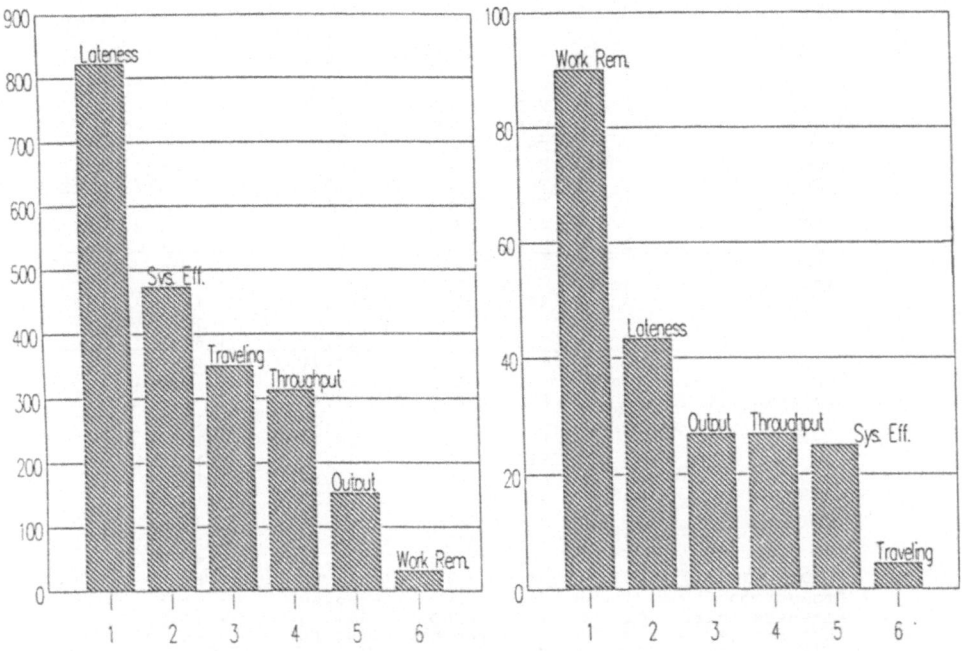

Fig. 1. F ratios of
         scheduling rules

Fig. 2. F ratios of
         queue capacities

over another when the difference between two performance values
was greater than the BLSD value.

Conclusion

Potential values of this paper include that a number of simple
scheduling heuristics were evaluated for a FMS, and efficient
statistical tools were adapted to determine most feasible rule
combinations. However, because this study handled only one
FMSconfiguration, the results cannot be assumed valid for all
configurations. What can be extended to other systems is the
methodology used in this study to evaluate FMS production
control parameters.

Bibliography
1. Choi, R. H. and E. M. Malstrom, "Physical Simulation of Work
   Scheduling Rules in a Flexible Manufacturing System,"
   Proceedings of the 8th Annual Conference on Computers and
   Industrial Engineering, Orlando, March, 1986.

2. Conway, R. W., W. L. Maxwell and L. W. Miller, Theory of
   Scheduling, Addison-Wesley, Massachusetts, 1967.

Table 1. Recommended scheduling rules by performance criteria

Performance Criteria	Rank	Rules	Mean	Difference	
System	1	VALUE/WINQ	0.4743	---	*
Effectivity	2	VALUE/NINQ	0.4742	- 0.0001	*
	3	VALUE/RANDOM	0.4723	- 0.0020	*
BLSD = 0.0039	4	VALUE/FMFS	0.4703	- 0.0040	
Traveling	1	FSFS/WINQ	2.026	---	*
Time (min.)	2	FSFS/NINQ	2.033	+ 0.007	*
BLSD = 0.0219	3	FSFS/FMFS	2.103	+ 0.077	
Production	1	FSFS/NINQ	670.01	---	*
Output (units)	2	FSFS/WINQ	669.90	- 0.110	*
BLSD = 5.1	3	S/PT/WINQ	644.30	- 25.710	
Throughput	1	FSFS/WINQ	158.10	---	*
Time (min.)	2	FSFS/NINQ	167.90	+ 9.8	*
	3	SPT/WINQ	176.30	+ 18.2	*
	4	SPT/NINQ	176.60	+ 18.5	*
BLSD = 30.5	5	SPT/RANDOM	220.00	+ 61.9	
	1	SLACK/WINQ	13.01	---	*
Work	2	DDATE/WINQ	15.65	+ 2.64	*
Remaining (min.)	3	FSFS/WINQ	22.44	+ 9.43	*
	4	SPT/NINQ	25.03	+ 12.02	*
	5	DDATE/NINQ	26.12	+ 13.11	*
	6	FSFS/NINQ	26.89	+ 13.88	*
	7	SPT/WINQ	27.72	+ 14.71	*
	8	SLACK/NINQ	29.23	+ 16.22	*
	9	S/PT/WINQ	43.24	+ 30.23	*
	10	S/PT/NINQ	51.45	+ 38.44	*
	11	VALUE/NINQ	59.77	+ 46.76	*
BLSD = 51.77	12	RANDOM/WINQ	64.89	+ 51.88	
Production	1	VALUE/NINQ	-336.60	---	*
Lateness (min.)	2	VALUE/WINQ	-317.30	+ 19.30	
BLSD = 11.5	3	VALUE/RANDOM	-252.80	+ 83.80	

*: Not significant at 95%

Table 2. Recommended queue capacity by performance criteria

Performance Criteria	Rank	Mean	Queue Capacity	Difference	
System	1	0.4205	4	---	*
Effectivity	2	0.4177	2	- 0.0028	
BLSD = 0.0013	3	0.4154	1	- 0.0051	
Traveling	1	2.348	1	---	*
Time (min.)	2	2.350	2	+ 0.002	*
BLSD = 0.008	3	2.359	4	+ 0.011	
Production	1	611.9	1	---	*
Output (units)	2	608.0	2	- 3.9	
BLSD = 1.7	3	605.0	4	- 6.9	
Throughput	1	330.1	1	---	*
Time (min.)	2	337.7	2	+ 7.6	
BLSD = 5.7	3	353.6	4	+ 23.5	
Work	1	63.36	1	---	*
Remaining (min.)	2	126.4	2	+ 63.04	
BLSD = 16.9	3	193.5	4	+130.14	
Production	1	- 109.2	1	---	*
Lateness (min.)	2	- 104.4	2	+ 4.8	
BLSD = 3.75	3	- 90.8	4	+ 18.4	

*: Not significant at 95%

3. Pegden, C. D., Introduction to SIMAN with Version 3.0 Enhancements, System Modeling Corporation, State College, Pennsylvania, 1986.

4. Peterson, R. G., "Design and Analysis of Experiments", Marcel Dekker, New York, N.Y., 1985.

Application of Simulation for Daily Production Planning and Control in Shop Production

Dipl.-Math Gabriele Schröder
Fraunhofer-Institute of Transport Engineering and Physical Distribution
Emil Figge Straße 75
4600 Dortmund 50
W.-Germany

1. Introduction

As a result of the increase in automation and the changing demands being placed on manufactoring industries, production control has taken a major significance within the field of manufactoring organization. Fig. 1 illustrates the tasks to be completed by production control and its position within the field of manufactoring organization:

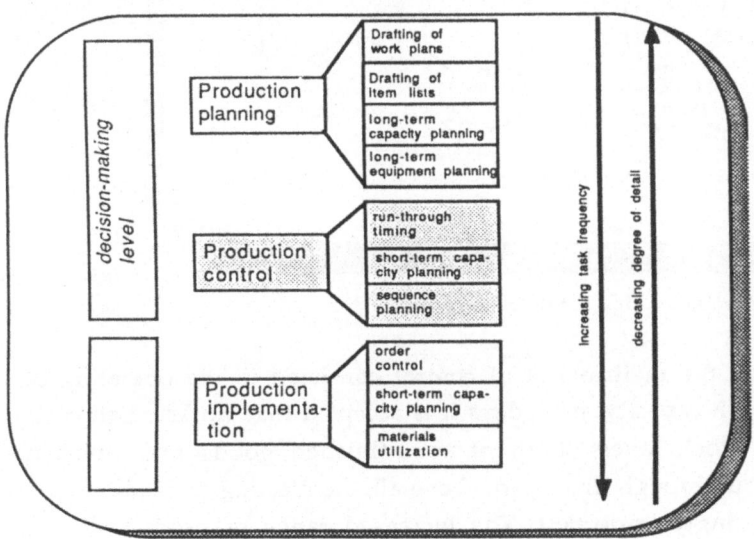

Fig. 1 Position of production control in manufactoring organization

Whereas production planning involves the medium and long term aspects of order processing (the determination of data relating to products and equipment), production control deals with the short term (run-through timing, short term capacity planning, sequence planning). This implies that the task frequency and the degree of detail involved are much greater with production control than with production planning. Thus it is not surprising that control operations in particular are becoming more and more dependent on computers.

2. Tasks and aims of production control

Production control seeks not only to retain a (uniformly) high utilization of capacity but to achieve short run through times, low stock levels and, above all, the accurate meeting of production deadlines (Just in Time). Fig. 2 /4/ illustrates the shift in emphasis which has taken place in recent years regarding these aims:

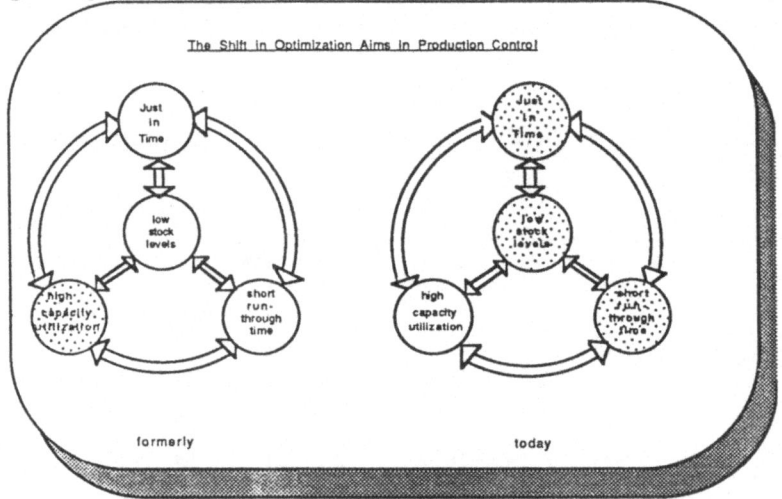

Fig. 2: The shift in optimisation aims in production control

Whereas formerly the main object of control consisted in the operating of installations at high capacity, nowadays it is accepted that stocks mean costs and, thus, low stock levels (both of semi-finished goods and finished products), short runthrough times and, above all, the meeting of deadlines are becoming increasingly important. The increased range of products being manufactored as a result of competition is accompanied by a transition to an increasing number of smaller and smaller lots. The complicated control mechanisms resulting from this leads to more and more complex demands being placed on production control, demands which cannot be met without the use of computer-assisted systems. This is also a result of the greater demands being placed on control quality due to automation.

Computer-assisted control systems operating on an analytical basis have been developed from various sources and are capable of meeting these demands for particular fields of application. Load-oriented production and OPT and the Kanban system serve as examples. In every case, the application of these systems depends on the particular marginal factors present. The Kanban system, for example, demands a limited range of articles low equipment levels, as constant a demand as possible and high production capcity flexibility. The major disadvantage of the analytical systems lies, however, in the fact that it is only possible to recognize a potential application requirement when it has arisen.(see /1/) The integration of simulation into control systems presents a decisive advantage in this case. With the aid of simulation, a future application requirement becomes apparent at the present point in time and thus renders it possible to bring the commencement of the application forward. However, the use of simulation for purposes of control offers further advantages of no lesser importance (see Fig. 3/3/)

Fig. 3: Advantages of simulation systems for the control of logistical process operating

There are various simulation systems today which incorporate one or more of the control tasks to be completed. The majority of these systems operate in two steps:

- The determination of the production plan using analytical procedures

- The simulated operation of the generated plan for the purposes of evaluation and   verification

At the Fraunhofer Institute in Dortmund a control system has been developed which uses simulation for the generation of the production plan.

3.The simulation system SIMON

The simulation system SIMON is a control system which has been developed for workshop production. It has the additional advantage of being able to determine the optimum lot size for any given situation. This is useful for situations in which several orders can be combined to form a lot. In this case, the total volume is not required for one particular delivery deadline and the lot can be divided up should bottlenecks arise.

An order-monitoring function has also been integrated into the system. This renders it possible to identify any deviations of the production process from the production plan immediately and to provide the simulator with data on the current level of production at any given time.

Whereas with analytical procedures it is easy to recognize the demarcation of individual control tasks one from the other, there is not the same degree of clarity with SIMON. This is because the different tasks are completed during simulation, dependent on the simulated system condition. Thus for example, capacity utilization and the capacity which the processing stations should possess in order to complete their tasks is a result which is obtained during a simulation run. A simulation run serves to arrive at an exact and feasible production plan which also provides information about its quality.

The results obtained from a simulation run include exact data on machine use with individual starting and finishing deadlines and an evaluation of the planned process operation regarding idle times and set up times as well as the holding times and any deviation from the order finishing times. In addition, a valued comparison with previous simulation runs is made. The comparitive facility enables the user to use his judgement and to select the best simulated run implemented.

The model of the production process to be simulated consists of processing stations and products.

A processing station is not only understood to be a machine but also includes manual workplaces at which defined operation may be completed as well as inflexibly linked installations. In the same way, machines or installations which can implement exactly the same operations (with the same output) are grouped together as a processing station. In this way, the processing station is clearly established on the basis of the operation to be performed.

The basic product data required per order unit are constituted by the pieces to be manufactored with the corresponding operations (including final assembly) and the corresponding reference times.

A production model constructed in this way is automatically provided with the dynamic data.

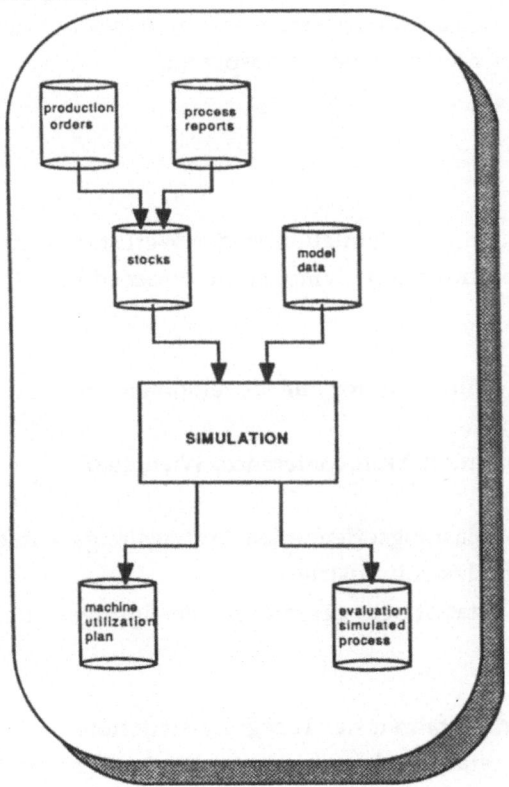

Fig. 4 Input and output data for SIMON

After the simulation has been started SIMON first of all generates a processing order for each operation to be completed per order unit and order lot. This processing order is fed into the order pool of the relevant station. The earliest and latest starting times are determined based on the processing status of the lot and the final deadline. In this way the room for manoeuvre is established for each individual operation. During simulation it is then attempted to allocate each and every process order within the scope available and at the same time to select the best order according to the situation at head. Thus, during simulation, on operable machine utilization plan is generated. This is accompanied by an evaluation of the planned process operation and provides the user with a list of data enabling him to intervene in the operation plan and weigh up the optimization criteria.

In an initial application of the system in the production plant of a supplier to the automobile industry the control systems were dramatically improved and the manufactoring process rendered more transparent.

4. References

/1/ Schmidt, R.:
Einsatzmöglichkeiten der Simulation in der Werkstattsteuerung
In: 21st Annual Simulation Symposium, Proceedings, Tampa 1988

/2/ Schürholz, A.:
Application of a Simulator for the development and check of controlling software
In: European Simulation Multiconference, Wien 1987

/3/ Ufer, H. A.:
Anwendung der belastungsorientierten Auftragsfreigabe in einem Unternehmen der Elektroindustrie
In: Fachseminar statistisch orientierte Fertigungssteuerung, Hannover 1984

/4/ Wiendahl, H.-P.:
Grundlagen neuer Verfahren der Fertigungssteuerung
In: Fachseminar statistisch orientierte Fertigungssteuerung, Hannover 1984

The White-Collar World: The Next Push for Productivity

A. K. SARRIS
Director of Systems Planning and Implementation
Ontek Corporation, Laguna Hills, California

Summary

There are a number of challenging issues which must be resolved
if white-collar productivity analysis is to come into the mainstream and if such analysis is to be applied to anything but
the most basic of clerical tasks. For white-collar productivity
analysis to be conducted on a regular, on-going basis,
management's perception of such analysis must be realigned to
focus on the positive aspects of working smarter. Even more
importantly, the techniques, approaches, methodologies and
tools for conducting white-collar productivity analysis must be
further developed and refined so that the analysis process itself is efficient, accurate and *productive* . The manufacturing
industry has reached the point where the white-collar world is
the major target for productivity improvement - and the push is
on.

Background

Computer-Integrated Manufacturing (CIM) has expressly had as
its goal the enhancement of manufacturing productivity through
the automation of the factory. While all sorts of high-technol-
ogy gadgetry, both hardware and software, has been utilized in
the pursuit of CIM and its associated goals, the majority of
that technology has been oriented toward automating *physical*
tasks. Whether the task takes place on the shop floor (e.g. a
machining or part assembly operation), next to the shop floor
(e.g. inspection), in the building next door (e.g. drafting) or
the suites upstairs (e.g. office automation), the focus has
been on mechanizing those activities that machines can do
equally as well or better than humans. The reason those machines were developed by technology vendors is because that is
exactly where industry said it needed help the most. The increasing cost of direct labor, coupled with more complex products and lower quality drove industry to seek a solution.

Today we see a different scenario. The cries for help from industry to technology vendors have a markedly different tone. Indirect costs, primarily the costs of professional, white-collar knowledgeworkers - the people who perform analytical tasks or who plan, manage and account for the activities of the enterprise - are growing rapidly. Given the declines in direct labor, indirect labor as a relative piece of the enterprise cost pie is *skyrocketing* . In fact, since much of the factory automation installed over the last few years requires skilled technicians to program and support it, the reduction in direct labor costs may have inadvertently contributed to the growth of indirect labor costs.

Clearly, reducing indirect costs, or even reducing their rate of growth, has become today's "hot button". The immediate reaction is to double the M.I.S budget - placing a personal computer on every desk and ordering those optional M.R.P. modules we thought we did not need when we first bought the system. Unfortunately, whereas a five-axis numerically-controlled multi-spindle mill, fed by a material handling robot and incorporating a vision inspection sensor, meets or exceeds the capabilities of a human using a pantograph left-over from 1945, today's management information systems do not "*manage.*" It is questionable whether all but a few even provide knowledgeworkers with the right information in a reasonably usable form with which to do their jobs. At best, then, today's information systems *support* white-collar workers, but the systems *themselves* certainly do not perform design or autonomously create production plans. Even expert systems are severely limited by the narrowness of the domains in which they can perform inferencing and the volume of rules and data which must be fed into them to do that.

So, everyone seems to agree that indirect costs, and the white-collar labor that comprises a significant portion of those costs, are a critical issue in today's manufacturing environment. Everyone also seems to have a certain "intuitive" feeling, based largely on the complaints of system users, that today's M.I.S. and engineering systems do not completely do the job they were intended to do. But where exactly are improvements needed? What aspects of white-collar work (e.g. analyzing, decision-making) do we need help with? How much improvement potential is there? If we had more intelligent systems available, systems that could: learn from their experiences; make predictions based on trend analysis and extrapolation;

automatically adjust their scope and scale to the preferences
and needs of individual users; and reconcile conflicting information about the enterprise - what would we do with such systems?

One good reason why vendors have not immediately offered better
productivity tools for white-collar workers is that industry is
not giving clear signals as to *where*, specifically, technology
is needed and of *what value* such technology would be. There
are a number of conditions that have lead to this situation.
First, there is a general reluctance on the part of manufacturing management to admit that white-collar productivity could be
better. Doing so seems to be viewed as admitting that we our-
selves, or Joe or Jane down the hall, are not working hard
enough or not doing our jobs right. Have we forgotten the old
adage that working *hard* is not the same as working *smart*?
Secondly, because there is such a void of technology for white-
collar workers, industry's expectations of potential solutions
seems to have been shamefully constrained. Word-processing is
helpful for documenting engineering changes, but that is not
the same, for example, as having a system present you with an
analysis of the *ramifications* of the E.C.O. you are about to
submit. But few people are willing to say that such advanced
capabilities are what we really need. Most people seem content
in asking for what we already have, and then complaining that
it does not really do the job.

However, even if management can overcome the "white-collar productivity as leper syndrome", a number of logistical issues
remain before white-collar productivity analysis can point the
way toward *real* productivity improvements.

Approach

To assess white-collar productivity and determine which technologies appear to be the most promising for achieving productivity improvements, one should perform the following analyses:

- identify the *activities* in the enterprise (emphasizing
those that white-collar workers currently perform), and the
resources (number and type of people, systems, etc.) that
are currently used to perform them; determine the inputs and
outputs from each activity and the *interrelationships* between and among activities,

- measure the current *performance* of those activities

against established standards, or against other activities,
to determine which are the largest resource users; determine
where the greatest *potential for improvement* lies,

• gain an *understanding of why* performance is the way it
is, and determine *what needs to be done* to improve it (independent of any particular technologies for achieving that
improvement),

• establish *improvement goals*, based on the importance of
each activity,

• determine the *ability of various technologies*, or potential technologies to achieve these goals,

• dollarize the data used for the analysis and perform
cost benefits analysis.

As straightforward as this process might seem, there are a number of methodological problems or gaps which must be addressed
before white-collar productivity analysis becomes science
rather than art. Two critical problems are summarized below.

Enterprise Modeling

The methodologies and tools in existence today for conducting
enterprise functional modeling are extremely limited in their
ability to represent analytical functions (i.e. functions in
which *logical* rather than *physical* processes are performed).
Additionally, today's information modeling tools do not address
many of the complex representational structures that are necessary to depict the interactions among functions and/or their
informational entities. For example, *cause and effect. relationships* are very difficult to represent.

Metrics and Measurements for White-Collar Productivity

Based on research recently conducted at Ontek Corporation, as
well as a a number of studies conducted within the last 2-3
years by government and private organizations such as The
American Productivity Center, there appears to be a major gap
in measurement techniques, metrics, and standards (standard
measurements) between the macro-level measures (i.e. country,
industry or company-wide models for econometric or parametric
comparisons to other countries, industries or companies) and
the shop floor-level measures (i.e. time and motion studies).
That is to say, existing measures tend to be either too high-
level (to the point of being so broad that they are not useful
in a practical sense) or too physically-oriented (meaning the

resource being measured must relate to the *physical* process of production). The gap then takes the form of two components: one strictly a leveling problem between macro and micro measures; and the second, related problem, of a lack of productivity measures specifically oriented to white-collar work.

Both the chasm between high and low-level measures and the lack of good, sound measures for white-collar productivity seem to be attributed to two causes. First, *white-collar productivity is hard to measure* . Unlike shop floor processes, one cannot easily count the resources (inputs) consumed by an activity such as "planning", "controlling" or "analyzing", or count the resultant outputs. Secondly, measurement of these processes also does not lend itself to the generalized parametric modeling used in macro-level productivity measurement. Specifically, while the physical movements a worker performs during an activity like "assemble part" may be standard across the industry, and while macro-models may allow company-specific parameters to be plugged into various general industry equations, the mid-world of *knowledgework seems to be viewed as being unique to each environment* , and often to each instance within that environment. It is encouraging that various individuals and organizations are now working to address these problems. The underlying approaches being advocated rely primarily on Normative Productivity Measurement, including the use of Nominal Group Techniques and the Delphi method.

Assuming one can model white-collar activities and measure their current productivity, there is still the issue of understanding *why* productivity is the way it is, as well as *where* and *how* productivity improvements might be achieved. This involves an integration and analysis of the functional models in conjunction with the results of the productivity measurement efforts. The interrelationships of the myriad of factors that come into play in the course of performing white-collar activities, principally issues such as cause and effect, are again the chief source of difficulties. This problem also has additional repercussions when identifying the operational impacts of technologies and determining dollarized cost benefits. Various methods of linking such factors and categorizing their behavior are being studied. The Productivity Network Approach developed by Dr. Bela Gold is an example of work conducted in this area. It is time to apply theoretical work like this to the *science* of productivity analysis.

Process Planning

Fifth (5th) Generation Group Technology

William B. Krag
Arthur Young & Company
Detroit, Michigan, USA

SUMMARY

As manufacturing organizations evolve through phases of capability, strategy, and emphasis, the organization of information has evolved as data manipulation technology has progressed and proliferated. The 1990's will usher in a transition to fifth (5th) generation management and, simultaneously, offer tremendous potentials for organizations to make large, competitively efficient jumps from existing phases to 5th generation platforms given detailed and careful prerequisite planning. This paper will review this fifth (5th) generation stage and describe two fundamental enhancements to company information needed to support this transition.

MASS SPECIALIZATION

"Toward Mass Specialization," (Figure 1) illustrates the five generations of manufacturing and lists several related trends. The trends noted in Figure 1 enhance the SME [1] definition and the title, "Toward Mass Specialization," has been contributed by this author. Of particular importance, is the implied necessity for competitive success both in the ability for an organization to be flexible and to organize their technical information well. Fifth Generation Group Technology addresses the organization of technical information and in so doing provides the means to act quickly -- to be flexible.

FLEXIBILITY

Flexibility has been discussed and defined in an excellent publication "Developing Flexibility for Excellence in Manufacturing" [2]. Broadly defined, it can be measured in the ability of a company to implement Just-In-Time (JIT) manufacturing which includes the capability to:

1. Rapidly design a new, quality product.
2. Rapidly launch a new, quality product.
3. Rapidly process an engineering change request.
4. Rapidly implement an engineering change notice.
5. Effectively establish JIT relationships with suppliers and customers.
6. Continuously improve manufacturing capabilities.
7. Continuously improve human capabilities.

Fifth Generation Group Technology, in large measure, enables the rapid processing of information, a major prerequisite in the above JIT requirements.

PARAMETRIC DESIGN

Concurrently, much emphasis in manufacturing is being placed upon parametric design. The concept of flexibility and parametric design, although usually discussed separately, are in fact mutually supportive and implemented using 5th generation group technology as illustrated in Figure 2 "Prerequisites to JIT/ Flexibility." Parametric design is typically associated with the design of finished products using predefined features (entities/ attributes) embedded in a computer graphics image library. Using such a library, product designers can now "assemble" a new piece part drawing from various predefined "features" from selection tables. This enables piece part designs to be made from predefined, standard (hopefully tooled/programmed) elements. This approach promotes flexibility because it enables the creation of an infinite number of end products from a finite number of inputs.

UNIVERSE OF PRODUCTION (U of P)

In traditional group technology perspectives the classification of elements of production spans product, process, support, and technology oriented categories of items as follows:

1 Product	11	Raw material
	12	Commercial items used in product
	13	Proprietary designed, finished, piece parts
	14	Features (attributes/entities)
	15	Sub-assemblies
	16	Models and options
2 Process	21	Machines
	22	Material handling equipment
	23	Measuring equipment
	24	Computer equipment
	25	Tools
	26	Gauges
	27	Fixtures
3 Support	31	Maintenance items
	32	Machine repair parts
	33	Suppliers (shop/medical) stationary, computer, etc.
4 Technology	41	Design rules
	42	Process rules
	43	Process capabilities including speeds/feeds
	44	Time standards
	45	Device control programs

Rarely have companies expanded their group technology classification and coded data bases beyond the first major category ("product") in the Universe of Production. Fifth generation manufacturing demands that accurate, descriptive, technical databases be created for all major categories in the U of P. Notice that this U of P expands well beyond just physical items and includes major categories for subjects such as features (#14) and technology (#40).

VISIBILITY AND POPULARITY

"The ability of an individual and an organization to draw upon the resources developed in the past and to 'see' their vision of the future is a powerful determinate of the 'quality' of management actions" [1]. Fourth (4th) generation group technology has provided the product-oriented visibility needed to view the past. This now needs to be married with a dynamic approach which tracks current trends in production volume and mix and overlays the results on the GT databases to provide "green, yellow and red" indicators of item preferences for future use. A simple algorithm for computer generated preference coding was advanced in SME technical paper MS 82-932 [3]. It remains for 5th generation manufacturing companies to overlay the important popularity dimension over the visibility gained with group technology.

IN-PROCESS STATES AND CONTROL BY REVISION LEVEL

Group technology has heretofore focused almost entirely on the organization of information dealing with the finished state of items (as acquired raw materials, commodities, finished piece parts, etc.). As dimensions and tolerances of finished parts change during the process of progressive manufacturing, it becomes important to differentiate between finished states and in-process states. Fifth generation, data driven companies should not only distinguish finished from in-process item states in the databases, but should also control design and especially manufacturability by item revision level to assure the production of reliable, durable, desirable, economic parts. A means for making these important distinctions was advanced in SME technical paper MS 85-881 "The CIM Descriptor" [4].

IN-PROCESS DIMENSIONS AND TOLERANCES

In June 1987 Pennsylvania State University hosted a very significant worldwide conference which proceedings documented the state-of-the-art in Computer Aided Process Planning (CAPP) [5]. Although a few papers suggested methodologies to calculate the number of cuts required, none advanced a rigorous, proven methodology for determining in-process dimensions and tolerances. This "Gap in CAPP" has fortunately been recognized by SME who:

1. Published the methodology in SME Tool & Manufacturing Engineers Handbook, Vol. 1, Chapter 2 (4th edition) by Mr. Oliver Wade [6].

2. Is currently in the process of generating a video tape lecture series in tolerance charting) likely publication in Fall 1988).

Given this basis of education and training, the tolerance chart methodology has been preserved and eventually will be converted from a manual to a graphic-based, computer assisted essential tool. The fifth (5th) generation manufacturing company will insist that tolerance stack analysis (charting) be performed to assure economic producibility of each component prior to launch. This then becomes an essential link in converting the "Voice of the Customer" to economic production technology as illustrated in Figure 3, "Quality Function Deployment."

SUMMARY

Fifth generation group technology for a 5th generation company will demand a quantum jump in the extent of item descriptive data bases needed to operate a truly flexible operation. The primary expansions in data bases required are from product-oriented items to process, supply, and technology categories as indicated by the expanding "Universe of Production." Using this approach, an organization can finally become competitive in a global market which increasingly demands the ability to provide "mass specialization."

BIBLIOGRAPHY

FIFTH (5th) GENERATION GROUP TECHNOLOGY

1. Savage, Dr. Charles M., "Fifth Generation Management for Fifth Generation Technology," published by SME/CASA, Dearborn, Michigan, 1987.

2. Hall, Dr. Robert W., Nakane, Dr. Jinichiro, "Developing Flexibility for Excellence in Manufacturing," published for Arthur Young/DEC conference "Manufacturing for the 90's," Dearborn, Michigan, March 11, 1988.

3. Krag, W. B., "Dynamic Standardization Using GT Classified and Coded Data Bases," SME technical paper MS 82-332.

4. Krag, W. B., "CIM Descriptor," SME technical paper MS 85-881.

5. "Computer Aided Process Planning," Penn State USA, 19th CIRP International Seminar in Manufacturing Systems, June 1-2, 1987.

6. Tool and Manufacturing Engineers Handbook, Volume 1, Chapter 2, by Oliver Wade, SME 4th Edition.

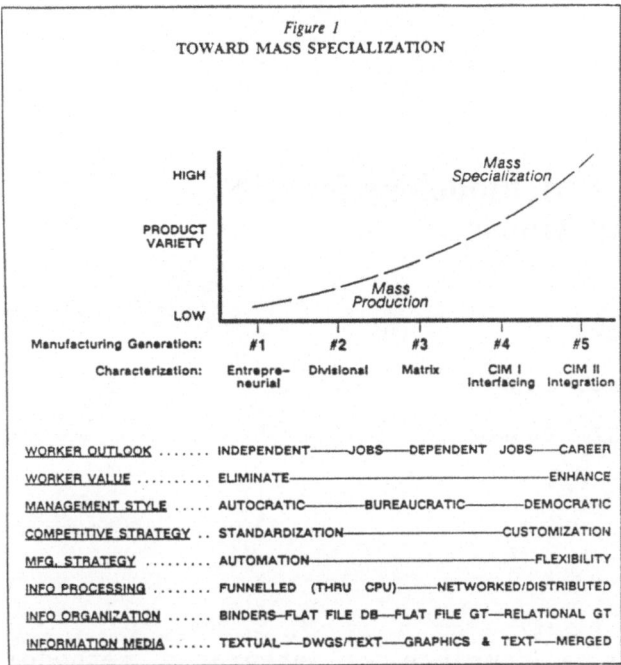

Figure 1
TOWARD MASS SPECIALIZATION

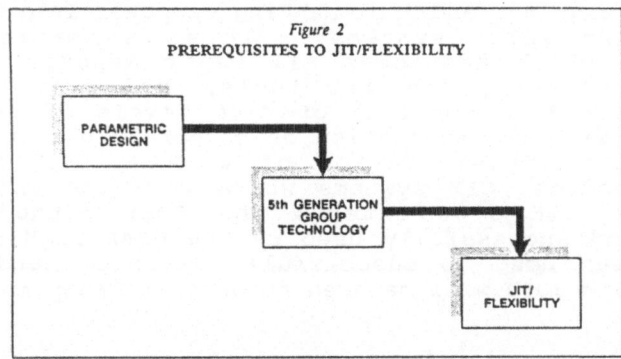

Figure 2
PREREQUISITES TO JIT/FLEXIBILITY

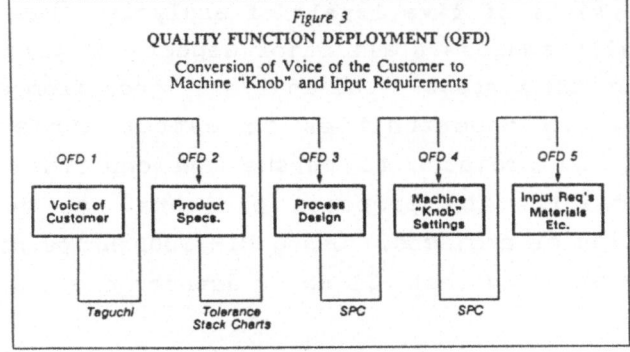

Figure 3
QUALITY FUNCTION DEPLOYMENT (QFD)

Conversion of Voice of the Customer to
Machine "Knob" and Input Requirements

CIM 3000: A Methodology for CIM Planning and Management

Donald R. Sloan

Wizdom Systems, Inc.
1260 Iroquois
Naperville, IL 60540

Summary

CIM 3000 is a complete, integrated methodology for planning, designing, and implementing Computer Integrated Manufacturing (CIM) systems. It is a comprehensive methodology which addresses all major aspects of CIM, provides supporting computer tools, and provides user training. CIM 3000 consists of five levels of analysis, which begin with the evaluation of corporate objectives and end with the implementation of working CIM systems. CIM 3000 produces CIM systems which work and which meet objectives. CIM 3000 extends the IDEF methodologies developed and successfully used by the USAF ICAM program. IDEF has been used to successfully automate hundreds of factories. CIM 3000 will be used to automate many more.

Introduction

CIM 3000 consists of **five levels** of analysis. This analysis systematically addresses all major aspects of planning and implementing CIM systems. The analysis goes from corporate objectives, to opportunities to better achieve these objectives, to projects to pursue the opportunities, to plans to realize the proposed projects, to successful implementation of projects. Using CIM 3000 helps manage CIM programs by assuring that all major aspects are addressed.

Level 1

Level 1 of CIM 3000 is called the **CIM Opportunities Survey.** In Level 1, the scope of the project is defined. Company documents are reviewed. Key people are interviewed.

The company objectives are determined and rated. Opportunities to better achieve the objectives are identified and assessed.

The outputs of Level 1 include the following:
o Prioritized list of business objectives.
o Cost matrix.
o High level cost benefit analysis of opportunities.
o Prioritized list of opportunities.
o List of short term, low cost, high return opportunities.
o Selection of longer term opportunities to be investigated.
o Plans for Level 2, 3, and 4 analysis of selected opportunities.

Level 1 concentrates on objectives and opportunities. It helps assure that the projects analyzed, proposed, and implemented meet the company objectives. Subsequent levels also use the objectives to focus their analysis.

Level 2
Level 2 of CIM 3000 includes **functional modeling** and analysis. In Level 2, an IDEF0 model is built of the current "AS-IS" operation, using the IDEFine-0 software modeling tool. An IDEF0 model describes all the functions (operations) in a factory or enterprise. It shows all the inputs, outputs, controls, and mechanisms for each function. Controls are inputs which influence outputs, but are not transformed into them. For example, parts are inputs and specifications are controls. Mechanisms provide the means of producing outputs. For example, machines, people, and computers are often mechanisms. An IDEF0 model describes operations by top-down decomposition and refinement. It starts with a high level model, then decomposes it into parts and refines it to add detail. The models are decomposed and refined until the appropriate level of detail is reached. This modeling provides a thorough understanding

of the AS-IS operation and pinpoints additional opportunities for improvement.

The cost/performance of the AS-IS operation is analyzed using time analysis, headcount analysis, and expense analysis. The IDEFcost software supports this cost analysis. The analysis identifies the high cost drivers of current operation and forms the basis for the cost / benefit analysis of TO-BE projects.

The State of the Art is reviewed in key technology areas. Potential projects are defined, evaluated, and ranked. Highly ranked projects are investigated further in Level 3.

Level 2 aids the management of CIM programs by providing a thorough understanding of the AS-IS environment. This understanding is necessary to determine what needs to be improved and how best to migrate to the improved environment.

Level 3

Level 3 of CIM 3000 builds **operation flow models** for selected potential projects. A flow model is built for the "AS-IS" operation and the proposed "TO-BE" operation. These models bridge between the Level 2 high level IDEF0 "AS-IS" model of current operations and the Level 4 detailed specifications for the "TO-BE" CIM system. The models combine functional, information, and dynamic modeling techniques. The information modeling is done using the IDEF-1X data modeling technique and is supported by the IDEFine-1X software product. IDEF-1X describes the main data entities in the system, their relationship to one another, and their data attributes (elements).

Level 4

Level 4 of CIM 3000 is the **engineering analysis and design.** It produces the strategic plan, the technology plan, the financial plan, and the system design. The strategic plan shows how strategic objectives are achieved by implementing specific projects. The technology plan assures that the enabling technology is in place to achieve the strategic objectives. The financial plans show the pattern of expenditure, provides cost / benefit analysis, and shows how the CIM program will impact the corporate financial statement. These plans explain the overall program and help get projects approved and funded.

Level 4 activities include defining potential manufacturing cells, simulating cell designs, simulating factory operations, and selecting potential vendors. It includes specifying computer hardware, software, and networks. It includes rank ordering projects by objectives and cost / benefit, and developing phased implementation plans.

Level 4 is supported by simulation packages and financial analysis software.

Level 4 aids the management of CIM programs by assuring the appropriate planning precedes implementation.

The complete analysis of an enterprise through Levels 1 to 4 can be completed in 6 to 9 months by a project team of 4 to 8 people. Automated tools, experienced staff, and good client training make these short intervals possible.

Level 5
Level 5 of CIM 3000 is the actual **implementation** of the CIM systems. It includes system, hardware, software, machine tool, and factory setup, installation, and operation.

Software design and implementation follow a rigorous design methodology consisting of the following steps.
1. Requirements analysis
2. Functional design
3. Data entity relation analysis
4. Logical data modeling

5. Detailed functional design
6. Software systems architecture
7. Test design and validation requirements
8. Physical data base design
9. Detailed software architecture
10. Prototype and user evaluation
11. Detailed software implementation plan
12. Module design and code
13. Module test
14. Software module integration
15. Software system integration
16. Hardware and software system integration and test
17. System performance tuning
18. User evaluation
19. Installation and training
20. Operation, maintenance, and support.

Software implementation is supported by popular Computer Aided Software Engineering (CASE) tools.

Conclusion

In conclusion, CIM 3000 systematically addresses all major aspects of planning and implementing CIM systems. Using CIM 3000 helps manage CIM programs by assuring that all major aspects are addressed. CIM 3000 produces CIM systems which work and meet objectives.

Just-in-Lead Time (JILT) Management: A Missing Link in CIM Strategy

Jay Nathan, CPIM, ME, PhD

University of Scranton

SUMMARY

This paper focuses on the importance of lead time management in a CIM environment. In most manufacturing companies, the handling of lead times and their variability had been poor--vendor lead times, processing lead times, and delivery lead times to market. Despite widespread use of micro and mini computers, the real-time linkages to equipment and processing technology (EPT) via Just In Time (JIT) management is not fully realized. This research describes "how" such a system can be developed and implemented.

INTRODUCTION

Computer Integrated Manufac.uring (CIM) has become a way of life for the productive manufacturing organizations. The use of JIT, total quality control system, and material requirements planning (MRP) systems have become a part of planning systems in a CIM environment. The essence of JIT porduction philosophy is its continual drive towards the streamlining of production capabilities that realize total quality control, lowest possible "total cost" products, and elimination of waste. Just-In-Time concepts focus on viewing the entire manufacturing process as a piece of constant flow, supported by numerous networks of work stations that attempt to maintain each dependent part for the good of the end product. It is a pull method which strives to get the material through the plant by only producing what is needed now, and balances the flow as opposed to balancing the capacity under MRP.

JUST IN LEAD TIME (JILT)

The growing popularity of MRP is due to its ability for time-phased planning and re-planning while interfacing with all levels of bill-of-material structure. One of the major short comings of MRP is that it works with one fixed lead time value. In a real-time CIM environment, the lead times or the just in lead times, if you will, have to be monitored and used in a real time to reflect the changing situations in the manufacturing process. The "push" method of releasing orders to the shop uses the fixed planning lead time. The components of fixed planning leadtime are: queue time, run time, set-up time, wait time, inspection, and move time. But the actual leadtime is a function of priority on the shop floor which gives us such varying degrees between planned and actual leadtimes.

JILT/CIM INTERFACE

The components of planning lead-time used in CIM, come under constant attack within the Just-In-Time philosophy. In fact, queues, set-ups, lot sizes, and safety stock are all inventory in one form or another. The JIT attitude towards inventory is that it is not necessary and should be minimized as far as possible. The JIT, rather than protecting against disruptions on the shop floor, as just-in-case inventory techniques tend to do, inventory hides problems and, in some cases, causes disruptions. The lead time embraced by JIT is only that time it actually takes to produce the required product by the given equipment and process technology (EPT). There is no safety lead time and should not be. Because reduction of set-up times, elimination of defects and rework, the minimization of handling and wait time, and the vast reduction of lot sizes produces the lowest cost product in the least amount of time with the best possible quality.

JILT MANAGEMENT WITH PERT

The program evaluation and review technique know as PERT, has been around with wide applications. The analogies between PERT and Bill of Materials are rather striking. PERT works with events, activities, and time. Events encompass start and completion times. Activities consume

resources like men and materials, and times can be expressed
in weeks like MRP. This network scheduling routine, much
like MRP, also has "precedence relationships". These
relationships, much like those in a bill of material, state
that certain tasks or activities cannot be started until
others are completed. Task independence and precedence
relationships can both be handled by the PERT technique.
The JILT management with PERT also identifies the bottleneck
work centers or machines where the efforts for reduced
setups and inventory are to be addressed. In addition, this
approach looks at the various inventory costs along the
critical path. The cost profile pits cost against activity
duration with direct and indirect dollars. As one expedites
the duration or shortens the activity time, direct costs
tend to increase while fixed, indirect costs decrease.

BILL OF MATERIALS AND PERT

A typical bill of material consists of sub-assemblies,
manufactured parts, and purchased raw materials (figure 1).
Each element of the BOM has a planning leadtime and the
number of each unit that is needed during that process.
Immediately, from looking at the BOM structure, one can
discern two distinct "paths" associated with producing an
end-item. Thus, with two paths, that of two sub-assemblies,
one can begin to break down these paths using the
similarities between BOM and PERT.

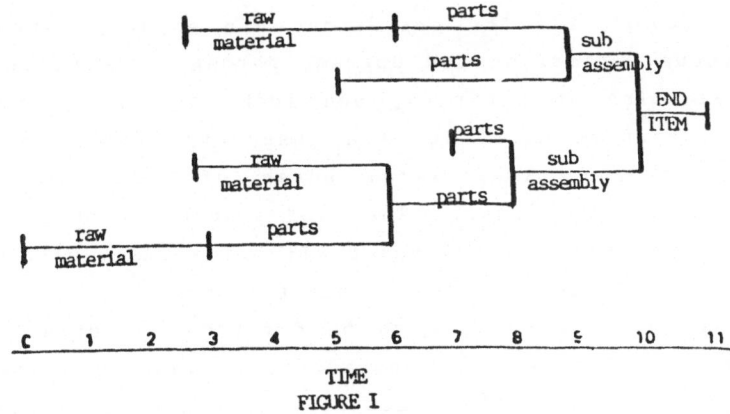

TIME

FIGURE I

JILT BENEFITS FOR CIM

The immediate impact of using JILT to evaluate planning leadtimes in a real-life CIM environment results in continual reduction in lead times. With the advent of using JILT to logically approach planning leadtime reduction in MRP, the leadtime reduction is not the only impact that this process has on CIM systems. Another important benefit is the opportunity to address many of the JIT zero principles (minimizing wastes of all types). Thus the impacts of using JILT to implement JIT principles within CIM systems are basically structured real-time linkages to equipment and processing technology with a prompting-type of reiterative analysis and evaluation criteria backed by a good networking CIM system.

As illustrated in figure 1, the bill of material information is used to develop a lead time evaluation and review which serves as a link in enhancing CIM's overall lead time effectiveness. Other benefits include: cycle time reduction, reduction in WIP cost, and increase in asset turnover. As WIP costs decrease, the ROI will increase which improves the cash flow.

Another benefit of JILT is the ability to identify bottleneck activities and helps in reconfiguring the plant layout among others. Alternate routings, reduction in set-ups and run times, and improved machine efficiencies are yet other major benefits.

CIM's network and its operations are greatly improved by continuous monitoring and upgrade through attacks on waste. By evaluating an activity, one looks for excessive queue time, excessive handling, and inspection time, inordinate set-up or run time, large scrap of rework levels of inventory on the floor. The JILT process becomes a medium for the CIM user to implement JIT principles which reflect changes in the shop floor and manufacturing process.

The last overall benefit to be derived from using JILT to implement JIT into the manufacturing planning system is a renewed effort to gain a competitive edge. By using JILT, one can begin to put a new focus on the factory. This is ultimately what Just-In-Time is all about -- producing quality products as fast as one can produce with little or no inventory!

REFERENCES

1. Monden, Y. "What Makes the Toyota Production System Really Tick?", Industrial Engineering, 13, no. 1 (January 1981), pp. 36-46.

2. Nellemann, D.O. and L.F. Smith, "Just-In-Time vs Just-In-Case", Production and Inventory Management, (Second Quarter 1982), pp. 13-21.

3. Schonberger, R.J. "Some Oberservations on the Advantages and Implementation Issues of Just-In-Time Production System", Journal of Operations Management, 1, no. 1 (November 1982), pp. 1-10.

4. Skinner, W. "The Focused Factory", Harvard Business Review (May-June 1974), pp. 1-13.

MRP in CIM for Selected Consumer Products

JAY NATHAN

University of Scranton and
Deborah L. Beaton
Nabisco Brands, Inc.

Summary

The materials requirements planning (MRP) in computer integrated manufacturing (CIM) has become a way of life for most
manufacturing companies. The benefits of MRP are many: inventory reduction, better scheduling, increased throughout,
control over lot sizes, improved problem visibility, among
others. This paper focuses on the application and use of MRP
for selected consumer products. By means of four different
scenarios, the reorder point system, just-in-time, MRP, and
the inventory system currently used are compared and results
are presented.

Introduction

Over the past several decades, manufacturing firms have
flourished. Consumer spending has exploded, resulting in an
extraordinary amount of gross sales. Break through in technological advances have occurred yearly. With all of these
advances, why aren't manufacturing firms reaping in an over-
abundance of profits? The answer is operating costs. Each
year, costs continue to rise. Building and equipment costs
are extraordinary and labor costs are flying through the roof.
Because of these reasons, manufacturing firms are stressing
cost reductions and increases in productivity to maintain
their profit margins. One area that corporations are focusing on to reduce manufacturing and distribution costs is
inventory.

Some firms are reducing their lead times so they can reduce
inventory levels, but at the same time, service their customers at the desired levels. They also may look for suppliers that are very reliable when it comes to on time
delivery and product quality. Others use inventory control

systems such as Reorder Point, Just-in-Time and Materials
Requirements Planning. These systems will be discussed
shortly.

This paper briefly reviews inventory control systems and
discusses the opportunities and problems of each. Then it
focuses on the application of one of these systems, material
requirements planning. It suggests ways to understand MRP
development, design and implementation by reviewing data from
selected products.

Reorder Point System
Some of the basic assumptions of a reorder point model are as
follows:

1. the demand rate is known and constant
2. lead time is constant
3. all items are delivered at the same time
4. no back orders or lost sales
5. order costs are fixed, they do not depend on order
   size
6. holding costs depend on the length of time items
   are on the floor and the average amount of inventory held
7. purchase costs are constant

Several problems exist in these assumptions: First, the assumption that demand is known and uniform is unrealistic.
This is particularly true with material and component demand
which tends to be lumpy because it depends on the production
decisions made for their parent items. Secondly, lead time
is rarely constant. Production bottlenecks or shortages from
suppliers often cause lead time to vary. Thirdly, to prevent
cancellations or back orders continuously which would be
caused by the unpredictability of demand, safety stock levels
would have to be increased tremendously. This is quite
costly and there is no definite guarantee that a stock out
would always be avoided. Fourth, quantity discounts often
cause order or purchase costs not to be fixed. Often a
discount for purchasing over a designated quantity is given
which lowers unit costs. Fifth, holding costs often are
calculated by the value of the item held, the number of times
it needs to be handled, and the type of storage required i.e.
dry, conditioned, or refrigerated. Sixth, reorder point

models are designed primarily for independent inventory items.
This assumption is not necessarily correct. Dependent demand
items are linked through the bill of materials, the "recipe"
of a parent item.

Reorder point systems fail to realize that the production
schedules of parent products become the requirements of
their components.

Seventh, items received at the beginning of low demand periods
cause high inventory costs and unnecessary levels of inventory
being carried. Lastly, reorder point systems do not look into
the future and thus, cannot be used effectively for planning
purposes.

Just-In-Time System

The ideal just-in-time system would be to have components
arrive at the same time they are needed so that there is no
inventory on hand. Just-in-time systems are pull systems that
are dependent on the accuracy of predicting demand for items.
It is best used in repetitive manufacturing environments where
similar products are manufactured continuously at high volumes. Even in some just-in-time systems the flow of work
through work stations is balanced so that each worker has
approximately the same amount of work.

Characteristics of a just-in-time system are as follows:
  1. The characteristics of a product that customers want
     are produced. This requires accurate forecasting.
  2. Products are produced only at customer demand rates.
     There is little or no inventory on hand.
  3. Production is triggered instantly and purchasing lead
     time is reduced to almost zero.
  4. Products are produced with perfect quality. The
     total quality control concept is maintained, making
     all employees responsible for product quality all
     along the production line.
  5. Lastly, production is achieved while focusing on
     the development and motivation of employees.

As you can see, there are several inconsistencies in the just-
-in-time system. Yes, it diminishes unnecessary inventory and
its related costs. However, many of the assumptions look

great on paper, but are difficult to achieve in a "real world"
production environment. First, many times it is difficult to
forecast customer demand absolutely 100 percent accurately.
Inaccurate forecasting would diminish the effectiveness of a
just-in-time system. Additionally, due to capacity constraints, production scheduling, raw material deliveries and
shortages, it is very unrealistic to have no lead time between production runs. Quality control recently appears to
be improving due to recent feats in quality circles, quality
incentives and machine technology. However, it continues
to be close to impossible to produce 100 percent quality
products on every production line day in and day out.

Materials Requirements Planning System
MRP is an information system for the planning and control of
manufacturing and purchasing. "It indicates priorities and
uses lead time estimates to determine start dates of each
component item in order to meet the desired completion date
of the end item."[9]   Materials Requirements Planning allows
management to look at the future and plan. It makes pro-
jections about what materials must be purchased and when.
Labor hours and capacity requirements may be developed and
projected on-hand balances of materials can be multiplied by
costs to develop inventory budgets. However, please note
that MRP does assume infinite manufacturing capacity.

Materials Requirements Planning performs three major functions:
 "1.   Order planning and control - when to release orders
       and for what quantity.
  2.   Priority planning and control - how the expected date
       of availability compares to the need of each item.
  3.   Provision of a basis for planning capacity requirements
       and development of broad business plans." [11]
MRP reviews the requirements of the master production schedule
and changes them into time phased net requirements.

In the past, MRP inventory methods were thought to work only
with the manufacturing of complex assembled products. However, recent developments has disproved this assumption.
Manufacturing firms that produce simple, one piece products

can and do use Materials Requirements Planning methods. MRP
inventory methods are an extremely effective way of managing
inventory for the following reasons:

1. The cost of inventory can be kept to a minimum.
2. MRP is flexible, reactive, and change sensitive.
3. MRP looks into the future and reviews the status of
    each item individually.
4. Inventory control is "action-oriented" rather than
    clerical "bookkeeping-oriented."
5. Order quantities are translated from requirements.
6. MRP takes into account the timing of requirements,
    coverage of net requirements, and order actions.

The Materials Requirements Planning system is an extremely effective, formalized inventory control system. It is much more
cost saving and efficient than the reorder point system, and
much more realistic and better covers stock outs than the
just-in-time system.

Review of Scenarios

The next facet of this paper will review several different
inventory scenarios using various inventory control methods.
Demand data was derived from actual forecasts of three consumer products. All scenarios contain data compiled from
1985 through 1987 for Brand XXA, Brand XXB, and Brand XXC.
All three products are relatively seasonal. (See Figure 1).
Inventory costs are obviously quite a bit higher than the
Scenario II, the "ideal" just-in-time system. Also notice,
however, that inventory costs are lower in Scenario III,
where inventory levels are minimized, than in Scenario I
where inventory levels vary from high to low throughout
several months. It appears also that Scenario III would have
less of a problem with stock outs than in Scenario I because
of Scenario I's inconsistent level of inventory. Undoubtedly,
Scenario III would have a better case fill performance than
the just-in-time model, Scenario II, for reasons previously
discussed. It is difficult to measure the cost of a stock
out, but one must weigh the price of a lost sale, product
cancellation, or loss of customer goodwill while evaluating
an inventory control method.

Figure 1

1987   Brand XXA   INVENTORY COSTS

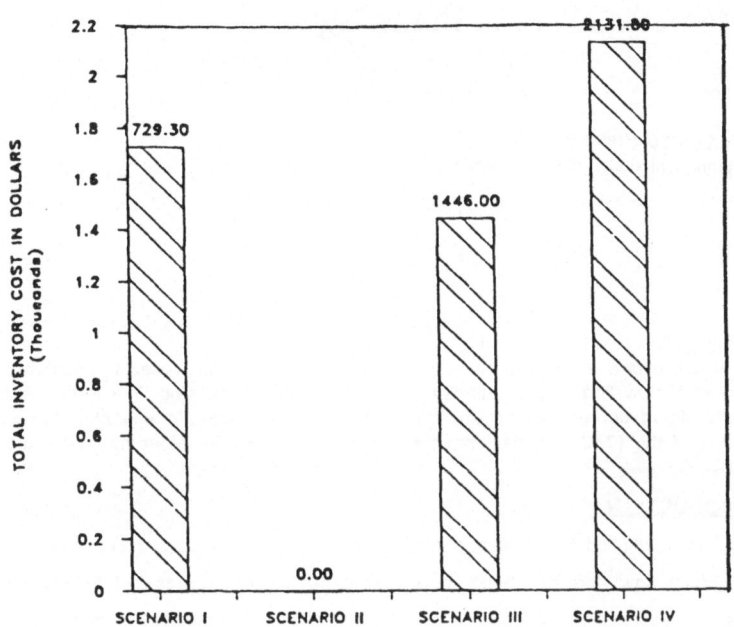

REFERENCES

1. Dilworth, James B., Production and Operations Management
    Manufacturing and Nonmanufacturing, New York, Random
    House, Inc., 1986.

2. Krajewski, Lee J., and Ritzman, Larry P., Operation
    Management Strategy and Analysis, Reading,
    Massachusetts, Addison-Wesley Publishing Company,
    Inc., 1987.

3. Orlicky, Joseph, Materials Requirements Planning The
    New Way of Life in Production and Inventory Management,
    New York, McGraw Hill Book Company, 1975.

4. Sawaya, William, and Giauque, William C., Production
    and Operations Management, New York, Harcourt Brace
    Jovanovich, Inc., 1986.

Dynamic Approach of Computer Automatic Process Planning

Y. Rong

Department of Precision Instruments
Tsinghua University, Beijing, China

ABSTRACT

In this paper, a dynamic process planning approach is proposed to improve the system efficiency of an FMC flow line. The relationship of tool life and processing time is first studied. As the values of tool life in some stations are adjusted, the processing time is changed so that the dynamic balance of the FMC flow line can be expected. A recursive algorithm is developed to optimize the process planning.

INTRODUCTION

In multi-product, multi-size production, a group technology (GT) based flexible manufacturing cell (FMC) flow line has became one solution for the purpose of increasing productivity.[1] To achieve a successful operation of an FMC flow line, the optimization of production should be made from process planning stage. Computer-aided process planning (CAPP) has been rapidly developed to produce a process plan, which is not only feasible but also with good quality.[7] CAPP is actually a two step process.[2] First, according to a coding system, a part is identified and assigned to an FMC flow line. Then operation details are generated for the processing in each station of the FMC flow line. In this paper, a dynamic process planning is studied to improve the quality of CAPP.

In an FMC flow line, the machines are usually divided into several workstations. After it is designed out, the station layout and machine number in each station are fixed. When a part passes through the FMC flow line, it is transformed from a raw material into a finished product. Fig. 1 is a sketch of an FMC flow line.

In a general CAPP system, process planning is made based on the information of parts and the capacity of machines. When a process plan is carried out, the optimum condition of each individual machine is emphasized. But the efficiency of the FMC flow line may not be optimum because of the unbalanced loading to each station, which is specially caused by the variation of part types.

The sysytem efficiency of an FMC flow line is a dynamic variable because variety of parts and machines have to be considered. For a long time, much research has addressed on the job scheduling problem in a flow shop environment. Many algorithms have been developed to increase the machine efficiency, in which some affecting factors are considered as variables, such as the job input sequence, batch size and processing path.[3] But the processing time in each station is usually treated as a constant, which is one of the main factors of the line balance, so that it is hard to achieve a best solution of the operation of the FMC flow line. Although in some loading model, the processing time is considered as a variable, the balance of an FMC flow line can not be achieved by simply accepting or rejecting a group of parts.[4]

In this paper, a dynamic CAPP approach is proposed to improve the efficiency of an FMC flow line. In this method, the part processing time in each station is adjusted by changing the values of tool life in some stations. Therefore, the line balance can be dynamically improved.

THE CONCEPT OF DYNAMIC CAPP

The purpose of dynamic CAPP is to reduce the influence of unbalanced loading to each workstation so that the system efficiency is improved at planning stage. From a conceptual view, the dynamic process planning difffers from a general process planning. In the general process planning, the objective of determining operation detailS of a part processed in a workstation is to optimize the operation in an individual machine, either minimize the unit production time or cost, or to maximize the unit profit. Once a part's processing sequence is chosen, the tool life is first selected as a "reasonable" constant. Then the machining parameters are such determined that the maximum available machine power is used to get the maximum material removal rate, or the shortest processing time in each station is purchased to reduce the throughput time of produced parts. Finally the processing time is calculated based on the operation parameters.[5][6] The process plan is only determined according to the geometric and precision requirement of parts and the capacity of available machines in the FMC flow line. There is no any relationship between the process plans of different parts, and different workstations. Therefore the efficiency of the FMC flow line may not be optimized because of the unbalanced loading to each workstation.

The dynamic process planning system emphasizes on the optimization of the FMC flow line. In such a system, the part processing time in each station will be determined with the consideration of the loading balance of the FMC flow line so that the maximum efficiency of can be expected. More concretely, the value of tool life can be changed in a certain range so that the processing time can be adjusted. For example, If a station has a heavy work load, the machining speed (or feedrate) could be set to a greater value (the tool life would be a little lower) to achieve a shorter processing time. In this way, the system efficiency can be controlled at planning stage.

The DESCRIPTION OF PROCESSING TIME

The processing time in each station of an FMC flow line basically depends on machining parameters (cutting speed v, feedrate f and cutting depth d), tool life and accuracy requirement. For example, in the case of processing a rotation shaft in an NC lathe, the processing time is:

Processing time = Σ Cutting-lenth / Feedrate + Σ Tool-reposition-time

$$= \Sigma \ L/(C_1 * \ f * \ v) + \Sigma \ t_1 \qquad (1)$$

where Feedrate = $f*N = C_1*f*v$ (in./min.)

and: $f =$ (in./rev.); $N =$ (rev./min.)$= C_1*v$ (rev./min.); C_1 is a constant;

Also, the tool life and machine's power should be considered:

$$\text{Tool-life} = C_k \ / \ (v^{k1} * \ d^{k2} * \ f^{k3}) \qquad (2.a)$$

$$\text{Horse-power} = C_2 * \ v * \ d * \ f^{k4} \qquad (2.b)$$

where C_k, C_2, k_1, k_2, k_3 and k_4 are constants.

In order to adjust the processing time without much lost of utilization of machine power, the tool life is taken as a basic variable. It is such changed that the machine power is kept as a constant and cutting parameters (v, d, f) are re-determined so that the processing time is adjusted.

In finish processing, cutting parameters are determined according to precision requirement. In rough machining, cutting parameters are limited by machine capacity (e.g. maximum power and cutting speed). The relationship between tool life and processing time can not be simply represented by an explicit function. It is related to cutting parameters, tool repositioning and precision requirement. When cutting parameters are determined, this relationship is actually set up. There is no any standard or optimal value of tool life which can be pre-determined before a process plan is carried out. Fig.2 shows this relationship in the case of different parts processed on an NC lathe.

THE OPTIMIZATION OF CAPP IN AN FMC FLOW LINE

Based on above discussion, an economic analysis is made as a criterion to optimize the operation of an FMC flow line. In this study, the maximum system efficiency is expressed as that the maximum unit profit is purchased by using existent facilities. First, the unit production time in station i is calculated:

$$U_i = a_i + t_i + b_i * t_i / T_i + K_i \qquad (3.a)$$

where a_i is the set up time (min./pc.); t_i is the machining time (min./pc.); b_i is the tool replace time (min./edge); T_i is the tool life (min./edge); K_i is an unit idle time (min./pc.) due to the phenomena of blocking and starvation, which is related to the inventory level and performance of material handling system, and can be estimated by empirical data or simulation technique.[10]

In a contiuous operation, the production time (ratio) to process a part is:

$$U = max \{U_i\}, \quad i = 1, 2, \ldots, n \qquad (3.b)$$

Then the unit production cost in station i is :[4]

$$q_i = \alpha * n_i * a_i + (\alpha + \beta_i) * t_i + (a * b_i + \gamma_i) * t_i / T_i \qquad (4.a)$$

The unit production cost in the FMC flow line is:

$$q = \Sigma q_i + \Sigma C_i * (U - U_i + K_i) \qquad (4.b)$$

where α is labor cost ($/min.); n_i is the number of operators attending to station i; β_i is machine overhead cost ($/min.); γ_i is tool cost per edge ($/edge); C_i is the idle machine cost ($/min.).

Finally the unit profit can be written as:

$$f = (e - q) / U \qquad (5.a)$$

where e is unit revenue, which is basically related to the material cost and selling price.

Therefore the objective of changing processing time is to maximize above function. i.e.:

$$max \ f; \qquad s.t.: \text{feasible machining parameters and tool life.} \qquad (5.b)$$

Fig. 3 shows the change of unit profit when the tool life is adjusted in a station.

THE PROCEDURE OF DYNAMIC CAPP

CAPP is a two step process. In the first step, according to a coding system (e.g. Opitz syetm), a part is identified and assigned to a manufacturing cell. The processing sequence is also determined based on the information of facility's capacity and production requirement. In the secod step of CAPP, operation details (machining parameters) in each workstation are generated such that the system efficiency of the FMC flow line is maximized.

In order to implement the optimal generation of operation parameters, a recursive algorithm is developed as shown as in Fig. 4. Basically, it contains a two level planning: cell level and station level. In station level the machining details are generated for a given tool life value. Then in cell level the work load and production cost are calculated so that the decision of changing the value of tool life is made to maximize the unit profit. After this changing, machining parameters are re-generated untill the optmal operation is achieved.

SUMMARY

Optimization of a production process is a main goal of computer applications to manufacturing system.[7] In order to achieve a good quality of CAPP, a dynamic process planning approach, in which the processing time is a variable, is proposed to improve the system efficiency of an FMC flow line, based on an analysis of production profit. The complex relationships between tool life and processing time, and production profit are studied. A recursive algorithm is developed to purchase the optimization of process planning.

REFERENCES

1. Steudel, H. J.; Park, T. *A Job Input Sequencing Algorithm for GT-based Flexibl Manufacturing Cell via Simulation*, ULTRTECH 86 Conference Proceeding, Vol.II, 1986

2. Wang, H. and Wysk, R. A. *Micro-GEPPS: A Microcomputer Based Process Planning System*, Computer-Aided/Intelligent Process Planning, ed. Liu, C. R. and Chang, T. C., 1986

3. Stecke, K.E. *Design, Planning, Scheduling and Control Problem of Flexible Manufacturing System*, Annals of Operations Research, Vol.3, 1985

4. Ham, I.; Hitomi, K. and Yoshida, T. *Group Technology*, Kluwer-Nijhoff publishing, 1985

5. Steudel, H. *Computer-Aided Process Planning: Past, Present and Futher*, International Journal of Production Research, Vol. 22, No. 2, 1984

6. Chang, T.C.; Wysk, R.A. *An Introduction to Automated Process Planning System*, 1985

7. Yellowley,I.; Kusiak,A. *Observations on the Use of Computers in the Process Planning of Machined Components*, Trans. of the CSME, Vol. 9, No. 2, 1985

8. Kusiak, A. *Integer Programming Approach to Process Planning*, International Journal of Advanced Manufacturing Technology, Vol. 1, No. 1, 1985

9. Chang, T.C.; Wysk, R.A.; Davis, R.P. and Choi, B. *Milling Parameter Optimization Throtgh a Discrete Variable Transformation*, Int. J. of Prod. Res., Vol.20, No.4, 1982

10. Rong, Y. and Tzou, H.S. *Simulation Designfor Batch Production: a Transaction Oriented Model*, Symposium on Advanced Manufacturing (SAM88), 1988, (to appear)

11. Wang,H. and Wysk, R.A. *Applications of Microcomputers in Automatic Process Planning*, Journal of Manufacturing System, Vol. 5, No. 2, 1986

12. Van Looveren, A.J.; Gelders, L.F. and Van Wassenhove, L.N. *A Review of FMS Planning Models*, Modeling and Design of Manufacturing Systems, ed. A. Kusiak, 1986

434

13. Kusiak,A. *Knowledge Engineering and Optimization in Automated Manufacturing Systems*, 2nd International Conference on Production System, Paris, 1987

14. Egbelu, P. J. *Planning for Machining in a Multijob, Multimachine Manufacturing Environment* , Journal of Manufacturing System, Vol. 5, No.1, 1986

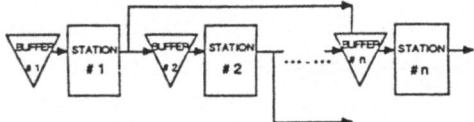

Figure 1. A Sketch of FMC Flow Line

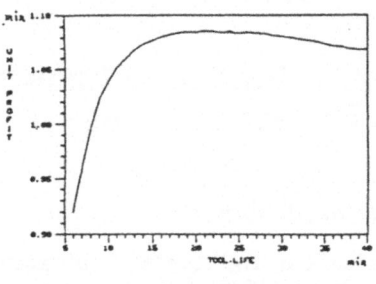

a. Processing Part #1

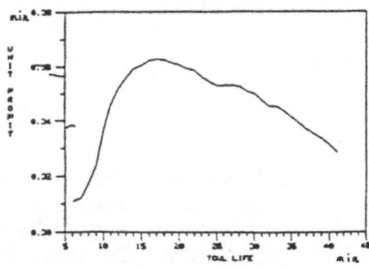

b. Processing Part #2

Figure 3. The change-of unit profit.

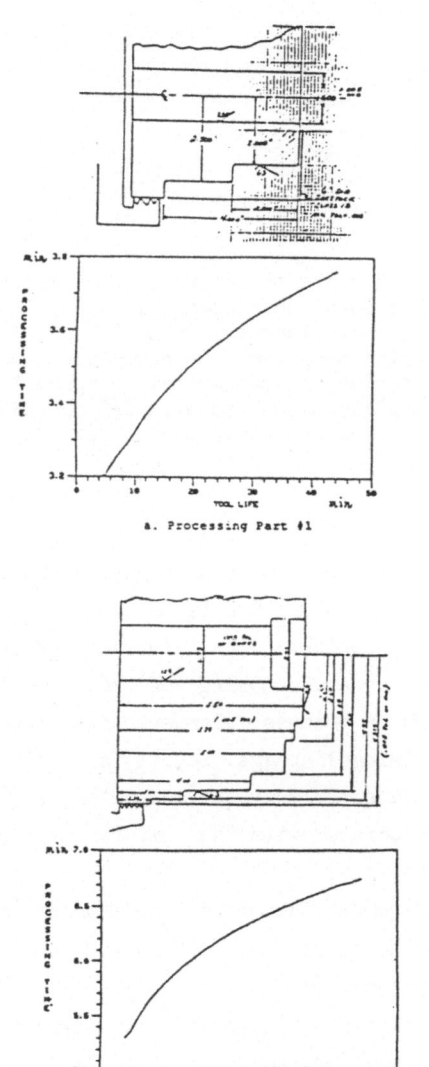

a. Processing Part #1

b. Processing Part #2

Figure 2.Relationship of tool life and processing time in a station

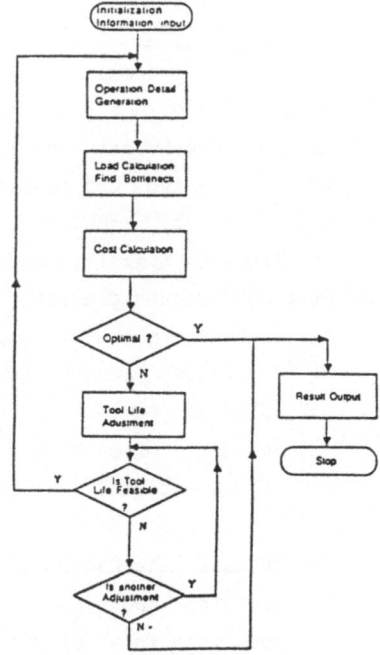

Figure 4 Program Flow Chart of Dynamic CAPP Model

An Application of Knowledge Engineering to Process Planning

Nobutaka Uemura

Production Engineering Laboratory
NEC Corporation
484,Tsukagoshi 3-chome, Saiwai-ku, Kawasaki,
Kanagawa 210, Japan

Summary
    To achieve the integration of CAD and CAM, a prototype of automated process planning system for machining has been developed. This system utilizes geometric processing and knowledge engineering methodology to systematize process planning. The rules which represent the expertise can reference geometric data directly. 16 samples of mechanical components were actually manufactured by the process planning system with CAD and CAM, and the process plans by the system were as good as those made by experts.

1. INTRODUCTION

    For process planning, which bridges design and manufacturing, computers have not been utilized sufficiently, while CAD is used in design and CAM in manufacturing individually. This obstructs the information flow between design and manufacturing. Also in manufacturing of mechanical components by machining, elaborate work by process planners has been necessary to translate CAD data into machining language. So, it is desirable for process planning to be computer-aided or computer-automated. [1][2]

    But, there are two major problems associated with the computerization of the process planning for machining. They are:

a) Advanced geometric processing is required in the process to extract a sequence of machining segments from the difference between the shape of raw material and the designed component. [3]

b) It is hard to incorporate the knowledge of experts into the system, because the knowledge is complicated and hard to systematize, and it contains knowledge about geometry.[4] One example of this kind of knowledge is about which tools and machining parameters are to be used to get the required shape and tolerances.

    We utilized knowledge-based techniques and geometric processing to deal with these problems, and have developed a prototype of fully automated production system for mechanical components.

In the next section, the utilized techniques are described. And, in section 3, the developed system is outlined.

2. THE BASIC APPROACH: GEOMETRIC PROCESSING AND INCORPORATION OF EXPERTISE

In this section, two basic approaches are described. At first, the geometric modeling for the geometric processing and the expert system to incorporate expertise for process planning are described. Then, the geometric manipulation in the extraction of the machining volume unit, and the inference with the reference of geometric information in the generation of the machining steps are described.

2.1 The Geometric Modeling

We use surface model to represent the shape of a designed component and a raw material, and various intermediate shapes generated in process planning. The design information consists of the shape (2.5-dimensional) and tolerances of the component using boundary representation with attribute data.

2.2 The Expert System

We developed the dedicated expert system. The inference engine can be invoked by process planning system, and applies IF-THEN type rule-based knowledge for process planning with forward reasoning. All data in process planning are handled by the "frame" system. Geometry data structure is handled as well as text data. Points, lines and faces are "instances" of the "frame" system. Their parameters are memorized in the "instances" as "slot" values. As the expert system can accept C language-like syntax and Japanese characters, rules are easy to understand.

2.3 The Extraction of Machining Volume Unit

A machining volume unit is a primitive volume which is one of divisions of the difference between the shape of raw material and the designed component.(Fig.1) Our system can handle face, slot, pocket and hole type of machining volume unit. The extraction process starts with the shape to be made. Extracted machining units are attached to the shape repeatedly until the shape is changed to the raw material. This simulates machining process backward. Each extraction consists of three steps: (1) to select a bottom face of a machining volume unit (the basic strategy is to select a lowest horizontal face of the work), (2) to define a top face of the machining unit (the basic strategy is to define a face which has same figure as the bottom face and same level as an nearest upper face of the bottom face), (3) to create a shape between the two faces as a machining volume unit.

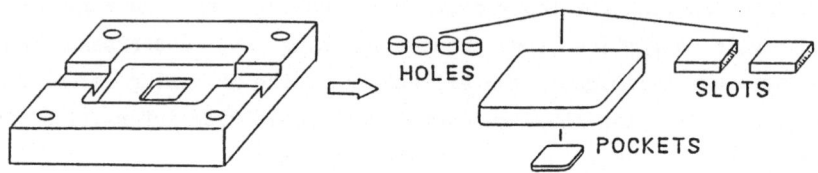

Fig.1 Extraction of machining volume units

2.4 The Generation of Machining Steps

A machining step is a sequence of operations of a tool between retraction or exchange of tool.(Fig.2) More than one machine steps are generated corresponding to the shape and tolerance of a machining volume unit utilizing the rule-based knowledge. The geometric information referenced by the rule are shape type, thickness, area of the bottom face, maximum and minimum radius of corner, open side face (number and position) etc. They are obtained by referencing a "slot" or calculation with built-in functions. Using a simple rule as an example (Fig.3), the reference of geometric information is explained. For a slot type of machining unit as shown in Fig.1, the "IF" part of the rule tests if the slot has two open side faces by the function "is_straight_slot", and if an end mill whose diameter is shorter than the slot width is available; and the "THEN" part decides that the end mill is to be used, and the cutter path of the machining step is shuttle type.

Fig.2 Generation of machining steps

```
IF
    is_straight_slot($unit)==TRUE
    exist_tool("endmill","diameter","<",straight_slot_width($unit))==TRUE
THEN
    select_tool("endmill","diameter","<",straight_slot_width($unit),$step)
    $step.cutter_pass_type = "shuttle"
```

Fig.3 Fragment of a rule

3. A PROTOTYPE SYSTEM USING THE BASIC APPROACH

Based on the techniques presented above, we have developed a prototype of production system which realizes the automation of whole manufacturing process after design to machining mechanical components. The system consists of three subsystems: a 3D-CAD, an automatic process planning system and a work cell.(Fig.4)

The process planning system, which is the key of the whole system, makes process plans automatically using generative method with a rule-based built-in type expert system. We implemented the system on two personal computers in order to realize a low cost system.

The system handles a mechanical component designed with 3-D CAD. The component has 2.5-dimensional shape, and can be made by machining metal blocks from one direction with a 3-axis machining center.(Fig.5) The system generates not only NC data for machining but also robot program for fixing the work.

In principle, the system simulates the process planning of experts. (Fig.6) In the first step, as was explained in sections 2.3 and 2.4, the system extracts the pieces of primitive volume to be machined, selects machining steps for each volume. Next, the system sorts the sequence of the machining steps considering the optimization of machining time, and finally generates the cutter motion. The system also decides the arrangement of fixtures to fasten the work on a pallet and generates the motion of a robot to handle the raw material, the machined component and fixtures. The final output of this process is NC data for the machining center and a program for the robot.

A rule-based expert system is invoked to solve the selection of machining steps , and the arrangement of fixture.

Sixteen samples of mechanical components were actually manufactured, and the process plan by the system was as good as the one plan made by experts. It took only a few minutes to make a process plan although using two personal computers instead of mainframe or mini computers.

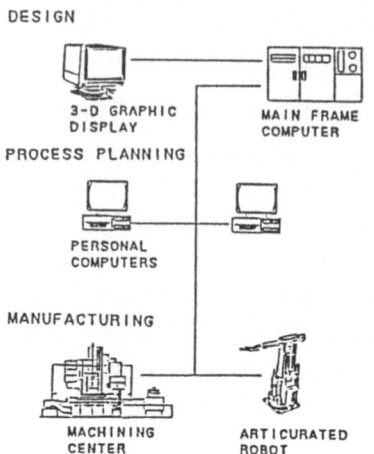

Fig.4 Hardware configuration

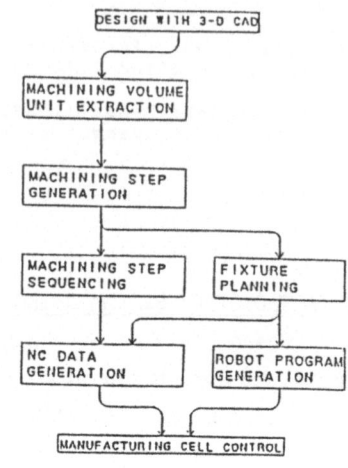

Fig.6 Software configuration

Fig.5 Sample workpieces

4. CONCLUSION

The development of the prototype of fully automated production system for mechanical component indicates that our approach, (knowledge-based techniques and geometric processing), is effective for dealing with the problems associated with the automatization of process planning.

Now, we have a plan to develop an process planning system that handles the expanded object such as 3-D shaped components, 6 directional machining and setup planning.

References
1. Chang, T.C.; Wysk, R.A.: An introduction to automated process planning systems. Prentice-Hall 1985.

2. Kochan, D.: CAM –developments in computer-integrated manufacturing. Springer-Verlag 1985.

3. Henderson, M.R.: Extraction and organization of form feature. Proc. 6th PROLAMAT (1985) 131-141

4. Davies, B.J.; Darbyshire, I.L.: The use of expert systems in process-planning. Annals of CIRP Vol.33 No.1 (1984) 303-306

An Expert Fuzzy Advisor Framework for Statistical Process Control

G. RODOLFO CHACON and DONALD H. LILES

Industrial Engineering Department
The University of Texas at Arlington

INTRODUCTION

Statistical Process Control is an effective problem solving tool used to assess and control the variability of the manufacturing process. Traditional implementation of SPC requires manual computation and the plotting of control charts. This, of course, can now be automated with the use of readily available SPC software. A software package, however, does not guarantee the success of an SPC implementation. Expert knowledge of the process and of SPC techniques is also required. Ongoing statistical process control entails a series of decisions, based upon human knowledge that is ill-defined, vague, and ambiguous. Among these decisions are the determination of the degree of instability of the process, the classification of patterns in the SPC signal, and the diagnosis of assignable cause variation. Zadeh's fuzzy set theory [5][6][7] helps to represent this type of ill-defined knowledge. This paper presents a framework for an expert system which uses a fuzzy set approach to assist in statistical process control decision making.

STATISTICAL PROCESS CONTROL EXPERT SYSTEMS

Statistical process control implies three basic activities or functions.

1. Monitoring: Data acquisition and the familiar computation of sample statistics for process centering and dispersion.

2. Assessment: Determination of control status (identification of non-random variation).

3. Diagnosis: Determination of assignable cause and necessary corrective action.

Implementation occurs in two phases. The first phase determines process capability and establishes appropriate statistical decision limits. The second phase is the ongoing statistical control of the manufacturing process.

Evans and Lindsay [2] have suggested the development of an expert system which addresses at least two of the above SPC functions. The proposed knowledge base is divided into three parts: rules for the analysis of instability (based in statistical inference);rules for the interpretation of patterns in the sample statistic;and rules for diagnosis. The rules for instability include rules for both "strong" conditions and rules for "weak" conditions. These are analogous to traditional runs tests at several levels of confidence. Pattern interpretation rules classify patterns into two main groups: simple and complex. Simple patterns are cycles, and changes in level (trends). Complex patterns include interactions and mixtures. Diagnosis rules are domain dependent. In their conclusion, Evans and Lindsay stress: the difficulty of "systematically addressing qualitative rules"; the need for an alternative approach to pattern recognition so that "the decision rules may be more objective and less analyst-dependent"; and the need for a consensus from SPC experts about the contents of an SPC knowledge base.

THE FUZZY APPROACH TO THE KNOWLEDGE REPRESENTATION AND MANIPULATION

Fuzzy set theory [5] is a generalization of classical set theory. In classical set theory an object is either a member of a set or is not. In fuzzy set theory an object may be partially in a set. This is accomplished by assigning a grade of membership between 0 and 1 to the object. In a classical set, the grade of membership is either 0 or 1. Fuzzy sets can be used to represent linguistic variables such as "strong", "weak", "near zero". These linguistic values can, in turn, be manipulated in a logical way using fuzzy operators. Fuzzy logic, which has potential for the management of uncertainty in expert systems, is defined by Zadeh as "a kind of logic using graded or qualified statements rather than ones that are strictly true or false" [6][7]. To quote Kandel: "Fuzzy-set theory provides a natural, conceptual framework for knowledge representation and manipulation in expert systems that have imprecise components, incomplete data, or unreliable sources of information" [3]. This is certainly descriptive of the process control environment. It is suggested that a fuzzy approach is appropriate to deal with the difficult interpretation problems inherent to SPC.

THE EXPERT FUZZY ADVISOR

The proposed expert fuzzy advisor is intended for use in a man-machine system composed of a process, a computer, and a human operator. The process includes a set of process parameters to be monitored and controlled, a measurable output or process result, and a set of process disturbances. The computer monitors the process via sensory input and presents a stimulative display to the operator. The human and the computer share the functions of analysis, diagnosis, and supervision. The human alone is charged with executing required corrective action and with performing ad-hoc statistical analysis.

The above structure gives the human operator direct access to information and an active role in decision making. This is desirable [4], especially in a dynamic process. Interactive decision making allows the operator to continue to develop knowledge based upon decisions that he/she has participated in.

The expert fuzzy advisor has two functional levels. The lower level performs process monitoring and preliminary analysis to detect process instability. The monitoring function is relatively domain independent and accomplishes data collection (sensory input), computation of sample

statistics for accuracy and precision, and presentation of a
stimulative control display to the operator. The analysis
function at this level performs the basic statistical
analysis and assesses process stability using the instability
rule base. These instability rules are expressed as fuzzy
sets. The overall instability of a process (m) is given by
the rule:

> "If a or b or c or... then m , where a, b, c,...
> are the current grades of membership relative to
> each of several instability tests and m is the
> maximum of them"

This measure of instability is multi-valued and becomes
part of the stimulus display presented to the operator.
Since this measure is displayed continuously, it acts as a
trend indicator and prompts corrective action. Not all cases
of instability can be detected by the rules. Therefore, the
operator is presented with the data and has the option of
making independent decisions.

The higher level analysis function is interactive
between the operator and the expert system. The objective
here is to identify and classify patterns in the sample
statistic. This is based, in part, upon information
provided by data analysis performed by the operator. Some of
this information is expressed in fuzzy terms. For example, a
correlation analysis would result in a correlation
coefficient, r, in the range −1 to +1. This index could be
used to indicate the presence of a linear trend in the sample
statistic. Fuzzy sets can be defined for the range of
possible correlation coefficients. The fuzzy set "near zero"
includes r values close to zero. The fuzzy set "strong
positive" ("strong negative") includes r values close to
+1(−1). Likewise, the fuzzy sets "weak positive" and "weak
negative" include r values not close to either +1, −1, or
zero. Fuzzy information of this type is used as input to
the expert system. Information about the characteristics of
a pattern that cannot be fuzzified from a quantitative
measure, is expressed in multi-valued terms. For example,
the distance between points can be expressed as "very small",

"small", "medium", "big", "very big". Using rules with multi-valued conditions provides an alternative approach to solve the problems expressed by Evans and Lindsay.

At this higher level also, the diagnosis function determines the cause of error as a function of the identified pattern. Fuzzy logic has been proposed by previous research as a promising approach for medical diagnosis [1]. The same approach would be appropriate for the diagnosis of the manufacturing process. Fuzzy logic allows multi-valued observations and more than one conclusion with a certain degree of uncertainty (based in fuzzy logic). Diagnosis rules are domain dependent, therefore the extent and complexity of the diagnosis knowledge base is dependent on the specific application.

SUMMARY

This research has the objective of developing a small expert system prototype for statistical process control. In order to alleviate the problems expressed by Evans and Lindsay the system uses multi-valued fuzzy logic to facilitate interactive SPC decision making.

REFERENCES

1. Anderson, J.; Bandler, W.; Kohout, L. J.; and Trayner, C. (1985). The Design of a Fuzzy Medical Expert System, Approximate Reasoning in Expert Systems, M. M. Gupta, A. Kandel, W. Bandler, J. B. Kiszka (editors), Elsevier Science Publishers B. V. (North- Holland).
2. Evans, J. R. and W. M. Lindsay (1987). Expert Systems for Statistical Quality Control, IIE, World Productivity Forum & 1987 International Industrial Engineering Conference Proceedings.
3. Kandel, A. (1986). Fuzzy Mathematical Techniques with Applications, Addison-Wesley, Mass.
4. Rasmussen, J. and Goodstein, L. P. (1985). Decision Support in Supervisory Control, IFAC Man-Machine Systems, Varese, Italy.
5. Zadeh, L. A. (1965). Fuzzy Sets, Information and Control, Vol. 8, 338-353.
6. Zadeh, L. A. (1984). Making computers think like people, IEEE Spectrum, August 1984.
7. Zadeh, L. A. (1985). The Role of Fuzzy Logic in the Management of Uncertainty in Expert Systems, Approximate Reasoning in Expert Systems, M. M. Gupta, A. Kandel, W. Bandler, J. B. Kiszka (editors), Elsevier Science Publishers B. V. (North- Holland).

Product Data Management

Sharad Sheth
Electronic Data Systems Corporation

INTRODUCTION

Engineering/manufacturing information takes many forms each of which is no
less important than the other in the overall process of creating a
product. The preparation, control and distribution of this information is
presently accomplished through labor intensive manual procedures. Design
processes which utilize manual drafting methods are still commonplace.
They result in the generation of paper, which is the main medium for
communication of the design and its related documentation to those
downstream that will convert that design into a product. The movement of
paper, in its varied forms, is a cumbersome process.

Where automated methods such as CAD/CAM are employed, "islands of
automation" frequently occur; applications are developed which fit the
needs of users and increase productivity within limited areas, but are not
flexible enough to interface to existing systems.

Present systems lack the management and control which alleviate
conditions which result in excessive delays in information transmission;
this limits the effectiveness of personnel by requiring them to perform
inappropriate or time-consuming clerical and/or administrative tasks.

A global solution is required, which encompasses all corporate data and
provides centralized control to ensure its accuracy and availability. This
paper will discuss the realistic possibilities of implementing a data
management system which recognizes the need for many forms of corporate
data and and adapts itself to that need.

DATA MANAGEMENT ISSUES

Data must be considered as a corporate resource. The various forms which
this data takes can be classified under the general title of Enterprise
Data. A fundamental corporate goal must be integration through management
of Enterprise Data, resulting in data which is shared, meaningful and
independent. The management of technical data, in particular, is critical.

Automation of the management, control, and distribution of product and
manufacturing engineering information will play a major róle in any CIM
strategy by providing the data management capabilities required to
optimize information flow within the technical and office environments.
Modular software design and standardized hardware platforms will ease this
integration effort with existing and future information processing
systems.

Each leap in technology within the CAD/CAM/CAE environment enables faster
data creation, analysis and modification. Data management systems must
keep pace to derive the full benefits of these advancing technologies. It
is essential to eliminate duplicative data bases and synchronize data, not
only from engineering and manufacturing but also from financial and
administrative systems.

Present methods of managing the massive flow of data within most
organizations are inadequate. Those who are aware of the inadequacies both
within and outside of the organizations are frequently frustrated by the
lack of an all encompassing solution. The wide variety of user practices
and procedures within various organizations has caused a reluctance to
develop systems which would have limited applicability. What is needed is
a solution that employs the best hardware and software technologies, and
is flexible enough to meet the requirements of users in various
disciplines.

CURRENT ENVIRONMENT

The creation of a new product or the revision of an existing product presently involves the manual generation of an enormous amount of paper documentation in addition to the engineering drawings. Even though computer systems may be utilized, they tend to be very paper-oriented when it comes to the output. The end users such as engineers and plant floor personnel utilize blueprints, aperture cards, microfiche, and plots generated by CAD systems --- all paper intensive.

It is also labor and time-intensive to rely on paper output and manual methods for distributing information. For example, engineers making product revisions deliver the appropriate paperwork to the reproduction area where the distribution process originates. In many cases, the inefficiencies involved in the reproduction and distribution process cause plant floor operations to suffer from outdated information.

The variety of CAD/CAM systems which have been implemented in various organizations have resulted in their own unique data management problems. How can files be efficiently moved from System X to System Y? Can we control the computer-aided engineering files created on one system so that they can be utilized for a computer-aided manufacturing operation on another system? It is obvious that the manual environment is not the only one of concern. The CAD/CAM environment does not currently present a unified, single point of control which the user truly needs for information access.

Another element of the current environment is the heterogeneous computer systems which can be found in many organizations. Certain computing platforms are more suited for specific applications; this is a fact which cannot be ignored. This situation is not conducive to the integration which is required for computer integrated manufacturing; it must be a high priority issue on which we must focus if we are to achieve the objectives of integration.

Further complicating matters is the fact that many businesses tend to be geographically dispersed. The advantages to be gained by utilizing the strengths of a specific state or country in providing a product or service are sometimes reduced by the inefficiencies encountered in managing the data critical to combining dispersed pieces into a quality end product.

GENERAL FEATURES REQUIRED

Easy Access to Data Through Common Devices

Control of Multiple Data Types

Document Conversion Capability

Portability

Interface Capability

Tailored User Interfaces

Multiple Security Levels

Cost Effective, Easily Accessible Archival Capability

Convenience/Batch Printing Capability

Multi-media Reproduction

Local/Remote Networking

STRATEGY

Applications such as CAE, CAD, CAM, and MRP/BOM generally provide their own file management functions; however when we have two or more applications together, bridges must be built between them. When one system changes, the software is upgraded and the bridges must be upgraded as well. Until now this has been a time-consuming and unreliable process. What is needed is a set of four independent, self sustaining modules with functions that control a variety of applications and interface with their computing platforms.

SOFTWARE FUNCTIONS

Extensive analysis of user requirements has resulted in identification of four major functions which a total data management system must incorporate. These functions must provide features which can be turned on or off according to the needs of the user.

The functions are:
    Document/File Management
    Engineering Change Management
    Project Management
    Document Distribution

Document/File Management

Document/File Management is the mechanism designed to provide control of all document activity including CAD and CAM files. This is the function with which a typical engineering/manufacturing user will have the most interaction. The frequency of access of engineering drawings will amplify the benefits to be derived.

Engineering Change Management

Engineering Change Management is a mechanism to control the process of modifications to engineering designs. It tracks the change process from the design initiation or revision through release, serving as a vehicle of notification and authorization.

Project Management

Project Management is the mechanism for organizing implemehtation of all engineering activities. It defines product life cycle events, completion dates, tasks, required resources, overlapping activities and associated budgets, with the end result being a "critical path" to project completion. Project Management is the organizational tool for controlling the product development and implementation life cycle in an efficient and cost effective manner.

Document Distribution

Document Distribution defines distribution patterns based on specific user needs. Physical and organizational hierarchies interact to define the distribution network. Once the network is defined, a trigger mechanism initiates the distribution process. The trigger can be either manual, automatic, or a combination of both. It might be activated as a function of time or as a function of project completion. For instance, distribution might be triggered automatically every six months for certain documents, or initiated by a project engineer at the completion of a project.

SOFTWARE INTERFACES

Today, due to the diversity of computing platforms and information
processing requirements, "islands of automation" are created that satisfy
niche requirements, without addressing data sharing at the global level.
The gaps between these islands must be bridged by providing a global
indexing scheme without disturbing individual applications. Typical
systems to which interfaces must be provided include:

    Bills of Material (BOM)
    Computer Aided Design (CAD)
    Computer Aided Manufacturing (CAM)
    Manufacturing Resource Planning (MRP)
    Business Systems
    Office Automation Systems

System implementation must be preceeded by a detailed requirements
analysis to determine both logical and physical interfaces to existing and
planned systems. A successful data management system will have the
capability to interface to installed systems. Equally important, the
personnel performing the requirements analysis must possess the knowledge
and experience to recommend logical, cost effective solutions.

HARDWARE COMPONENTS

The hardware architecture is based on the use of various components
connected through open systems architecture. To provide a wide range of
functions and capabilities, components should be selected to satisfy
varying requirements for performance and cost. Typical hardware components
include:

Input Devices

    - Paper Scanners (page and large format)
    - Aperture Card Scanners
    - CAD Files (vector and plot)

Storage Devices

    - Magnetic Tape
    - Magnetic Disk
    - Optical Disk with Autochanger and Drives

Display Devices

    - Engineering Workstations
    - Dedicated Video Display Terminals (VDTs)
    - Enhanced Personal Computers

Output Devices

    - Electrostatic Printers
    - Laser Printers
    - Dot Matrix Printers
    - Film Printers
    - Aperture Card Reproducers

Processors

    - Mainframes
    - Minicomputers
    - File Servers
    - Engineering Workstations
    - Personal Computers

BENEFITS

An effective data management system enhances the utilization of present and planned automation systems by providing a global data management tool. Among the expected benefits are:

- Increased productivity through improved information flow.

- Fast and reliable access to data through a centralized data base.

- Effective management and control of information.

- Reduction in project lead times by insuring that the right information gets to the right people at the right time.

- Reduction in distribution costs through elimination of the need for couriers or mail delivery.

- Reduction in "downstream" errors due to unreliable data. For example, parts might be ordered or manufactured to the wrong revision level due to unreliable release and distribution mechanisms.

- Reduction in reproduction costs through elimination or reduction of the need for blueprints, microfilm copies or photocopies.

- Reduction in floorspace requirements through elimination of hardware such as blueprint machines and large storage cabinets for drawings.

  -Elimination of "satellite" files in various departments and on the shop floor since data can be accessed from strategically located terminals and printers and access a centralized data base.

- Elimination of lost documents resulting from an inefficient distribution process.

- Reduction of redundant information by achieving integration of applications under a single data management and control system.

CONCLUSION

All of the discussion to this point has been centered on the fact that there is an obvious need for data management within a company which hopes to remain competitive in the products produced or services offered. It has been demonstrated that the technology is in place to implement data management systems on a scale suitable to any application.

The one factor which has not been discussed yet is the primary factor which will ensure that a data management implementation plan will become a reality--management commitment to the project. This commitment is not something which can be sold to a company or developed in a research lab. It is generated from within the corporate hierarchy by visionary people who understand the needs of the company and are prepared to take the steps necessary to insure its survival.

Automated Extraction and Recognition of Internal Defect Configurations in Solid Log Modeling

L. G. Occeña [*] and Jose M. A. Tanchoco [**]

[*]Department of Industrial Engineering
University of Missouri-Columbia
Columbia, Missouri USA 65211

[**]School of Industrial Engineering
Purdue University
West Lafayette, Indiana USA 47907

Abstract

The objective of this paper is to describe a procedure developed for the automated extraction and recognition of internal defect profiles in hardwood log processing. The procedure uses a five-stage filtering approach to extract the defect hull, and a three-stage characterization approach to recognize the defect configuration type. The problem of computer recognition of internal log defects will find parallel work in the interface problem of CAD and CAM.

Introduction

Research is currently being done on the application of non-invasive internal examination technology, such as computed axial tomography scanning and nuclear magnetic resonance scanning, to the detection of internal defects in hardwood logs [1,2,3]. In the hardwood lumber industry, where the manufacturing process hinges upon the production of clear-faced lumber, this capability to detect internal defects and thus better plan the log breakdown process is viewed with great interest.

A concomitant problem to this new capability is the issue of how the resulting scan data will be used. Imaging and preprocessing methods to convert the scan data to a graphic image for interpretation by a human sawyer is one alternative. A more economical alternative process-wise is to route the scan data directly to the computer for analysis. This alternative crosses into the realm of automated processing and machine understanding.

The problem of computer recognition of internal log defects will find parallel work in the interface problem of CAD and CAM, where automatic recognition of machinable features leading towards process plan formation is the objective [4,5,6]. Unlike features in machinable parts which are geometrically and topologically regular and reproducible, however, biological features in wood are irregular and unpredictable. While a slightly different approach may be required, underlying principles are related. Sculptured surface and fractal geometry applications are good examples.

The objective of this paper is to give an overview of a procedure developed for the automated extraction and recognition of internal defect profiles in hardwood logs. A more detailed discussion is given in [7].

Domain Definition

Trees are classified as either hardwood or softwood. Hardwoods, such as maple, oak, or poplar, are distinguished by a deciduous foliage that falls off in autumn, and a tree form marked by multiple rebranching. Hardwoods are generally used to make furniture and fine-finish woodwork because of their rich and colorful heartwood. Owing to this end use, particular attention is given to the presence of degrading defects and blemishes on the wood surface. In fact the value of hardwood lumber is measured by a grading procedure which places a high value on defect-free lumber faces. Knots, splits, decay, etc. are considered degrading defects.

Hardwood logs are processed with the objective of obtaining the most number of clear or defect-free wood pieces. In hardwood log breakdown, the biggest limitation is the inability of the human sawyer to see the location and orientation of internal defects inside the log. By the time internal defects are revealed by the opening sawcuts, the remaining log is already constrained to a particular sawing pattern.

Studies on the use of non-invasive imaging techniques [1,2,3] provided the needed information for a better planning of the log breakdown process. Follow up research on the use of the information on internal defects resulted in the development of a pattern directed sawing policy for hardwood logs [8]. The policy involves the examination of the defect configuration inside the log, a search for a description of the configuration, the characterization of the configuration, and the prescription of an appropriate breakdown pattern based on the configuration detected. This policy was later implemented in a logic-based computer model. The C-Prolog model can automatically generate a set of sawing instructions from the log and internal defect information [9].

A critical component of the model was the capability to automatically extract and recognize the internal defect configuration. The input consists of log and defect information from the imaging operation. The output is an edge-list representing the connected vertices of the defect configuration hull. This edge-list becomes the input for the log breakdown module of the model. The information flow in the model is depicted in

Figure 1 in the appendix, showing the extraction as a distinct process, and the recognition as part of the pattern directed breakdown process.

Automated Defect Configuration Extraction

The approach for extracting the defect configuration is to first map the three-dimensional mass of defects inside the log to two-dimensional space, then to envelope the boundary of the resulting mass of defects with a concatenation of boundary points. The result is called a defect hull. The hull is not convex, which is desirable because this permits the accumulation of defects to appear in the form of major axes. These axes are used in determining the breakdown pattern. There are five stages to the hull extraction: log profile extraction, initial defect filtration, measurement of defect density, density filtering, and defect hull extraction.

The first stage extracts the log profile using a recursive sweep of the x, y, and z directions of the log exterior. This extraction is done because only the log boundary is pertinent for referencing purposes. Furthermore, only the starting and ending points of the log boundary in a vertical sweep are required in each longitudinal slice. Figure 2 in the appendix illustrates the extraction.

In the course of the log breakdown process, defects located near the log surface or the log ends are removed by the finishing operations of edging and trimming. These defects are filtered out in the second stage as noise in the data. To implement the filtering, an imaginary prismatic boundary based on the log length and cross-section of the log small end is created, using the minimum defect size of 1-1/4 inches as tolerance.

As a measure of the defect concentration inside the log, a defect density is obtained for each defect in the third stage. The result is a two-dimensional mapping of the defect densities along the z-axis. This procedure can be justified by the lengthwise sawing of the log which makes defects stringed along the length of the log more significant than defects stringed along the cross-section.

The fourth stage is a refinement of the data, involving the elimination of defects with density measures less than the minimum degrading defect size of 1-1/4 inches. The remaining defects are then summed along the z-axis to complete the two-dimensional mapping. The final map represents the mass of concentrated defects that will degrade the lumber if not contained. In the fifth stage, the defect hull bounding the aggregation of defects is extracted. The defect hull is obtained by concatenating the bounding points in a counterclockwise direction. The bounding points consist of the maximum and minimum points in a range of points, obtained by scanning first along the x axis, and then along the y axis of the two-dimensional map [10]. Figure 3 in the appendix illustrates the defect hull formation.

Defect Configuration Characterization

The automated defect hull recognition is a characterization of the hull relative to the number of major axes and other axial relations required as input to the log breakdown module. If there is only one major axis, that axis is considered as dominant. The challenge occurs when there are multiple axes, which then have to be isolated by partionining. Recursive partitioning of the log sections eventually brings the problem down to the level of a single major axis for which the solution is well-defined. There are also extreme cases where the log is free of defects, or completely inundated with defects.

If the boundary points in the defect hull are evaluated counterclockwise in pairs, a change in the sign of the slope will occur about the inflection points. These inflection points are candidates for vertices of major axes. Figure 4 in the appendix illustrates the inflection points.

An inflection point is considered a vertex of a major axis if it satisfies two necessary conditions: (a) in a counterclockwise sweep, the inflection point must be convex, and (b) in a counterclockwise sweep, the height of an inflection point should be greater than its base width. A sample output is given in Figure 5.

Bibliography

[1] Taylor,F.W.,   F.G.Wagner,Jr.,   C.W.McMillin,   I.L.Morgan,   and F.F.Hopkins. Locating Knots by Industrial Tomography - A Feasibility Study. Forest Products Journal, 34(5):42-46, 1984.
[2] Funt,B.V. and E.C.Bryant. Detection of Internal Log Defects by Automatic Interpretation of Computer Tomograpy Images. Forest Products Journal, 37(1):56-62. 1987.
[3] Chang,S.J., P.C.Wang, and J.R.Olson. NMR Imaging of White Oak Logs. Technical Forum, 41st Annual Mtg. of the FPRS. Louisville,Ky. 1987.
[4] Staley,S.M., M.R.Henderson, D.C.Anderson. Using Syntactic Pattern Recognition to Extract Feature Information from a Solid Geometric Data Base. Computers in Mechanical Engineering. September 1983.
[5] Henderson,M.R. and D.C.Anderson. Computer Recognition and Extraction of Form Features: A CAD/CAM Link. Computers in Industry, 5:329-339. 1984.
[6] Joshi,S.B. Automated Understanding of Part Features for Automated Process Planning. Ph.D. Dissertation. School of Industrial Engineering. Purdue University, West Lafayette, In. 1987.
[7] Occeña,L.G. and J.M.A.Tanchoco. Pattern Directed Extraction and Recognition of Defect Configurations in Solid Log Models. Technical Paper. Dept of IE, Univ. of Missouri, Columbia, Mo. 1988.

[8] Occeña,L.G. Pattern Directed Sawing in the Presence of Internal Defect Detection. Technical Paper. Dept of IE, Univ. of Missouri, Columbia, Mo. 1988.

[9] Occeña,L.G. and J.M.A.Tanchoco. PDIM - A Logic Based Model for Automated Log Breakdown Process Planning. Technical Paper. Dept of IE, Univ. of Missouri, Columbia, Mo. 1988.

[10] Occeña,L.G. and J.M.A.Tanchoco. Automated Extraction of Star-Shaped Hulls. Technical Paper. Dept of IE, Univ. of Missouri, Columbia, Mo. 1988.

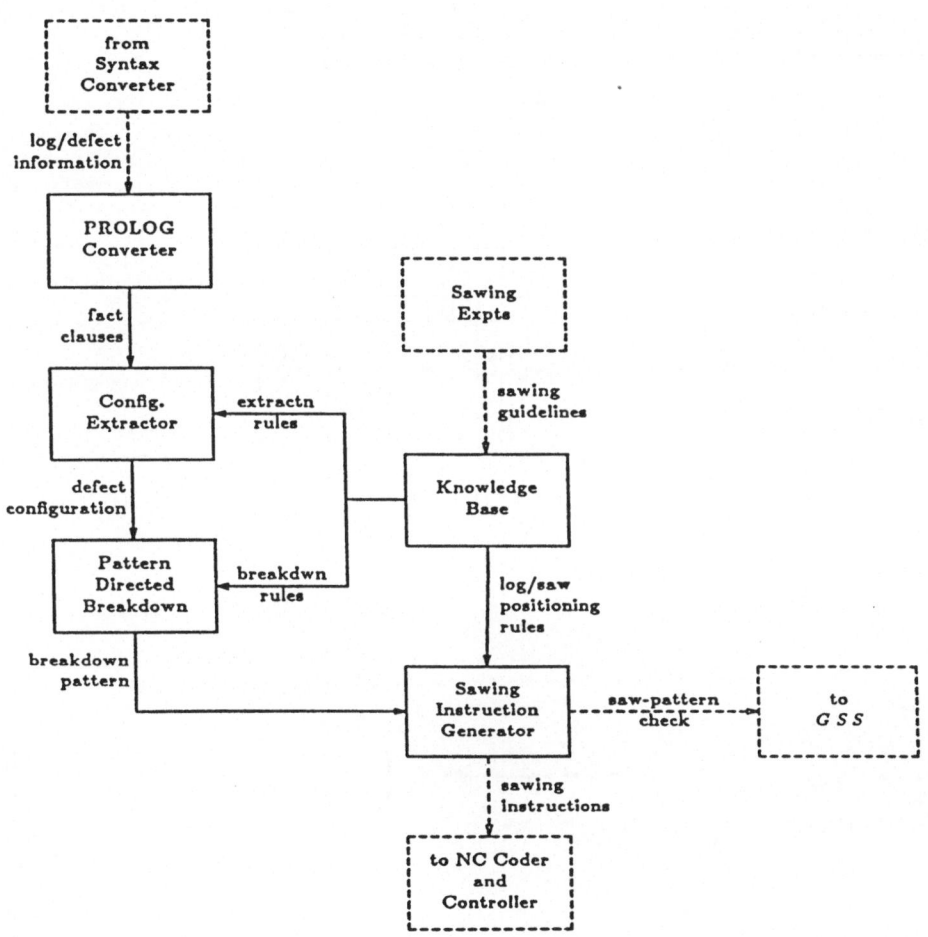

Figure 1. PDIM Information Flowchart

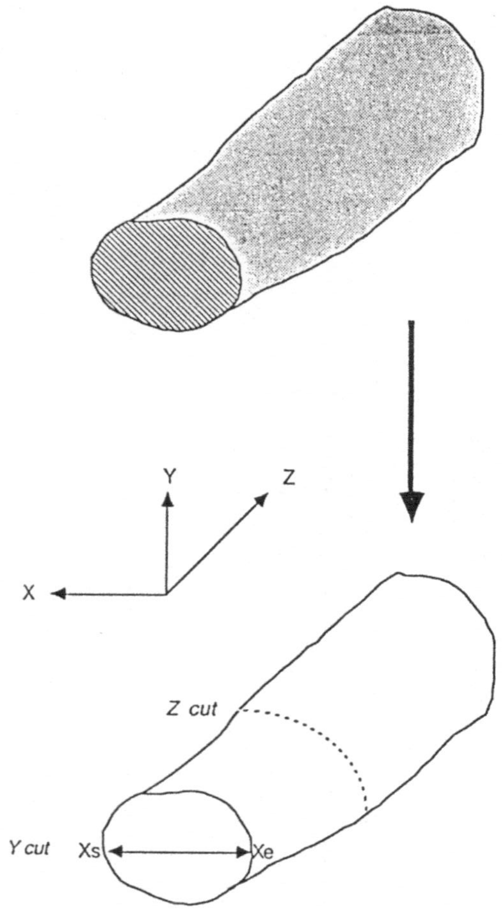

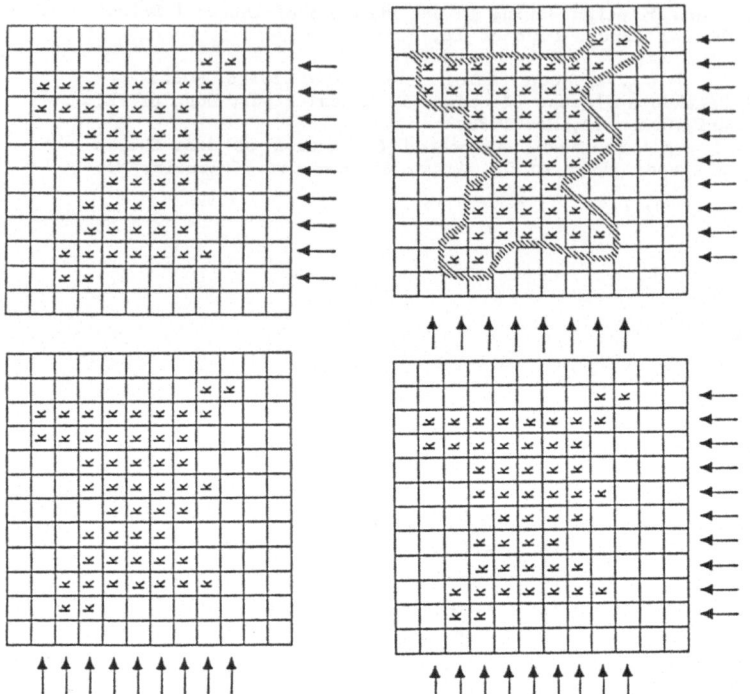

Figure 3 · Defect Hull Formation

Figure 2. Log Boundary Extraction

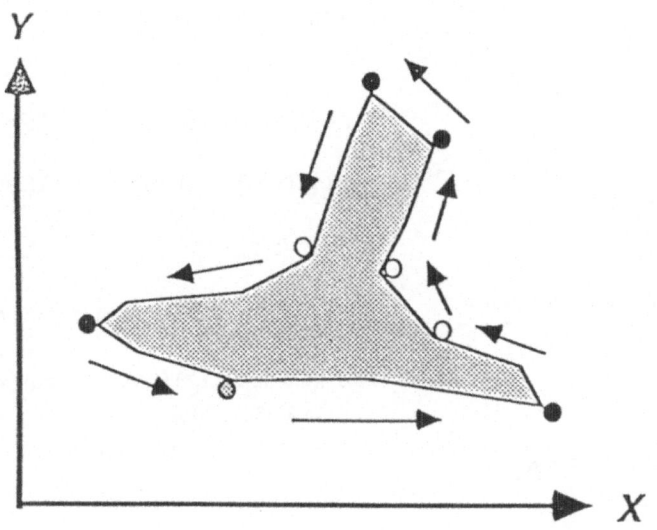

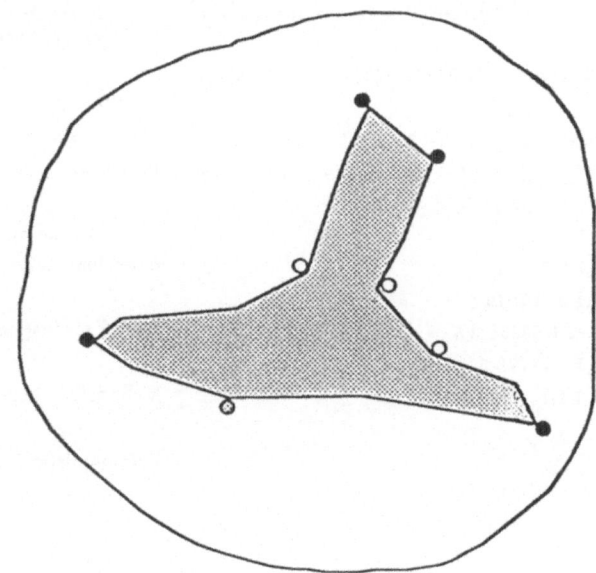

- ◉ Axis Vertex
  Convex
  Inflection Point

- ⊗ Convex
  Inflection Point

- ○ Inflection Point

Figure 4. Inflection Point Illustration

HULL FORMATION

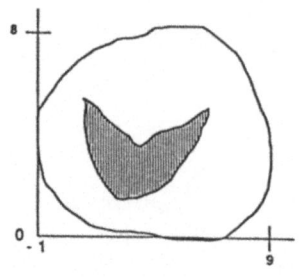

x [(3,3),(1,3),(1,2),(1,1),(2,0),(2,-1),(2,-1),(3,0),(4,1),(3,2)]

y [(4,1),(3,3),(2,2),(1,3),(1,1),(2,-1),(3,0),(4,1)]

singlex [(3,3),(1,3),(1,2),(1,1),(2,0),(2,-1),(3,0),(4,1),(3,2)]

singley [(3,3),(2,2),(1,3),(1,1),(2,-1),(3,0),(4,1)]

hull [(3,3),(2,2),(1,3),(1,2),(1,1),(2,0),(2,-1),(3,0),(4,1),
(3,2),(3,3),(2,2)]

HULL CHARACTERIZATION

inflection_pts [[(3,3),(2,2),(1,3)],[(2,2),(1,3),(1,2)],
[(1,2),(1,1),(2,0)],[(1,1),(2,0),(2,-1)],
[(2,0),(2,-1),(3,0)],[(3,0),(4,1),(3,2)],
[(4,1),(3,2),(3,3)],[(3,2),(3,3),(2,2)]]

convex_pts [(1,3),(1,1),(2,-1),(4,1),(3,3)]

relatives [[(2,2),(1,3),(1,1)],[(1,3),(1,1),(2,0)],
[(2,0),(2,-1),(4,1)],[(2,-1),(4,1),(3,2)],
[(3,2),(3,3),(2,2)]]

axes [[(1,3),(1.58579,1.58579),2.34314],
[(3,3),(2.58578,2),2.34316]]

number of axes = 2

major_axis5 above major_axis6 by 44.9999 degrees
major_axis6 below major_axis5 by 44.9999 degrees

direction(major_axis6,(3',3),(2.58578','2),2.34316).
direction(major_axis5,(1','3),(1.58579','1.58579),2.34314).

1,3 2.34314 1
1,1 0.402019 5
2,-1 1.86354 2.5
4,1 1.77778 5
3,3 1.17158 0.5

logdef_status(log_boundary1,multiple_axes).

logbnds(log_boundary1,[-1,0,9,8]).

hullbounds(log_boundary1,[1,-1,4,3]).

Figure 5. Sample Case

Contents of Volume II

460

462

Contents of Volume III

464

Chapter V: Object Recognition, Imaging and Sensors

466

Chapter VI: Control

Chapter VII: Motion Coordination

Chapter VIII: Communication and Networking

Chapter IX: Factories of the Future - Plant Organization

Chapter X: Future Trends

Author Index

Scientific Fundamentals of Robotics

1

M. Vukobratović, V. Potkonjak,

Dynamics of Manipulation Robots

Theory and Application

1982. XIII, 303 pp. 149 figs. (Communications and Control Engineering Series). ISBN 3-540-11628-1

2

M. Vukobratović, D. Stokić

Control of Manipulation Robots

Theory and Application

1982. XIII, 363 pp. 111 figs. (Communications and Control Engineering Series) ISBN 3-540-11629-X

3

M. Vukobratović, N. Kirćanski

Kinematics and Trajectories Synthesis of Manipulation Robots

1986. X, 267 pp. 66 figs. (Communications and Control Engineering Series) ISBN 3-540-13071-3

Contents: Kinematic Equations. – Computer-Aided Generation of Kinematic Equations in Symbolic Form. – Appendices I–III. – Inverse Kinematic Problem. – Kinematic Approach to Motion Generation. – Dynamic Approach to Motion Generation. – Motion Generation for Redundant Manipulators. – References. – Subject Index.

4

M. Vukobratović, N. Kirćanski

Real-Time Dynamics of Manipulation Robots

1985. XII, 239 pp. 43 figs. (Communications and Control Engineering Series) ISBN 3-540-13072-1

Contents: Introduction. – Survey of Computer-Aided Robot Modelling Methods. – Computer-Aided Method for Closed-Form Dynamic Robot Model Construction. – Computer-Aided Generation of Numeric-Symbolic Robot Model. – Model Optimization and Real-Time Program-Code Generation. – Examples. – Appendix 5.1. – References. – Subject Index.

5

M. Vukobratović, D. Stokić, N. Kirćanski

Non-Adaptive and Adaptive Control of Manipulation Robots

1985. X, 383 pp. 111 figs. (Communications and Control Engineering Series) ISBN 3-540-13073-X

Contents: Computer-Assisted Generation of Robot Dynamic Models in Analytical Form. – Non-Adaptive Control Manipulation Robots with Variable Parameters. – Adaptive Control Algorithms. – Computer-Aided Control Synthesis. – Implementation of Control Algorithms. – Subject Index.

6

M. Vukobratović, V. Potkonjak

Applied Dynamics and CAD of Manipulation Robots

1985. XII, 305 pp. 187 figs. (Communications and Control Engineering Series) ISBN 3-540-13074-8

Contents: General About Manipulation Robots and Computer-Aided Design of Machines. – Dynamic Analysis of Manipulator Motion. – Theory of Appel's Equations. – Closed Chain Dynamics. – Computer-Aided Design of Manipulation Robots. – References. – Subject Index.

Springer-Verlag
Berlin Heidelberg New York London Paris Tokyo Hong Kong

M. Vukobratović, D. Stokić

Applied Control of Manipulation Robots

Analysis, Synthesis and Exercises

1989. Approx. 495 pp. 100 figs. ISBN 3-540-51469-4

Contents: Concepts of Manipulation Robot Control. – Kinematic Control Level. – Synthesis of Servo Systems for Robot Control. – Local Optimal Regulator. – Control of Simultaneous Motions of Robot Joints. – Stability Analysis of Nonlinear Model of Robot. – Synthesis of Robot Dynamic Control. – Variable Parameters and Concept of Adaptive Robot Control. – Control of Constrained Motion of Robots. – Software Package for Synthesis of Robot Control. – Subject Index.

M. Vukobratović

Applied Dynamics of Manipulation Robots

Modelling, Analysis and Examples

1989. Approx. 495 pp. 176 figs. ISBN 3-540-51468-6

Contents: General About Robots. – Computer Forming of Mathematical Model of Manipulation Robots Dynamics. – Computer Method for Linearization and Parameter Sensitivity of Manipulation Robots Dynamic Models. – Connection Between the Moving and Fixed System. – Manipulator Kinematical Model. – Determining Velocities and Accelerations. – Momentum of Rigid Body with Respect to a Fixed Pole. – Specifities of Lever-Mechanisms Dynamics. – Mathematical Models of Driving Units. – Automatic Forming of Dynamic Models. – Dynamics of „ASEA" Mechanism. – Programme Support for Dynamics Modelling of Manipulation Robots. – Subject Index.

M. Vukobratović (Ed.)

Introduction to Robotics

With contributions by M. Djurović, D. Hristić, B. Karan, M. Kirćanski, N. Kirćanski, D. Stokić, D. Vujijić, M. Vukobratović

1988. Approx. 350 pp. 228 figs. ISBN 3-540-17452-4

Springer-Verlag
Berlin Heidelberg
New York London
Paris Tokyo Hong Kong

Contents: Preface. – General Introduction to Robotics. – Manipulator Kinematic Model. – Dynamics and Dynamic Analysis of Manipulation Robots. – Hierarchical Control of Robots. – Microprocessor Implementation of Control Algorithms. – Industrial Robot Programming Systems. – Sensors in Robotics. – Elements, Structures and Application of Industrial Robots. – Robotics and Flexible Automation Systems. – Appendix.

Springer